Control Theory for Humans

Quantitative Approaches
to Modeling Performance

Control Theory for Humans

Quantitative Approaches to Modeling Performance

Richard J. Jagacinski
Ohio State University

John M. Flach
Wright State University

LEA 2003
LAWRENCE ERLBAUM ASSOCIATES, PUBLISHERS
Mahwah, New Jersey
London

Lawrence Erlbaum Associates, Inc., Publishers
10 Industrial Avenue
Mahwah, New Jersey 07430

Cover design by Kathryn Houghtaling Lacey

Library of Congress Cataloging-in-Publication Data

Jagacinski, Richard J.
    Control theory for humans : quantitative approaches to modeling performance /
Richard J. Jagacinski, John M. Flach.
       p.   cm.
    Includes bibliographical references and indexes.
    ISBN 0-8058-2292-5 (cloth : alk. paper) — ISBN 0-8058-2293-3 (pbk. : alk. paper)
       1. Human information processing.   2. Perception.   3. Human behavior.   4. Control theory.
    I. Flach, John.   II. Title.

BF444 .J34   2002
153—dc21   2002024443
                                                                                 CIP

Books published by Lawrence Erlbaum Associates are printed on acid-free paper,
and their bindings are chosen for strength and durability.

Printed in the United States of America
10 9 8 7 6 5 4 3 2 1

To our advisors and teachers,
and to their advisors and teachers,

To our students, and to their students,

And to the Spirit that unites them

Control Theory for Humans

The feedback principle introduces an important new idea in nerve physiology. The central nervous system appears to be a self-contained organ receiving signals from the senses and discharging into the muscles. On the contrary, some of its most characteristic activities are explainable only as circular processes, traveling from the nervous system into the muscles and re-entering the nervous system through the sense organs. This finding seems to mark a step forward in the study of the nervous system as an integrated whole. (p. 40)

— *Norbert Wiener* (1948)

Indeed, studies of the behavior of automatic control systems give us new insight into a wide variety of happenings in nature and human affairs. The notions that engineers have evolved from these studies are useful aids in understanding how a man stands upright without toppling over, how the human heart beats, why our economic system suffers slumps and booms, why the rabbit population in parts of Canada regularly fluctuates between scarcity and abundance (p. 66). . . . Man is far from understanding himself, but it may turn out that his understanding of automatic control is one small further step toward that end. (p. 73)

— *Arnold Tustin* (1952)

. . . some of the most interesting control problems arise in fields such as economics, biology, and psychology (p. 74) . . . for those who want to understand both modern science and modern society, there is no better place to start than control theory. (p. 82)

— *Richard Bellman* (1964)

All behavior involves strong feedback effects, whether one is considering spinal reflexes or self-actualization. Feedback is such an all-pervasive and fundamental aspect of behavior that it is as invisible as the air we breathe. Quite literally it is behavior — we know nothing of our own behavior but the feedback effects of our own outputs. To behave is to control perception. (p. 351)

— William T. Powers (1973)

By *manual control* we mean that a person is receiving through his senses (visual, vestibular, tactile, etc.) information about the ideal states of some variables of a given system, as well as the output states of those variables, separately or in combination. His task is to manipulate mechanical devices — handles or knobs or buttons or control sticks or even pedals — in order to minimize error or some more complex function, whatever is appropriate. (p. 171)

— Thomas B. Sheridan and William R. Ferrell (1974)

REFERENCES

Bellman, R. (1964/1968). Control theory. Reprinted in D. M. Messick (Ed.), *Mathematical thinking in behavioral sciences: Readings from Scientific American* (pp. 74–82). San Francisco: W. H. Freeman.

Powers, W. T. (1973). Feedback: Beyond behaviorism. *Science, 179,* 351–356.

Sheridan, T. B. & Ferrell, W. R. (1974). *Man-machine systems: Information, control, and decision models of human performance.* Cambridge, MA: MIT Press.

Tustin, A. (1952/1968). Feedback. Reprinted in D. M. Messick (Ed.), *Mathematical thinking in behavioral sciences: Readings from Scientific American* (pp. 66–73). San Francisco: W. H. Freeman.

Wiener, N. (1948/1968). Cybernetics. Reprinted in D. M. Messick (Ed.), *Mathematical thinking in behavioral sciences: Readings from Scientific American* (pp. 40–46). San Francisco: W. H. Freeman.

Contents

About the Authors

Richard J. Jagacinski received a bachelor's degree from Princeton University in electrical engineering and a doctoral degree in experimental psychology from the University of Michigan. He is a professor in the Department of Psychology at The Ohio State University, where he has taught since 1973. His research interests include perceptual-motor coordination, decision making in dynamic contexts, human factors, aging, and human interaction with the natural environment.

John M. Flach received his doctoral degree in experimental psychology from The Ohio State University in 1984. From 1984 to 1990, he held joint appointments in the Department of Mechanical and Industrial Engineering, the Institute of Aviation, and the Psychology Department of the University of Illinois. He is currently a professor in the Department of Psychology at Wright State University where he teaches graduate and undergraduate courses. His research is directed at human perceptual/cognitive/motor skills with particular interest in generalizations to the design of human–machine systems.

Preface

This book provides a tutorial introduction to behavioral applications of control theory. It primarily deals with manual control, both as a substantive area of study and as a useful perspective for approaching control theory. It is the experience of the authors that by imagining themselves as part of a manual control system, students are better able to learn numerous concepts in this field. The intended reader is expected to have a good background in algebra and geometry and some slight familiarity with the concepts of an integral and derivative. Some familiarity with statistical regression and correlation would be helpful, but is not necessary. The text should be suitable for advanced undergraduates as well as graduate students in the behavioral sciences, engineering, and design. Topics include varieties of control theory such as classical control, optimal control, fuzzy control, adaptive control, learning control, and perception and decision making in dynamic contexts. We have additionally discussed some of the implications of control theory for how experiments can be conducted in the behavioral sciences. In each of these areas we have provided brief essays that are intended to convey a few key concepts that will enable the reader to more easily pursue additional readings should they find the topic of interest.

Although the overall text was a collaborative effort that would not have been completed otherwise, each of the authors took primary responsibility for different chapters. JF was primarily responsible for chapters 1–10, 13, 14, 17, and 22; RJ was primarily responsible for chapters 11, 12, 15, 16, 18–21, and 23–26; chapter 27 was a joint denouement.

ACKNOWLEDGMENTS

The authors are grateful to Richard W. Pew, Louis Tijerina, Hiroyuki Umemuro, and William Levison for providing comments on substantial portions of the manuscript.

James Townsend, Greg Zacharias, and Fred Voorhorst provided comments on selected chapters. Countless students have also provided comments on early drafts of
some chapters that have been used in seminars at Wright State University. We highly
value their efforts to improve the manuscript, and hold only ourselves responsible
for any shortcomings. We also thank Shane Ruland, Tom Lydon, Kate Charlesworth-
Miller, Donald Tillman, Pieter Jan Stappers, and David Hughley for technical support. The Cornell University Department of Psychology was a gracious host to RJ
during two sabbaticals that facilitated this work. Finally, we thank our colleagues,
friends, and families for graciously enduring our frequent unavailability over the last
half dozen years as we worked on this project, and for their faithful encouragement
of our efforts.

– Richard J. Jagacinski
– John M. Flach

Perception/Action:
A Systems Approach

What is a system? As any poet knows, a system is a way of looking at the world.
— Weinberg (1975)

Today we preach that science is not science unless it is quantitative. We substitute correlation for causal studies, and physical equations for organic reasoning. Measurements and equations are supposed to sharpen thinking but . . . they more often tend to make the thinking non-causal and fuzzy. They tend to become the object of scientific manipulation instead of auxiliary tests of crucial inferences.

Many — perhaps most — of the great issues in science are qualitative, not quantitative, even in physics and chemistry. Equations and measurements are useful when and only when they are related to proof; but proof or disproof comes first and is in fact strongest when it is absolutely convincing without any quantitative measurement.

Or to say it another way, you can catch phenomena in a logical box or in a mathematical box. The logical box is coarse but strong. The mathematical box is fine grained but flimsy. The mathematical box is a beautiful way of wrapping up a problem, but it will not hold the phenomena unless they have been caught in a logical box to begin with.
— Platt (1964; cited in Weinberg, 1975)

As Weinberg noted, a system is not a physical thing that exists in the world independent of an observer. Rather, a system is a scientific construct to help people to understand the world. In particular, the "systems" approach was developed to help address the complex, multivariate problems characteristic of the behavioral and biological sciences. These problems were largely ignored by conventional physical sciences. As Bertalanffy (1968) noted, "Concepts like wholeness, organization, teleology, and directiveness appeared in mechanistic science to be unscientific or metaphysical" (p. 14). Yet, these constructs seem to be characteristic of the behavior of living things. General systems theory was developed in an attempt to frame a quantitative scientific theory addressing these attributes of nature. This quantitative theory is intended as a means (not an end) to a qualitative understanding of nature. This book intends to introduce some quantitative tools of control theory that might be used to build a stronger logical box for capturing the phenomena of human performance.

The term *system* is sometimes used in contrast to *environment*. The system typically refers to the phenomenon of interest (e.g., the banking system or the circulatory system), and the environment typically refers to everything else (e.g., other political and social systems, other aspects of the physiology). Again, as the opening quote from Weinberg indicated, the observer determines the dividing line between system and environment. For example, in studying chess, the system might include both players and the board. Here the phenomenon of interest might be the dynamic coupling of the two players through the board and the rules of the game. In this case, the environment would include all the social, psychological, and physical processes assumed to be peripheral to the game. Alternatively, the system might be a particular player (and the board). From this perspective, the opponent is part of the environment. Thus, the observer is not primarily interested in the motives and strategies of the opponent. The primary goal is to determine how the player of interest responds to different configurations of the board. Finally, the system might be the game of chess. That is, neither opponent is considered in the system description, which might be a listing of the rules of the game or an enumeration of all possible moves. These are not the only choices that might be made.

Another aspect of the "way of looking at the world" is how clean the break is between the system and the environment. For example, Allport (1968), in discussing the systems approach to personality, wrote:

> Why Western thought makes such a razor-sharp distinction between the person and all else is an interesting problem. . . . Shinto philosophy by contrast, regards the individual, society, and nature as forming the tripod of human existence. The individual as such does not stick out like a raw digit. He blends with nature and he blends with society. It is only the merger that can be profitably studied. (p. 347)

Western science, in general, tends to have a preference for a clean, "razor-sharp" break between the system and the environment. So, for example, physicists typically go to great lengths to isolate the systems of interest from environmental influences (e.g., vacuums and huge concrete containments). In psychology, Ebbinghaus' choice of the nonsense syllable was an attempt to isolate human memory from the influence of the environment. As a result of his success, generations of psychologists studied human memory as a system sharply isolated from environmental factors, such as meaning and context. Descriptions of systems where the break between system and environment is sharp (i.e., where the environment is considered irrelevant) are generally referred to as *closed systems*.

The information-processing approach to cognition tends to break up the cognitive system into isolated components that can be studied independently from each other (e.g., sensation, perception, memory, decision making, and motor control). Each process is treated as a distinct "box" that is only loosely coupled to the other components. Laboratory tasks are designed to isolate component processes, and researchers tend to identify their research with one box or another (e.g., "I study perception" or "I study decision making"). Although the links between components are explicitly represented in information-processing models, research programs tend to focus on one component or another, and the other components tend to be treated as part of the environment. Thus, research programs have traditionally been formulated as if the component of interest (e.g., perception or decision making) is an effectively closed system. For example, each chapter in a standard cognitive psychology text is relatively independent from the other chapters. In other words, there are relatively few cross-references from one chapter to another.

However, appreciation for the coupling between system and environment is growing in all fields of science. Psychology is beginning to appreciate the importance of interactions between components within distributed cognitive systems. Also, there is a growing appreciation that the boundaries between human and environment are not so sharp. For example, researchers are beginning to realize that the sharp boundaries between human memory and environment created by the exclusive reliance on nonsense syllables was a very narrow perspective on the phenomenon of remembering. For example, Kintsch (1985) explained:

> What a terrible struggle our field has had to overcome the nonsense syllable. Decades to discover the "meaningfulness" of nonsense syllables, and decades more to finally turn away from the seductions of this chimera. Instead of the simplification that Ebbinghaus had hoped for, the nonsense syllable, for generations of researchers, merely screened the central problems of memory from inspection with the methods that Ebbinghaus had bequeathed us. (p. 461)

In some respects, the growth of general systems theory reflects recognition of the rich coupling that generally exists among natural phenomena. Thus, no matter how a system is defined, there will almost never be a razor-sharp break between system and environment. The flow of matter, energy, and information from environment into systems is considered to be fundamental to the self-organization exhibited by living systems (e.g., biological, cognitive, and social systems). Descriptions of systems where the flow between system and environment are recognized as fundamental to the phenomenon are referred to as *open systems*.

Some open system approaches to human performance are referred to as "ecological." These approaches tend to emphasize the dynamic interaction between humans and environments (e.g., Brunswik, 1955; Gibson, 1966, 1979). The term *ecology* is often used in place of *environment* to emphasize the relational properties that are most important to the coupling of human and environment. Von Uexküll (1957) used the term *umwelt* to refer to the world with respect to the functional abilities of an animal (i.e., the ecology). For example, a person confined to a wheelchair may live in the same environment as a person with normal locomotor abilities, but the functional significance of objects in that environment (e.g., stairs) will be very different. Thus, these people live in different "ecologies" or different "umwelten." The stairway is a passage for one, but an obstacle to the other. Shelves that can be easily reached by one are impossible for the other to reach. A hallway that allows one to turn easily will be too narrow for the other. Radical ecological approaches tend to approach the Shinto philosophy in which humans as distinct systems disappear and the ecology (or umwelt) becomes the system of interest.

The study of perception and action, in particular, seems to be well suited to an open systems or ecological perspective. Action reflects a "hard" (i.e., force) coupling and perception reflects a "soft" (i.e., information) coupling between a human (or human–machine) system and its environment (Kugler & Turvey, 1987). The idea of coupling is not particularly radical. Many images of human performance systems include both forward loops (action) and feedback loops (information) to represent the coupling between human and environment. However, the language, the analytic techniques, and the experimental logic employed to study human performance often reflect a simple stimulus–response logic that greatly underestimates the richness of the coupling. In particular, causality is sometimes viewed as unidirectional (behavior is a response to a stimulus) and the creative aspects of behavior that shape the environment and that seek out stimulation are often ignored. This stimulus–response framework leads to a reductionistic approach to human performance in which perception and action are isolated (both in theories and in laboratories) as distinct stages in a linear sequence. Perception and action tend to be treated as distinct systems (and only marginally open systems).

Figure 1.1 illustrates multiple ways to look at a cognitive system. At the top, the system is treated as a "black box." From this behaviorist perspective, the focus is on relations between stimuli and responses. Alternatively, the black box can be pictured as a sequence of information-processing stages. From this perspective, researchers attempt to describe the transfer functions for each stage. An implicit assumption of this information-processing perspective is that behavior can be understood as a concatenation of the transformations at each stage of processing. That is, the output from one stage is thought to be the input to the next stage of processing. The presence of feedback is often acknowledged as a component of the information-processing system. However, this feedback loop is typically treated as a peripheral aspect of the process and the implications of closing-the-loop are generally not reflected in the experimental logic or the computational models of information processing. One implication of closing-the-loop is that the cause–effect relation that is typically assumed between stimulus (cause) and response (effect) breaks down. In a closed-loop system, the stimulus and response are locked in a circular dynamic in which neither is clearly cause or effect. The stimuli are as much determined by the actions of the observers as the actions are determined by the stimuli. The intimacy of stimuli and response is better illustrated when the boxes are shifted so that action is to the left of perception in the diagram. Note that this change in how the processing loop is pictured does not alter the sequential relations among the stages. With respect to the logic of block diagrams, the third and fourth images in Fig. 1.1 are isomorphic. Within this circular dynamic, the boundaries among the component information-processing stages become blurred and emergent properties of the global dynamic become the focus of interest. This ecological perspective tends to focus on higher order properties of the perception–action dynamic, rather than on the local transfer functions of component stages.

This book introduces some of the quantitative and analytical techniques that have been developed to describe the coupling of perception and action. The language of control theory will help researchers to move past simple stimulus–response descrip-

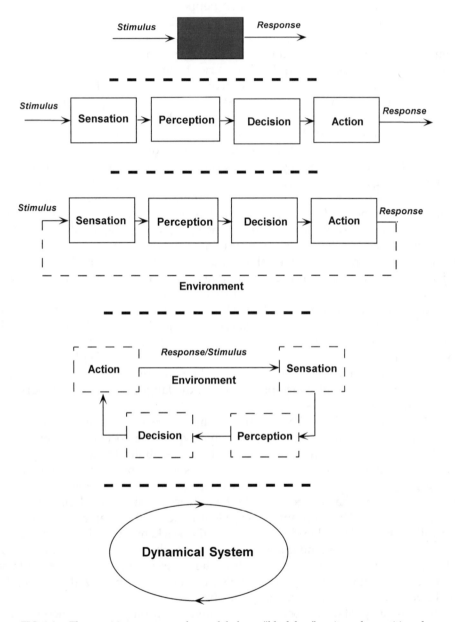

FIG. 1.1. The cognitive system can be modeled as a "black box"; or it can be partitioned into information-processing stages; or it can be viewed as an abstract dynamical system in which the distinctions among stages and between actor and environment become submerged within the overall dynamic.

tions of behavior without falling into the trap of mentalism. Many mental constructs (e.g., percept, memory, schema, etc.) that pepper the field of cognitive psychology might be better understood as emergent properties of complex dynamic systems. These are real phenomena (e.g., perceiving, remembering, knowing) that reflect dynamic interactions between humans and ecologies. The language of control theory will play an important role in the discovery of the psychophysical basis for these emergent properties of the cognitive system. This language will provide a perspective for looking at the world in a way that will enhance our appreciation and understanding of the dynamic coupling of perception and action within natural ecologies.

It is important to keep Weinberg's view of a system as a "way of looking at the world" in mind while reading this book. Control theory can be both a metaphor for human information processing (e.g., the cybernetic hypothesis) and an analytic tool for partitioning and modeling human performance data (e.g., Bode analysis). In this respect, it is similar to other tools that are familiar to behavioral researchers (e.g., analysis of variance and the theory of signal detectability). The utility of the analytic tools does not necessarily depend on embracing a particular associated metaphor or a particular theory of behavior. For example, analysis of variance provides a useful analytic tool, independent of the value of additive factors logic for identifying stages of information processing from reaction time data. Examining performance in terms of relative operating characteristics can be useful, independent of the value of the ideal observer metaphor for signal detection. Also, it is important to appreciate that the analytic tools often provide the means for testing the limits of the metaphors. This book focuses on control theory as a tool for evaluating human performance in the context of dynamic tasks. The analytic tools of control theory provide a valuable perspective on perception and action. Further, the value of these tools greatly exceeds the value of any particular metaphor. That is, the language of control theory spans the multiple perspectives shown in Fig. 1.1.

One area of research that has benefited from the use of a control theoretic perspective is the study of "manual control," which is typically associated with engineering psychology. This research focuses on the human's ability to close-the-loop as the "driver," or "pilot," in control of a vehicle, or as an "operator" managing an industrial process. For example, there has been a great investment to develop models of the human pilot so that designers can better anticipate the stability limits of high performance aircraft. There are several excellent reviews of this area of research (e.g., Frost, 1972; Hess, 1997; Sheridan & Ferrell, 1974; Wickens, 1986). Examples from this literature are used throughout this book. A goal of this book is to provide a tutorial introduction to make the manual control literature more accessible. Many lessons learned in the study of human–machine systems can be generalized to help inform the basic understanding of human performance.

The ultimate goal of this book is not to advocate for a particular theoretical perspective, although the theoretical biases are surely apparent. It does not push the "cybernetic hypothesis." It does not argue that the system is linear or nonlinear. Rather, it introduces behavioral scientists to a quantitative language that has evolved for describing dynamic control systems. In most universities, introductory courses on control theory are only offered in the context of electrical or mechanical engineering. Thus, the courses are taught in a context that makes sense for engineers (e.g., electrical circuits). This makes it very difficult for nonengineers who have an interest in dynamical aspects of behavior to learn this language. This book presents the language of dynamic control systems in a context that is more friendly to nonengineers and perhaps to beginning engineers as well. Although everyone might not share an interest in perceptual-motor skill, it is expected that readers have experiences with control or regulation of movement at some level. These experiences may provide a fertile ground for nurturing the skills and intuitions offered by a control theoretic language.

This book introduces the mathematical box of control theory in the context of human perceptual-motor skill. The goal is to help students to stand on the mathematical box so that they can appreciate the logical box of control theory. Hopefully, this appreciation will foster qualitative insights about human performance and cognition.

REFERENCES

Allport, G. W. (1968). The open system in personality theory. In W. Buckley (Ed.), *Modern systems research for the behavioral scientist* (pp. 343–350). Chicago: Aldine.

Bertalanffy, L. von (1968). General system theory — A critical review. In W. Buckley (Ed.), *Modern systems research for the behavioral scientist* (pp. 11–30). Chicago: Aldine.

Brunswik, E. (1955). *Perception and representative design of psychological experiments* (2nd ed.). Berkeley: University of California Press.

Frost, G. (1972). Man–machine dynamics. In H. P. VanCott & R. G. Kinkade (Eds.), *Human engineering guide to equipment design* (pp. 227–309). Washington, DC: U.S. Government Printing Office.

Gibson, J. J. (1966). *The senses considered as perceptual systems*. Boston: Houghton Mifflin.

Gibson, J. J. (1979). *The ecological approach to visual perception*. Boston: Houghton Mifflin.

Hess, R. A. (1997). Feedback control models: Manual control and tracking. In G. Salvendy (Ed.), *Handbook of human factors and ergonomics* (pp. 1249–1294). New York: Wiley.

Kintsch, W. (1985). Reflections on Ebbinghaus. *Journal of Experimental Psychology: Learning, Memory, and Cognition, 11*, 461–463.

Kugler, P. N., & Turvey, M. T. (1987). *Information, natural law, and the self-assembly of rhythmic movement*. Hillsdale, NJ: Lawrence Erlbaum Associates.

Platt, J. R. (1964). Strong inference. *Science, 146*, 351.

Sheridan, T. B., & Ferrell, W. R. (1974). *Man–machine systems*. Cambridge, MA. MIT Press.

von Uexküll, J. (1957). A stroll through the worlds of animals and man. In C. H. Schiller (Ed.), *Instinctive behavior* (pp. 5–80). New York: International Universities Press.

Weinberg, G. M. (1975). *An introduction to general systems thinking*. New York: Wiley.

Wickens, C. D. (1986). The effects of control dynamics on performance. In K. R. Boff, L. Kaufman, & J. P. Thomas (Eds.), *Handbook of perception and human performance* (Vol. 2, pp. 39.1–39.60). New York: Wiley.

2

Closing the Loop

Give me a dozen healthy infants, well-formed, and my own specified world to bring them up in and I'll guarantee to take any one at random and train him to become any type of specialist I might select – doctor, lawyer, artist, merchant-chief, and, yes, even beggar-man thief, regardless of his talents, penchants, tendencies, abilities, vocations, and race of his ancestors.

—Watson (1925; cited in Gleitman, Fridlund, & Reisberg, 1999, p. 709)

George Miller – together with his colleagues Karl Pribram, a neuroscientist, and Eugene Galanter, a mathematically oriented psychologist – opened the decade with a book that had a tremendous impact on psychology and allied fields – a slim volume entitled Plans and the Structure of Behavior (1960). In it the authors sounded the death knell for standard behaviorism with its discredited reflex arc and, instead, called for a cybernetic approach to behavior in terms of actions, feedback loops, and readjustments of action in the light of feedback. To replace the reflex arc, they proposed a unit of activity called a "TOTE unit" (for "Test-Operate-Test-Exit").

—Gardner (1985)

A very simple system is one whose output is a simple integration of all its input. In such a system, it might be said that the input causes the output. Behaviorism was an attempt to explain human behavior using the logic of such a simple system. This is sometimes referred to as *stimulus–response* (S–R) psychology, because all behaviors are seen as the response to external stimuli. As Watson (1925) suggested, it was believed that if the stimuli were orchestrated appropriately, then the resulting behavior was totally determined. The complexity seen in human behavior was thought to reflect the complexity of stimulation.

In control theory, such a system would be called *open-loop*. It is important not to confuse open-loop with the term *open system* used in chapter 1. Figure 2.1 shows two

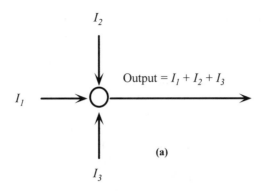

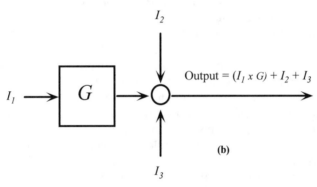

FIG. 2.1.   Two examples of open-loop systems.

schematic examples of open-loop systems. The system in Fig. 2.1(a) is a simple adding machine. The output (i.e., behavior) of this system is the simple addition of all the inputs. The system in Fig. 2.1(b) is a little more interesting. It weights one subset of inputs differently from the others. Suppose G is a multiplicative constant. It might be considered an "attention filter." Those inputs that enter to the left of the attention filter might be magnified so that they make a greater contribution to the output than those inputs that enter to the right; or they may be attenuated so that they have a smaller effect on the output depending, respectively, on whether G is greater than or less than 1.0. Thus, although the behavior of this system is affected by all inputs, the system has some power to selectively enhance some of the inputs. If the weighting given to stimuli entering to the left of the attention filter was great enough, then the contribution from those stimuli entering to the right might be insignificant in terms of predicting the output. To predict the behavior of this system, observers would have to know more than just the inputs. They would also have to know something about the mechanism of attention. For example, if the attention filter magnifies the inputs that it operates on and if observers failed to take into account stimuli on the right of the attention filter, then fairly accurate predictions of behavior would still be possible. However, if observers failed to account for stimuli to the left of the attender, then predictions of output will not be accurate. Likewise, if the attention filter attenuated the inputs that it operated on, then those inputs would have very little impact on the output in comparison to the impact of inputs not operated on. Of course, although only three inputs are shown in the figures, the nervous system could be a vast array of input channels with associated attention mechanisms on each channel to select those channels that will impact the performance output.

Do open-loop systems make good controllers? They do in situations in which there is not a significant disturbance affecting the system output. For example, the vestibular-ocular reflex is a much cited physiological example of open-loop control (e.g., Jordan, 1996; Robinson, 1968). The input is the velocity of a person's head movement, and the output is eye position. The eye position changes to approximately null out the effects of head movement. Therefore, if a person is gazing at a stationary object, then head movements will not significantly disrupt the gaze. If there were forces introduced that would impede the eye movements, then the open-loop controller has no way to compensate or adjust to this disturbance or to correct the "error" in eye position. However, such disturbances are rare, so this mechanism works well. An advantage of this open-loop system is that it works more quickly than a system relying on visual feedback, that is, that waited for an indication that the image of the object was moving across the retina in order to adjust the eye position (e.g., Robinson, 1968).

Consider an example of a situation in which an open-loop control system would not work well, a controller for regulating temperature. To accomplish this feat open loop would require that designers know precisely the amount of heat required to raise the room temperature the specified amount. They would have to take into account changes in efficiency of the heating element over time, changes in the filtering system (e.g., dirty vs. clean air filters), changing amounts of heat loss from the room as a function of insulation and outside temperatures, the number of people in the room, and so on. Thus, running the furnace at a fixed level for a fixed duration would lead to different room temperatures, depending on the impact of these other factors. Unless there were a means for taking all these factors into account, an open-loop controller would not be a satisfactory controller for this application.

A solution to this control problem is to use a closed-loop controller. A closed-loop controller is one that monitors its own behavior (output) and responds not to the input, per se, but to the relation between the reference input (e.g., desired temperature) and the output. This capacity to respond to the relation between reference input and output is called *feedback control*. Figure 2.2 shows a negative feedback control system. The output in this system is fed back and subtracted (hence negative) from the input. The attender then operates on the difference between one set of inputs and the output. This difference is called the *error* and the inputs that are compared to the output are generally referred to as the *reference*, or *command*, *inputs*. Those inputs that enter to the right of the attender are referred to as *disturbances*.

The behavior of this negative feedback controller can be derived from two constraints that must be satisfied. The first constraint is that the error is the difference between the current output and the reference:

$$\text{Error} = \text{Reference} - \text{Output} \qquad (1)$$

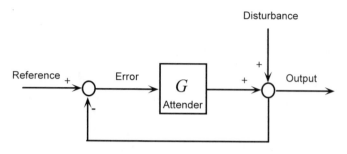

FIG. 2.2.   A simple closed-loop system.

The second constraint is that the output is a joint function of the error and the disturbance:

$$\text{Output} = (\text{Error} \times G) + \text{Disturbance} \tag{2}$$

The joint impact of these two constraints can be seen by substituting the first equation
into the second equation:

$$\text{Output} = [(\text{Reference} - \text{Output}) \times G] + \text{Disturbance} \tag{3}$$

It is convenient to rearrange the terms so that all terms involving the output are on
the same side of the equation:

$$\text{Output} = (\text{Reference} \times G) - (\text{Output} \times G) + \text{Disturbance}$$

$$\text{Output} + (\text{Output} \times G) = (\text{Reference} \times G) + \text{Disturbance}$$

$$\text{Output} (1 + G) = (\text{Reference} \times G) + \text{Disturbance}$$

Further rearranging the terms, the following equation shows the output (OUT) as a
function of the disturbance (DIST) and reference (REF) inputs:

$$\left[\frac{G}{1+G}\right]\text{REF} + \left[\frac{1}{1+G}\right]\text{DIST} = \text{OUT} \tag{4}$$

G represents the transfer function of the attention filter. Note that if G is a simple
magnification factor (i.e., gain), as this gain increases the first term $\left[\frac{G}{1+G}\right]$ approaches
1 and the second term $\left[\frac{1}{1+G}\right]$ approaches zero. The result is that the equation approaches REF = OUT. That is, the output matches the reference signal. Another way
to think about the behavior of the negative feedback system is that the system behaves so as to make the error signal (the difference between the reference and the output) to go nearly to zero. In this arrangement, the disturbance input loses its causal

potency. In other words, the system behaves as it does, in spite of the disturbance input.

More specifically, for the system in Fig. 2.2, if G is large (e.g., 99), then the output will be approximately equal to the reference input $[99/(1 + 99) = .99]$. The effective weighting of the disturbance will be quite small $[1/(1 + 99) = .01]$. In contrast, in Fig. 2.1, if Input 2 were a disturbance, then its weighting would be 1.0. Thus, a closed-loop control system can diminish the effects of disturbances in a more efficient fashion than open-loop control systems. Additionally, if the value of G were to vary by about 10% (e.g., $G = 89$), then that would not strongly affect the output of the closed-loop system. The weighting of the reference input would still be approximately unity $[89/(1 + 89) = .99]$. However, in the open-loop system, the value of G would typically be much smaller to start with, and a 10% variation in G would result in a 10% variation in the component of the output corresponding to Input 1. Thus, variations in the open-loop gain have a larger effect than variations of the closed-loop gain.

The thermostat controls for regulating room temperature are typically designed as closed-loop, negative feedback controllers. These controllers run the heating (or cooling) plants until the room temperature matches the reference temperature specified for the room. These controllers do not need information about changes in the efficiency of the heating element, changes in the filtering system, or changes in the amount of heat loss from the room. Although the well-designed thermostat will consistently meet the temperature goal that has been set, the actions that it takes to meet the same goal will vary depending on the values of the disturbance factors. It might be said that the actions of the temperature regulation system adjust to the specific circumstances created by the confluence of disturbance factors.

Part of the attraction of the cybernetic model of human behavior that Miller, Galanter, and Pribram (1960) provided in the TOTE Unit (Fig. 2.3) is the ability of this negative feedback controller to survive situations with significant disturbances. The environment in which people behave is replete with disturbances or unexpected events to which they need to adjust. Survival often depends on the ability to cope with the unexpected. The capacity to utilize negative feedback, that is, the ability to perceive actions in the context of intentions, is an important skill that humans and other biological organisms have for coping with the environment. Thus, perhaps by understanding the behavior of negative feedback systems, people may be able to better understand their own behavior.

The diagram and the equations in Fig. 2.2 are gross simplifications. An important characteristic of all physical control systems is the presence of time delays. Delays result in constraints on the efficiency of the negative feedback controller. The interaction between the gain of the attention filter (i.e., sensitivity to error) and delay in feedback will determine the stability of the control system. A *stable* control system will converge to a fixed value or narrow region (normally in the vicinity of the reference or goal). An *unstable* system will diverge from the goal (i.e., error will grow over time). Because of the presence of delays, actual negative feedback controllers will always exhibit speed–accuracy trade-offs. If gain is too high (i.e., error is responded to faster than appropriate given time delays), then the response will be unstable, and large error will be the consequence. An important aspect in designing negative feedback controllers is choosing the appropriate balance between speed and accuracy. If

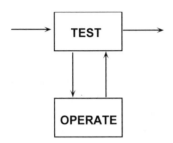

FIG. 2.3. Miller, Galanter, and Pribram's (1960) TOTE Unit (for Test-Operate-Test-Exit). From *Plans and the Structure of Behavior* (p. 26) by G. A. Miller, E. Galanter, & K. Pribram, 1960, New York: Holt, Rinehart, & Winston. Adapted by permission.

the balance between speed and accuracy is not chosen correctly, then either sluggish or oscillatory behavior will result. It is interesting to note that dysfunctional behaviors, like the ataxia described by Wiener (1961), resemble the behavior of a poorly tuned negative feedback controller. Thus, the negative feedback system as a model for human behavior provides some interesting hypotheses for investigating both abilities and disabilities in human performance. According to Wiener, the

> patient comes in. While he sits at rest in his chair, there seems to be nothing wrong with him. However, offer him a cigarette, and he will swing his hand past it in trying to pick it up. This will be followed by an equally futile swing in the other direction, and this by still a third swing back, until his motion becomes nothing but a futile and violent oscillation. Give him a glass of water, and he will empty it in these swings before he is able to bring it to his mouth. . . . He is suffering from what is known as cerebellar tremor or purpose tremor. It seems likely that the cerebellum has some function of proportioning the muscular response to the proprioceptive input, and if this proportioning is disturbed, a tremor may be one of the results. (p. 95)

This proportioning of the muscular response to the proprioceptive feedback may be roughly analogous to the function of the box that has been called the *attention filter*, although the full details of the control mechanism may be far more complex (e.g., Hore & Flament, 1986). One of the central issues here is a consideration of how the properties of the attention filter together with other dynamic constraints on response speed and feedback delay all contribute to determining whether the negative feedback controller is a good, stable controller, or a poor unstable controller.

The proportioning of muscular response can also be critical as human's close-the-loop in many complex sociotechnical systems. A classic example is the pilot of an aircraft. A recent National Research Council (NRC, 1997) report reviewed "aircraft-pilot coupling (APC) events." These are "inadvertent, unwanted aircraft attitude and flight path motions that originate in anomalous interactions between the aircraft and the pilot" (p. 14). One of the most common forms of APC events are pilot-induced, or pilot-involved, oscillations (PIOs) — the NRC prefers the second terminology because it does not ascribe blame to the pilot. The report described a PIO event that was recorded in an early T-38 trainer:

> The initial oscillation was an instability of the SAS-aircraft [stability augmentation system] combination with no pilot involvement. To eliminate the oscillation, the pilot disengaged the pitch SAS and entered the control loop in an attempt to counter the resulting upset. Triggered by these events, at the pilot's intervention a 1.2 Hz (7.4 rad/sec) oscilla-

tion developed very rapidly. In just a cycle or so, the oscillation had achieved an amplitude of ±5 g, increasing gradually to ±8 g, perilously near aircraft design limits. Recovery occurred when the pilot removed himself from the control loop. (p. 16)

The arm tremor and the pilot-involved oscillations are prototypical examples of how a negative feedback control system will behave if the gain is too high relative to the time delays and other dynamic constraints on the system. This book aims to help the reader to understand and appreciate the factors that determine the stability of closed-loop systems.

It is important to note that there can be both open-loop and closed-loop descriptions of the same system, depending on the level of detail used. For example, Fig. 2.2 shows the feedback mechanism that underlies the behavior of this system. However, Equation 4 can be considered an open-loop description of this system. Namely, the error variable is not explicitly represented in this equation, and only the effective relation between the inputs and the output is represented. Similarly, the response of the system in the stretch reflex that a patient's leg exhibits when a doctor taps the knee with a mallet can be described in different ways. As noted by Miller et al. (1960), for many years psychologists described this reflex behavior as a function of a stimulus delivered by the mallet and a subsequent response of the leg. It was regarded as an elementary stimulus–response unit of behavior, which was an open-loop description. However, a more detailed physiological analysis of this behavior reveals feedback loops that act to stabilize the position of the limb in response to various disturbances (e.g., McMahon, 1984). Thus, at a more detailed level of description, the behavior is closed-loop. As noted by Miller et al. (1960), acknowledging and analyzing the roles of feedback control mechanisms is an important step in understanding human behavior. The description that includes the feedback loops is richer than the simple SR models. See Robertson and Powers (1990) for a detailed discussion of why a control theoretic orientation might provide a better foundation for psychology than the more traditional behaviorist (S–R) model.

Finally, it is important to recognize that the servomechanism in Fig. 2.2 is a very simple abstraction of the regulation problem. In most any physical system, there will be dynamic constraints in addition to G in the forward loop. These dynamic constraints are properties of the *controlled system*, or *plant*, as illustrated in Fig. 2.4. For ex-

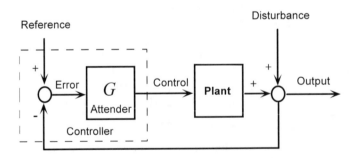

FIG. 2.4. The forward loop normally includes both the control function (G) and the dynamics of the controlled system or plant. The plant typically represents the physical constraints on action (e.g., limb or vehicle dynamics).

ample, delays associated with information processing and inertial dynamics of an arm would be properties of the plant that would need to be controlled in a simple arm movement, and the aerodynamics of the aircraft would be properties of the plant in the pilot–machine system. The dynamics of the plant are not at the discretion of the controller. They constitute part of the control problem that must be solved. A controller must be carefully tuned by its designer to the dynamics of the controlled system (e.g., time delays) and to the temporal structure of the disturbances and references, if it is to function properly. For example, the dynamic properties of aircraft change across altitudes. Thus, an autopilot that is stable at lower altitudes may become unstable at higher altitudes. Control system designers can anticipate the changing dynamic properties and can program adjustments to the autopilot so that it behaves as a different servomechanism (e.g., different gain) depending on its current altitude. Thus, much of the "intelligence" of the autopilot rests with the designer. It is not intrinsic to the servomechanism. For this reason, the servomechanism is probably not the best metaphor for modeling the adaptive, creative aspects of human perceptual motor skill. Thus, Bertalanffy (1968) pointed out:

> The concept of homeostasis also retains a third aspect of the mechanistic view. The organism is essentially considered to be a reactive system. Outside stimuli are answered by proper responses in such a way as to maintain the system. The feedback model is essentially the classical stimulus–response scheme, only the feedback loop being added. However, an overwhelming amount of facts shows that primary organic behavior as, for example, the first movements of the fetus, are not reflex responses to external stimuli, but rather spontaneous mass activities of the whole embryo or larger areas. Reflex reactions answering external stimuli and following a structured path appear to be superimposed upon primitive automatisms, ontogenetically and phylogenetically, as secondary regulatory mechanisms. These considerations win particular importance in the theory of behavior, as we shall see later on. (p. 19)

The discussion begins with simple servomodels not because they are comprehensive metaphors for human perceptual-motor skill, but because they provide the best place to begin learning the language of control theory. Understanding the behavior of simple servomechanisms is a baby step toward understanding the complex, nested coupling of perception and action that gives rise to the emergent skills of human perceptual-motor control.

REFERENCES

Bertalanffy, L. von (1968). General system theory—A critical review. In W. Buckley (Ed.), *Modern systems research for the behavioral scientist* (pp. 11–30). Chicago: Aldine.

Gardner, H. (1985). *The mind's new science: A history of the cognitive revolution.* New York: Basic Books.

Gleitman, H., Fridlund, A. J., & Reisberg, D. (1999). *Psychology* (5th ed.). New York: Norton.

Hore, J., & Flament, D. (1986). Evidence that a disordered servo-like mechanism contributes to tremor in movements during cerebellar dysfunction. *Journal of Neurophysiology, 56,* 123–136.

Jordan, M. I. (1996). Computational aspects of motor control and motor learning. In H. Heur & S. W. Keele (Eds.), *Handbook of motor control* (pp. 71–120). San Diego, CA: Academic Press.

McMahon, T. A. (1984). *Muscles, reflexes, and locomotion.* Princeton, NJ: Princeton University Press.

Miller, G. A., Galanter, E., & Pribram, K. (1960). *Plans and the structure of behavior*. New York: Holt, Rinehart & Winston.

National Research Council (1997). *Aviation safety and pilot control: Understanding and preventing unfavorable pilot–vehicle interactions*. Washington, DC: National Academy Press.

Robertson, R. J., & Powers, W. T. (1990). *Introduction to modern psychology: The control-theory view*. Gravel Switch, KY: The Control Systems Group, Inc.

Robinson, D. A. (1968). The oculomotor control system: A review. *Proceedings of the IEEE, 56*, 1032–1049.

Watson, J. B. (1925). *Behaviorism*. New York: Norton.

Wiener, N. (1961). *Cybernetics, or control and communication in the animal and the machine* (2nd ed.). Cambridge, MA: MIT Press. (Original work published 1948)

Information Theory and Fitts' Law

The information capacity of the motor system is specified by its ability to produce consistently one class of movement from among several alternative movement classes. The greater the number of alternative classes, the greater is the information capacity of a particular type of response. Since measurable aspects of motor responses such as their force, direction, and amplitude, are continuous variables, their information capacity is limited only by the amount of statistical variability, or noise, that is characteristic of repeated efforts to produce the same response. The information capacity of the motor system, therefore, can be inferred from measures of the variability of successive responses that S attempts to make uniform.

—Fitts (1954)

In the decades since Fitts' original publication, his relationship, or "law," has proven one of the most robust, highly cited, and widely adopted models to emerge from experimental psychology. Psychomotor studies in diverse settings—from under a microscope to under water—have consistently shown high correlations between Fitts' index of difficulty and the time to complete a movement task.

—MacKenzie (1992)

Wiener's (1961) book on cybernetics was subtitled "control and communication in the animal and the machine." Whereas the focus of this book is on control, this particular chapter focuses on communication. In particular, this chapter explores the metaphor that the human information-processing system is a kind of communication channel that can be described using information theory. It begins with this metaphor because it has dominated research in psychology over the last few decades. Associated with the communication metaphor is the research strategy of chronometric analysis, in which reaction time is used as an index of the complexity of underlying mental processes (e.g., Posner, 1978).

The next few chapters focus on the simple task of moving an arm to a fixed target—for example, reaching to pick up a paper clip. This simple act has been described using both the language of information theory and the language of control theory. This provides a good context for comparing and contrasting different ways that information theory and control theory can be used to model human performance. An introduction to information theory, however, starts with another simple task, *choice reaction time*.

A choice reaction time task is a task where several possible signals are each assigned a different response. For example, the possible stimuli could be the numbers 1 to 5 and the possible responses could be pressing a key under each of the 5 fingers of one hand. Thus, if the number 1 is presented, press the key under the thumb, whereas if the number 3 is presented, press the key under the middle finger. Merkel (1885, cited in Keele, 1973) found that as the number of equally likely possible alternatives increased from 1 (i.e., 1 number and 1 key; this is called simple reaction time) to 10 (i.e., 10 numbers and 10 keys), the reaction time increased. The relation between reaction time (RT) and number of alternatives found by Merkel showed that, for every doubling of the number of alternatives, there was a roughly constant increase in RT. These results are plotted in Fig. 3.1. Figure 3.1(a) shows reaction time as a function of the number of alternatives and Fig. 3.1(b) shows reaction time as a function of the base two logarithm of the number of alternatives. [The base two logarithm of a number is the number expressed in terms of powers of two. For example, the $\log_2 (1) = 0$, because $2^0 = 1$; the $\log_2 (2) = 1$; the $\log_2 (4) = 2$; the $\log_2 (8) = 3$; and the $\log_2 (64) = 6$, because $2^6 = 64$.]

The base two logarithm of the number of alternatives is a measure of information called a *bit* (see Shannon & Weaver, 1949/1963). The number of bits corresponds to the average number of yes–no questions that would be required to identify the correct stimulus if all alternatives are equally likely. For example, if the number of alternatives is 2 (i.e., the numbers 1 or 2), then a single question is sufficient to identify the correct alternative (Is the number 1?). If the number of alternatives is 4 (i.e., the numbers 1, 2, 3, and 4), then two questions are required (i.e., Is the number greater than 2? Yes! Is the number 3?). Figure 3.1b shows that over the range from about 1.5 to 3 bits there is a linear increase in response time with increase in information as measured in bits. This suggests that, at least over a limited range, the rate [slope of the line in Fig. 3.1(b)] at which the human can transmit information is constant. From these data, the slope can be estimated to be on the order of 140 milliseconds per bit (Keele, 1973).

The information statistic does more than simply reflect the number of alternatives; it is influenced by the probability of alternatives. The information statistic calculated for Merkel's data reflected the fact that each alternative was equally likely on every trial. If one alternative is more probable than another, then the information measure will be less. For example, the flip of a fair coin provides more information than the flip of a biased coin. The following equations show the information for the flip of a fair coin and for a biased coin that shows heads on 8 out of 10 tosses. The biased coin is more predictable. Thus, less information is communicated in the flip of the biased coin:

$$H(x) = \sum_i p(x_i) \log_2 \frac{1}{p(x_i)} \tag{1}$$

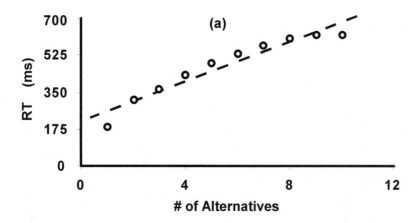

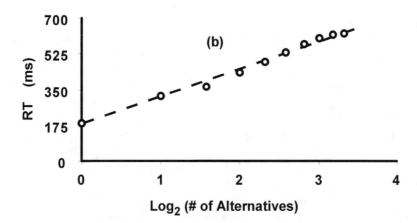

FIG. 3.1. Choice reaction time in a key pressing task as a function of (a) the number of alternatives, and (b) the base two logarithm of the number of alternatives (bits). Data from Merkel (1885; cited in Keele, 1973).

where $H(x)$ is the information in x, which has several alternatives indexed by i, and $p(x_i)$ is the probability of a particular alternative x_i.

$$H(fair\ coin) = p(heads) \times \log_2 \frac{1}{p(heads)} + p(tails) \times \log_2 \frac{1}{p(tails)}$$

$$= .5 \times \log_2 \frac{1}{.5} + .5 \times \log_2 \frac{1}{.5}$$

$$= \log_2 2 = 1$$

$$H(biased\ coin) = .2 \times \log_2 \frac{1}{.2} + .8 \times \log_2 \frac{1}{.8}$$
$$= .2 \times \log_2 5 + .8 \times \log_2 1.25$$
$$= .72 \tag{2}$$

Hyman (1953) changed the information-processing requirements in a choice reaction time task by manipulating stimulus probabilities. He found that the average reaction time, as in Merkel's experiment, was a linearly increasing function of the amount of information. Thus, reaction time depends on what could have occurred, not what actually did occur. An identical stimulus (e.g., a particular light) and an identical response (e.g., a particular key) will take shorter or longer depending on the context of other possibilities (e.g., the number and probabilities of other events).

Accuracy was stressed in the Merkel and Hyman experiments. Suppose people are asked to go faster? As people go faster, the number of errors increases. Thus, there is a speed–accuracy trade-off. If a person performs perfectly, then the information transmitted is equal to the information in the stimulus (e.g., as a function of number of alternatives and probability of alternatives). However, if a person responds randomly, then no information is transmitted. That is, an observer who knows only the subject's response will be no better at predicting what the stimulus was than an observer who does not know the subject's response. By looking at the correspondence between the subject's responses and the stimuli, it is possible to calculate the information transmitted. If there is a perfect one-to-one correspondence (i.e., correlation of 1) between responses and stimuli, then the information transmitted is equal to the information in the stimulation. If there is a zero correlation between stimuli and responses, then no information is transmitted. If the correlation is less than one but greater than zero, then the information will be less than the information in the stimulation, but greater than zero. Hick (1952) demonstrated that when people were motivated to emphasize speed, the information transmission rate remained constant. That is, the increase in errors with increased speed of responding could be accounted for by assuming that the information transmission rate for the human was constant.

Woodworth (1899) was one of the first researchers to demonstrate a similar speed–accuracy trade-off for continuous movements. He showed that for continuous movements, variable error increased with both the distance (amplitude) and the speed of the movement. Fitts (1954) was able to link the speed–accuracy trade-off, observed for continuous movements, with the speed–accuracy trade-off in choice reaction time tasks by using the information statistic.

To understand the way this link was made, the information statistic must again be considered. The information transmitted by an event (e.g., a response) is a measure of the reduction in uncertainty. If there are eight alternative numbers, each equally likely, then the uncertainty is the base two log of eight, which is equal to three. If a person is told the number, now the uncertainty is the base two log of one, which is equal to zero. Thus, the naming of the number reduces uncertainty from three bits to zero bits. Three bits of information have been transmitted. If a person is told that the number is greater than four, then uncertainty has been reduced from three bits [$\log_2$ (8)] to two bits [$\log_2$ (4)]. Thus, one bit (3-2) of information has been transmitted. This can be expressed in the following formula:

$$-\log_2\left(\frac{\#\ of\ possibilities\ after\ event}{\#\ of\ possibilities\ prior\ to\ event}\right) = \frac{Information}{Transmitted} \qquad (3)$$

For the aforementioned examples, this would give:

$$-\log_2\left(\frac{1}{8}\right) = \log_2(8) = 3\ \ bits$$

$$-\log_2\left(\frac{4}{8}\right) = \log_2(2) = 1\ \ bit$$

Now consider the problem of movement control. Fitts (1954) examined the movement time in three tasks. One task was a reciprocal tapping task in which the subject moved a stylus from one contact plate to another. Fitts varied the width of the contact plate (W) and the distance between contact plates (A) (Fig. 3.2). Subjects were instructed to move as fast as possible ("score as many hits as you can"), but were told to "emphasize accuracy rather than speed." The second task was a disk transfer task in which subjects had to move washers, one at a time, from one post to another. In this task, the size of the hole in the disk and the distance between posts was varied. The third task was a pin transfer task. Subjects moved pins from one set of holes to a second set of holes. The diameter of the pins and the distance between holes were varied. The data was the average time to accomplish the movement (i.e., from one contact to the other, from one post to the other, or from one hole to the other).

How can these tasks be characterized in terms of their information transfer requirements? The uncertainty after the movement is the width of the target. That is, if the subject responds accurately, then the position of the stylus will be inside the width of the contact plate. Uncertainty about where the washer will be depends on the difference between the size of the hole in the washer and the size of the post. If it is a tight fit, then there will be very little uncertainty about washer position; if it is a loose fit, then there will be more uncertainty. Similar logic can be applied to the pin transfer task. Thus, the uncertainty after a successful movement can be specified as the target width (W) or the tolerance in fit.

How can the uncertainty prior to the movement be specified? This is not so obvious. What is the number of possible positions the stylus could end up in? As Fitts (1954) noted, "This . . . is arbitrary since the range of possible amplitudes must be inferred" (p. 388). For both practical and logical reasons, Fitts chose twice the ampli-

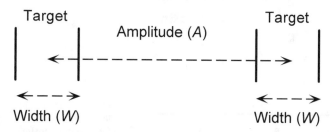

FIG. 3.2.   In the reciprocal tapping task, Fitts varied the distance or amplitude (A) between two target contact plates and the width (W) of the target area.

tude (2A) as the uncertainty prior to the movement. He explained that "the use of 2A makes the index correspond rationally to the number of successive fractionations required to specify the tolerance range out of a total range extending from the point of initiation of a movement to a point equidistant on the opposite side" (p. 388).

Using this logic, Fitts (1954) defined the binary index of difficulty for a movement as:

$$-\log_2\left(\frac{W}{2A}\right) = \log_2\left(\frac{2A}{W}\right) = ID(bits) \tag{4}$$

Figure 3.3 shows the movement times that were found for each task plotted as a function of *ID*. Again, as with the Merkel data, there is an approximately linear increase in movement time with increases in the amount of information transferred as measured using the index of difficulty (*ID*). Thus, Fitts' Law is as follows:

$$Movement\ Time = a + b\log_2\left(\frac{2A}{W}\right) \tag{5}$$

where *a* and *b* are empirically derived constants. The inverse of *b* is an index of the information-processing rate of the human (bits/s). Fitts' data suggested that the information-processing rate was approximately 10 bits/s.

A number of alternatives to Fitts' model have been proposed. For example, Welford (1960, 1968) suggested making movement time "dependent on a kind of Weber fraction in that the subject is called upon to distinguish between the distances to the far and the near edges of the target. To put it another way, he is called upon to

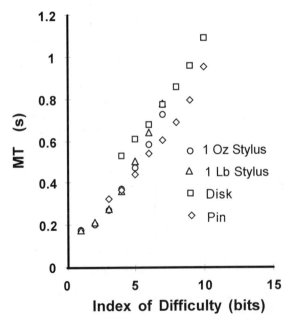

FIG. 3.3. Movement times for three tasks (reciprocal tapping, disk transfer, and pin transfer) from Fitts (1954) plotted as a function of Index of Difficulty.

choose a distance W out of a total distance extending from his starting point to the far edge of the target" (p. 147). Based on this logic, Welford proposed the following alternative to Fitts' Law:

$$MT = K \log\left[\frac{A + (\frac{1}{2})W}{W}\right] = K \log\left(\frac{A}{W} + .5\right) \tag{6}$$

MacKenzie (1989, 1992) showed how a model can be more closely tied to the logic of information theory. Whereas Fitts' model was derived through analogy to information theory, MacKenzie's model is more faithful to Shannon's Theorem 17, which gives the information capacity for a communication channel (C) as a function of signal power (S) and noise power (N):

$$C = B \log_2\left(\frac{S + N}{N}\right) \tag{7}$$

where B is the bandwidth of the channel. An oversimplified, but intuitively appealing, way to understand this theorem follows. Before a signal is sent, the number of possibilities includes all possible signals (S) plus all possible noises (N). Thus, $S + N$ represents a measure of the possibilities before a signal is sent. Once a signal is sent, the remaining variation is only that resulting from the noise (N). Thus, N represents a measure of the possibilities after a particular signal s has been sent over the channel with noise N. So ($S + N$) represents the set of possibilities before a particular signal s has been sent and N represents the set of possibilities after signal s has been sent. N is variation around the specific signal s.

MacKenzie's model that uses the logic of Theorem 17 is:

$$MT = a + b \log_2\left(\frac{A + W}{W}\right) \tag{8}$$

Figure 3.4 shows a comparison of the Fitts and MacKenzie models. Note that the two functions are more or less parallel for small widths. The result of the parallel functions is that there is very little practical difference between these two models in terms of the ability to predict movement times (e.g., Fitts' original equation yields an r of .9831 for the tapping task with the 1 oz stylus, whereas MacKenzie's equation yields an r of .9936). The differences are greatest when W is large relative to A. This is primarily at the lower indexes of difficulty. An important theoretical limitation of both of these models concerns the relative effects of target amplitude and width on movement time. Both models predict that the contributions of A and W to movement time should be equivalent, but in opposite directions. Thus, doubling the amplitude should be equivalent to halving the width. However, Sheridan (1979) showed that reductions in target width cause a disproportionate increase in movement time relative to similar increases in target amplitude.

Crossman (1957; also see Welford, 1968) suggested a further modification to Fitts' original equation that derives from "the very heart of the information-theoretic meta-

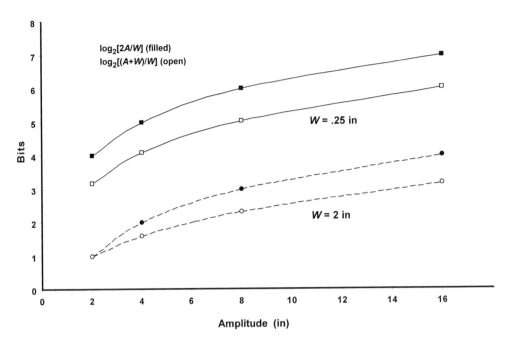

FIG. 3.4.   A comparison of Fitts' and MacKenzie's equations for computing index of difficulty for the amplitudes and two of the widths used by Fitts (1954).

phor" (MacKenzie, 1992, p. 106). This modification is useful because it permits analytically considering the impact of error in the model of movement time. In Fitts'
model and in many other models of movement time, error rates are controlled experimentally. That is, the subjects are instructed to keep errors at a minimum. Generally,
statistical tests are used to examine whether the error rate varies as a function of the
experimental conditions (i.e., combinations of amplitude and width). If there is no
statistically significant differences in error rates, then error is ignored and further
analyses concentrate on movement time. Problems arise, as MacKenzie pointed out,
when attempts are made to compare results across movement conditions and experiments that have different error rates.

The adjustment comes from the assumption that the signal is "perturbed by white
thermal noise" (Shannon & Weaver, 1949/1963, p. 100; cited in MacKenzie, 1992, p.
106). For the motor task, this assumption requires that the endpoints of the movements be normally distributed about the target. The uncertainty in a normal distribu-
tion is $\log_2(\sqrt{2\pi e}\sigma)$, where σ is the standard deviation. This simplifies to $\log_2(4.133\sigma)$.
If this assumption is accepted, then the noise term in the movement time model
$[\log_2(W)]$ should equal $\log_2(4.133\sigma)$. In simpler terms, this means that the endpoints
of the movements should be normally distributed with mean at the target center and
with 4.133 standard deviations (or ±2.066 standard deviations) of the distribution
within the target boundaries. This means that 96% of the movements will hit the target and 4% of the movements will terminate outside the target boundaries. If the experimental error is not equal to 4%, then the W defined by the experimental task is
not an appropriate index of the channel noise. If the error rate is less than 4%, then W

is an overestimation of channel noise. If the error rate is greater than 4%, then W underestimates the channel noise.

If the error rate is known, then the channel noise or effective target width, W_e, can be computed using a table of z scores:

$$W_e = \frac{2.066}{z} \times W \qquad (9)$$

where z is the z score, such that $\pm z$ contains the percentage of the area under the unit-normal curve equivalent to the obtained hit rate (100 − error %). For example, if the experimental width is 1 and the obtained error rate is 1%, then:

$$W_e = \frac{2.066}{2.58} \times 1 = .80$$

In this case, the effective target width is smaller than the experimentally defined target width. In other words, the subject was conservative (generated less noise than allowed by the experimental conditions). If the experimental width is 1 and the obtained error rate is 10%, then:

$$W_e = \frac{2.066}{1.65} \times 1 = 1.25$$

In this case, the effective target width is larger than the experimentally defined target width. This subject generated greater noise than allowed by the experimental conditions so the effective target width is greater than the experimentally specified target width.

Thus, MacKenzie's complete model for movement time is:

$$MT = a + b \log_2\left(\frac{A + W_e}{W_e}\right) \qquad (10)$$

This model uses the logic of information theory to account for both movement time and error rates.

Whether or not one accepts Fitts' or MacKenzie's model, it is clear that the speed–accuracy trade-off seen for continuous movements is qualitatively very similar to that seen for choice reaction times. This suggests that using the information statistic taps into an invariant property of the human controller. How can this invariant property be described? One tempting metaphor that motivated the use of information statistics is that of a limited capacity (bandwidth) communication channel. However, the next chapter explains that movement times can also be modeled using a control system metaphor. A control theory analysis is able to model movement time with the same accuracy as information theory. However, control theory allows predictions about the space–time properties of movements that cannot be addressed using information statistics.

Whereas this chapter has considered discrete movements to stationary targets, the general approach of characterizing information transmission can also be applied to continuous tracking of moving targets (e.g., Elkind & Sprague, 1961). Information theoretic differences in the movement capabilities of different joints can be quantified in continuous tracking tasks (e.g., Mesplay & Childress, 1988) and they complement similar descriptions of discrete movement capabilities (e.g., Langolf, Chaffin, & Foulke, 1976).

REFERENCES

Crossman, E.R.F.W. (1957). The speed and accuracy of simple hand movements. In E.R.F.W. Crossman & W. D. Seymour, *The nature and acquisition of industrial skills.* Report to M.R.C. and D.S.I.R. Joint Committee on Individual Efficiency in Industry.

Elkind, J. I., & Sprague, I. T. (1961). *IRE Transactions on Human Factors in Electronics, HEF-2,* 58–60.

Fitts, P. M. (1954). The information capacity of the human motor system in controlling the amplitude of movement. *Journal of Experimental Psychology, 47,* 381–391.

Hick, W. E. (1952). On the rate of gain of information. *Quarterly Journal of Experimental Psychology, 4,* 11–26.

Hyman, R. (1953). Stimulus information as a determinant of reaction time. *Journal of Experimental Psychology, 45,* 188–196.

Keele, S. W. (1973). *Attention and human performance.* Pacific Palisades, CA: Goodyear.

Langolf, G. D., Chaffin, D. B., & Foulke, J. A. (1976). An investigation of Fitts' Law using a wide range of movement amplitudes. *Journal of Motor Behavior, 8,* 113–128.

MacKenzie, I. S. (1989). A note on the information theoretic basis for Fitts' Law. *Journal of Motor Behavior, 21,* 323–330.

MacKenzie, I. S. (1992). Fitts' Law as a research and design tool in human–computer interaction. *Human Computer Interaction, 7,* 91–139.

Merkle, J. (1885). Die zeitlichen Verhaltnisse der Willensthatigheit. *Philosophische Studiën, 2,* 73–127. Cited in S. W. Keele (1973). *Attention and human performance.* Pacific Palisades, CA: Goodyear.

Mesplay, K. P., & Childress, D. S. (1988). Capacity of the human operator to move joints as control inputs to protheses. In *Modeling and control issues in biomedical systems* (DSC Vol. 12/ BED Vol. 11, pp. 17–25). ASME Winter Annual Meeting, Chicago, IL.

Posner, M. I. (1978). *Chronometric explanations of mind.* Hillsdale, NJ: Lawrence Erlbaum Associates.

Shannon, C., & Weaver, W. (1949/1963). *The mathematical theory of communication.* Urbana: University of Illinois Press.

Sheridan, M. R. (1979). A reappraisal of Fitts' Law. *Journal of Motor Behavior, 11,* 179–188.

Welford, A. T. (1960). The measurement of sensory-motor performance: Survey and appraisal of twelve years' progress. *Ergonomics, 3,* 189–230.

Welford, A. T. (1968). *Fundamentals of skill.* London: Methuen.

Woodworth, R. S. (1899). The accuracy of voluntary movement. *Psychological Review, 3*(3), 1–114.

Wiener, N. (1961). *Cybernetics, or control and communication in the animal and the machine* (2nd ed.). Cambridge, MA: MIT Press. (Original)

4

The Step Response: First-Order Lag

Strictly speaking, we cannot study a man's motor system at the behavioral level in isolation from its associated sensory mechanisms.

—Fitts (1954)

The term dynamic refers to phenomena that produce time-changing patterns, the characteristic pattern at one time being interrelated with those at other times. The term is nearly synonymous with time-evolution or pattern of change. It refers to the unfolding of events in a continuing evolutionary process.

—Luenberger (1979)

Movement times and error rates provide important information about movement, but certainly not a complete picture. Information theory has been an important tool for psychology, but it has also constrained thinking such that response time and accuracy have become a kind of procrustean bed where questions are almost exclusively framed in terms of these dependent variables. There seems to be an implicit attitude that if response duration and accuracy have been accounted for, then there is a complete and satisfactory basis for understanding behavior. In reality, response time and accuracy provide a very impoverished description of the movement involved in target acquisition. This section introduces some tools for building more detailed descriptions of the time evolution of the movement. First, consider a task used by Fitts and Peterson (1964). In this task, the subject holds a stylus on a home position until a target appears. As soon as a target appears, the subject is to move the stylus from the home position to the target position as quickly and accurately as possible. Movement time in this task turns out, as in the reciprocal tapping task, to be an approximately linear, increasing function of the index of difficulty.

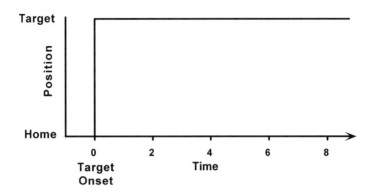

FIG. 4.1. A unit step input.

The home position and target can be described as a reference signal to the controller. Figure 4.1 shows what this would look like if this reference signal was plotted as a function of space and time. The absence of a target is a signal to stay in the home position. The onset of the target is an instantaneous change of reference from the home position to the target position.

Figure 4.2 shows a block diagram for a control system. This system is called a *first-order lag*. The symbol with the triangle shape represents integration. Whereas the output in Fig. 2.2 was proportional to error, the output for the first-order lag is proportional to the integral of error; or (because differentiation is the inverse of integration) error is proportional to the derivative of the output. To say it another way, the rate of change (i.e., velocity) of the output of the first-order lag is proportional to error. If error is large, then this system will respond with high velocity; as error is reduced, the rate of change of the output will be reduced. When the error goes to zero, the velocity of the output will go to zero. The response for a first-order lag to a step input as in the Fitts and Peterson (1964) experiment is shown in Fig. 4.3. Note that when the rate of change (i.e., slope) is high the output is far from the target (i.e., large error), but the slope becomes less steep as the output approaches the target (i.e., error diminishes). As the output reaches the target, the slope goes to zero. Thus, the response comes to rest (i.e., zero rate of change or zero velocity) at the target. Mathematically, it could be said that the function asymptotes at the target value.

It is also important to note that the first-order lag responds immediately to the step response. The lag is reflected in the fact that it takes time for the system to achieve a *steady state* output. Steady state refers to the asymptotic portion of the response curve, where the output is approximately constant (approximately zero rate of

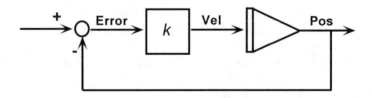

FIG. 4.2. A block diagram showing a first-order lag with time constant of $1/k$.

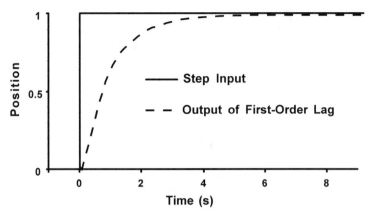

FIG. 4.3.   The time domain response of a first-order lag with $k = 1.0$ to a step input.

change). Thus, whereas the step input instantaneously shifts from one steady state to another, the lagged output reflects a more gradual movement from one steady state to another. This gradual response to a sudden input is common in many physical systems. For example, turning a knob on an electric stove is comparable to a step input. The temperature change at the burner is a lagged response to this step input. That is, the temperature gradually increases to reach a steady state level corresponding to the level specified by the control setting. In common language, *lag* and *time delay* are often used synonymously. However, in control theory, these terms refer to distinctive patterns of response. Lag refers to the gradual approach to a steady state following a step input. As discussed later, the lag involves the loss of high frequency information. Time delay refers to a gap between the time of the input and the initiation of a response to that input. Some have made the analogy to the delay between the read and write heads of a tape recorder. A time delay does not involve the loss of any information. Most biological control systems will have characteristics of both lags and time delays in their response to external stimuli.

Could the first-order lag be a model for human movement? Crossman and Goodeve (1983) were among the first to consider control theoretic descriptions as an alternative to Fitts' information theoretic model of human movements:

> If limb position is adjusted by making limb velocity proportional to error, the limb's response to a step function command input, i.e. a sudden change of desired position, will be an exponential approach to the new target with time constant dependent on loop gain. If the motion is terminated at the edge of a tolerance band, this predicts a total time proportional to the logarithm of motion amplitude divided by width of tolerance band. (p. 256)

The response of the first-order lag to a step input of amplitude A can be shown to be:

$$Output = A - Ae^{-kt} \tag{1}$$

Using the previous equation, it is possible to analytically solve for the time that it would take the first-order lag to cross into the target region (i.e., come within one half of a target width, W, of the step input with amplitude, A):

$$A - \frac{1}{2}W = A - Ae^{-kt} \tag{2}$$

$$\frac{1}{2}W = Ae^{-kt}$$

$$\frac{W}{2A} = e^{-kt}$$

$$\ln\left(\frac{W}{2A}\right) = -kt$$

$$\frac{1}{k}\ln\left(\frac{2A}{W}\right) = t \tag{3}$$

Using the following property of logarithms:

$$\log_a x = \frac{\ln x}{\ln a} \tag{4}$$

The result is:

$$\frac{\ln 2}{k}\log_2\left(\frac{2A}{W}\right) = t \tag{5}$$

This relation is very similar to what Fitts found — a logarithmic relation in which movement time is directly proportional to amplitude and inversely proportional to the target width.

The gain factor k determines the speed at which the target is acquired. For large k movement time will be short. For small k movement time will be longer. The term "$1/k$" is called the time constant for the first-order lag. In $1/k$ units of time, the output of the first-order lag will reach 63% of its steady state (asymptotic) value in response to a step input. Figure 4.4 shows how the response of the first-order lag varies as a function of k.

From a psychological standpoint, the gain of the forward loop (k) is an index of the sensitivity of the system to error. It determines the proportionality between the perceived error (perception) and the speed of the response to that error (action). If k is large, then the system is very sensitive to error. Thus, errors are reduced very quickly. If k is small, then the system is relatively insensitive to error. Thus, the system is sluggish in responding to the error.

When a communication channel was used as a metaphor through which to understand human movement, the critical parameter was the channel bandwidth or information-processing rate in bits per second. If a control system is the metaphor used, then the time constant becomes the key parameter. In the Fitts and Peterson (1964) experiment, the information-processing rate was approximately 13.5 bits/s. Compa-

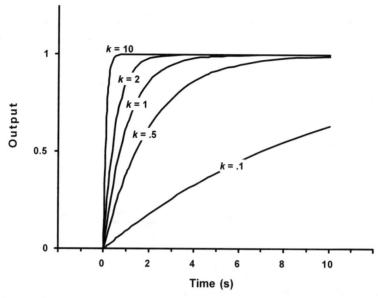

FIG. 4.4. The response of a first-order lag to a step input at various values of forward-loop gain (k) or time constant ($1/k$). Note that as gain increases, the time to approach the asymptotic steady state decreases. The output achieves 63% of steady state in one time constant.

rable performance would be expected for a first-order lag with a time constant ($1/k$) equal to .107 s. This is derived in the following equations, where IP refers to the information-processing rate, which is the inverse of the slope parameter in Fitts' model (see chap. 3):

$$\frac{\ln(2)}{k} = \frac{1}{IP} \quad \text{Note that } IP = \frac{1}{b}$$

$$\frac{.693}{k} = \frac{1}{13.5}$$

$$\frac{1}{k} = .107 \tag{6}$$

Figure 4.5 shows the time history of the response of a first-order lag for one of the amplitudes (3 in) used in the Fitts and Peterson (1964) experiment. The dotted lines show the edges of the targets for the four widths used in the experiment. The predicted movement times for these four different experimental conditions can be read from the graph as the time coordinates for the intersections of the target edges and the movement time history. The function that results will be consistent with Fitts' Law. With respect to movement time, both the limited capacity communication channel and the first-order lag metaphors are equivalent in providing models for the data from Fitts' experiments. However, the first-order lag is a stronger model, because there are more ways to falsify it. That is, in addition to predicting movement time, the first-order lag predicts the time history of the movement. Thus, this model can be re-

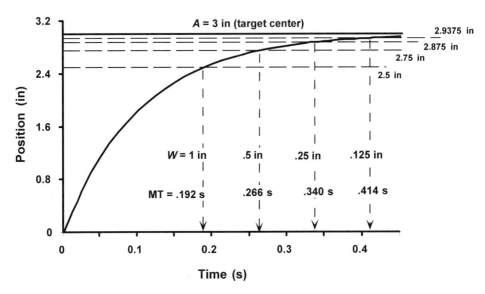

FIG. 4.5. The time history for a first-order lag with time constant .107 s for a movement of 3 in to targets of 1, .5, .25, .125 in width. These parameters were used in an experiment conducted by Fitts and Peterson (1964).

jected if the time histories for human movements differ in significant ways from those that would be generated by the first-order lag. The information-processing channel metaphor does not make any predictions about time histories. Chapters 6 and 7 take a closer look at movement time histories to see whether the predictions of a first-order lag are consistent with human performance.

It could be argued that the first-order lag does not directly account for the variability in the distribution of movement endpoints within the target. However, a stochastic element could be added to the model, such that the "effective step input" is probabilistically related to the actual experimental constraints (the amplitude and target width). However, there is no basis within information theory to account for the movement trajectories.

REFERENCES

Crossman, E.R.F.W., & Goodeve, P. J. (1983). Feedback control of hand-movement and Fitts' Law. *Quarterly Journal of Experimental Psychology, 35A*, 251–278. (Original work presented at the meeting of the Experimental Psychology Society, Oxford, England, July 1963.)

Fitts, P. M. (1954). The information capacity of the human motor system in controlling the amplitude of movement. *Journal of Experimental Psychology, 47*, 381–391.

Fitts, P. M., & Peterson, J. R. (1964). Information capacity of discrete motor responses. *Journal of Experimental Psychology, 67*, 103–112.

Luenberger, D. G. (1979). *Introduction to dynamic systems.* New York: Wiley.

5

Linear Systems: Block Diagrams and Laplace Transforms

> *There are many real-world problems which defy solution through formal mathematical analysis if one demands perfection in terms of exact models. The best model may be one which is a gross approximation if viewed in terms of total system fidelity but which is easily manipulated, easily understood, and offers insight into selected aspects of system structure. The prime demand we will place upon a mathematical technique is that it be useful in contributing to our understanding of system behavior.*
>
> *—Giffin (1975)*

Chapter 4 presented the response of a first-order lag as:

$$Output = A - Ae^{-kt} \qquad (1)$$

How was this determined? What is the relation between the block diagram shown in Fig. 5.1 and this equation?

This chapter considers some of the properties of linear systems. These properties, which are simplifying assumptions with regard to real-world systems and especially with respect to human performance, are nonetheless very useful assumptions in many instances. The assumption of linearity will greatly facilitate the ability to manipulate block diagrams and differential equations. Manipulating block diagrams and differential equations will, in turn, lead to greater understanding and insight. Assumptions are dangerous only when researchers forget that they have been made and then mistake the constraints due to these assumptions as invariant properties of the phenomenon they hope to understand.

The block diagram is an important tool for visualizing the elements and the relations among elements within dynamic systems. Differential equations are critical analytic tools for deriving the response of dynamic systems. Laplace transforms are

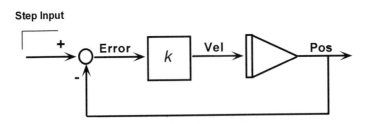

FIG. 5.1. A block diagram showing a first-order lag with time constant of $1/k$.

tools that make it easier to manipulate and solve linear differential equations. These tools are important for any scientists who want to describe and predict the behavior of dynamic systems.

LINEARITY

A system is linear if a particular input, I_1, to the system produces the output, O_1, and a different input, I_2, results in the output, O_2, and a third input that is a weighted sum of the first two inputs $[(k_1 \times I_1) + (k_2 \times I_2)]$ leads to the output $[(k_1 \times O_1) + (k_2 \times O_2)]$; ($k_1$ & k_2 are constants). This property is called *superposition*. The assumption of linearity is fundamental to many reductionistic approaches in science. If systems are linear, then it is possible to predict the response to complex stimulus situations from knowledge about the responses to more elemental stimuli that are linear components of the complex situation. *Reductionism* is a strategy of science that attempts to discover the fundamental stimuli (or particles or elements) from which all complex stimuli can be constructed. To the extent that the world is linear, then knowledge of the system's response to these fundamental stimuli is sufficient to predict responses to all other stimuli, which can be described in terms of these basic stimuli. The linear assumption will be a powerful tool for building models of dynamic systems. However, the validity of this assumption is questionable. For example, the existence of perceptual thresholds and the existence of rapid transitions between oscillatory movement patterns (chapter 21) are not consistent with assumptions of linearity. The problems of nonlinear systems are addressed later. For now, the assumption of linearity as a powerful heuristic that can lead to important insights and intuitions about dynamic systems is accepted.

CONVOLUTION

Block diagrams are important conventions for visualizing the structure of dynamic systems. Using this convention, signals (e.g., input and output) are represented as directed lines (arrows) and boxes are used to represent operations performed on those signals or transfer functions. Up to this point, these transfer functions have been treated as constant multipliers. That is, the output from a box has been represented as the input multiplied by the transfer function. This is not generally correct if the signals are thought of as functions of time. The operation performed on the signal by a transfer function is actually *convolution*. The word "convolve" means literally to "roll or fold together." That is a good description of what is accomplished by the mathematical operation of convolution—the signal is folded together with the *impulse response* of the system element (or box).

The impulse response is the response to a very brief signal. Because the signal is brief, the output of the system element reflects the dynamics of the element (i.e., the transfer function) only. Mathematically, the impulse is a limiting case in which the signal is infinitely brief, and has integrated area equal to one. Thus, the impulse response reflects only the constraints represented in the transfer function. For linear elements, the impulse response is a complete description of their dynamics.

To introduce the mathematical operation of convolution, it is useful to begin with an example in discrete time. In discrete time, the impulse is a single input that has magnitude equal to one and duration equal to one unit. This example is taken from Giffin (1975) using an economic model of research investments. The impulse response of the research system is such that for each dollar spent on research this year, there will be no profit this year or in Year 1, but a $2 profit in Year 2, a $3 profit in Year 3, a $1 profit in Year 4, and no profit after Year 4. This impulse response characterizes the dynamics of the research environment. It takes 2 years before today's research results in profits and the research done today will be obsolete after 5 years. There will be a peak return on the research investment in Year 3, with lower returns in Years 2 and 4. This impulse response is illustrated in the top graph of Fig. 5.2.

Suppose that someone invests $10,000 this year, $15,000 next year, $20,000 the following year, $25,000 the next year, and no dollars after that. This pattern of investment is the input signal to the dynamic research system. This input signal is shown in the bottom half of Fig. 5.2. What output can be predicted? The output can be analytically computed as the convolution of the input with the impulse response of the system. The convolution operation is illustrated in Fig. 5.3. The top graph in Fig. 5.3 shows the output for each investment. Thus, for the investment of $10,000, there is $0 return in Year 0 and Year 1; a return of $20,000 ($2 for every dollar invested) in Year 2; $30,000 in Year 3; $10,000 in Year 4; and $0 after Year 4. Similarly, the investment of $15,000 dollars in Year 1 shows its first return of $30,000 in Year 3, $45,000 in Year 4, $15,000 in Year 5, and $0 return after Year 5. Note that despite differences in scale, the bars corresponding to each input show the same qualitative pattern (reflecting the system dynamics or impulse response) with a peak output in the third year following the input. The differences in scale reflect the differences in the magnitude of each particular input.

The bottom graph of Fig. 5.3 shows the total response of the system. This represents the convolution or the folding together of the responses from each individual input. Thus, the total output at Year 5 represents the combined results of the $15,000 invested in Year 1, the $20,000 invested in Year 2, and the $25,000 invested in Year 3. This convolution operation is represented in the following equation:

$$g(Year) = \sum_{k=0}^{Year} f(k)h(Year - k) \tag{2}$$

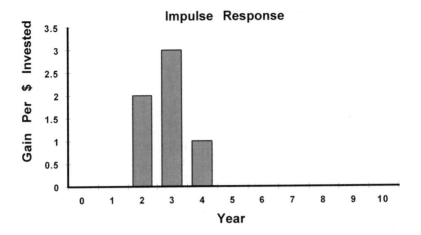

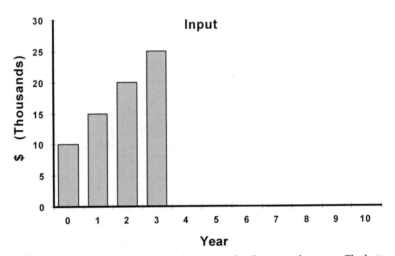

FIG. 5.2.   The top graph shows the *impulse response* for the research system. The bottom graph shows the investment or input presented to the system.

where $h(Year - k)$ is the impulse response at time $(Year - k)$, $f(k)$ is the input at time (k), and $g(Year)$ is the total output. This calculation is shown in Table 5.1.

Thus, for linear systems, the operation of convolution allows an analytical calculation of the behavior (i.e., output) of a system from knowledge of the inputs and the system dynamics. This calculation involves a folding together of the input to the system with the dynamic constraints (impulse function) of the system itself.

The logic for continuous time systems is analogous to that for discrete time systems. The convolution operation in continuous time is represented by the following equation:

$$g(t) = \int_0^t f(u)h(t - u)du \qquad (3)$$

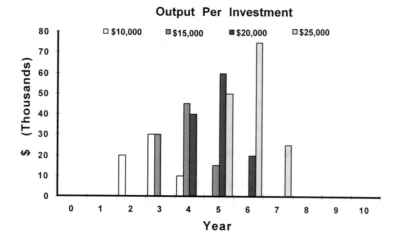

Output Per Investment

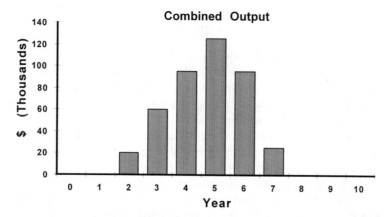

Combined Output

FIG. 5.3. The top graph shows the contribution of each investment to the total output for each year. The bottom graph shows the total system output. This output represents the convolution of the input and the impulse response shown in Fig. 5.2.

TABLE 5.1
Convolution Example

Year	$g(Year) = \displaystyle\sum_{k=0}^{Year} f(k)h(Year - k)$
0	$(10)(0) = 0$
1	$(10)(0) + (15)(0) = 0$
2	$(10)(2) + (15)(0) + (20)(0) = 20$
3	$(10)(3) + (15)(2) + (20)(0) + (25)(0) = 60$
4	$(10)(1) + (15)(3) + (20)(2) + (25)(0) = 95$
5	$(10)(0) + (15)(1) + (20)(3) + (25)(2) = 125$
6	$(10)(0) + (15)(0) + (20)(1) + (25)(3) = 95$
7	$(10)(0) + (15)(0) + (20)(0) + (25)(1) = 25$
8	$(10)(0) + (15)(0) + (20)(0) + (25)(0) = 0$
9	0
10	0

where $h(\ )$ is the impulse response, $f(\ )$ is the input, and $g(\ )$ is the output. The integral for continuous time is analogous to summation for the discrete case. For motor control and control systems in general, the systems are often modeled as continuous time systems. However, analysts generally finesse the convolution operation through the use of Laplace transforms.

THE LAPLACE TRANSFORM

Transforms can often be used to simplify computations. For example, transforming numbers into their logarithms is a way to accomplish multiplication through the simpler operation of addition. For example, consider the following problem:

$$
\begin{array}{ll}
1953 & \log_{10}(1953) = 3.290702243 \\
\times\ 1995 & \log_{10}(1995) = \underline{3.299942900} \\
3{,}896{,}235 & \qquad\qquad\quad 6.590645143 \\
& 10^{6.590645143} = \quad 3{,}896{,}235
\end{array}
$$

In this example, the transform method is to first look up the logarithm for each number in a table of logarithms. Then the logarithms are added. Then return to the table to look up the inverse of the logarithm, which is raising 10 to the calculated power. This inverse is the answer to the multiplication problem. However, the answer was arrived at using addition, not multiplication.

The Laplace transform will work in an analogous fashion, but the benefits of this transform are that the mathematically formidable calculations involving differentiation and integration can be accomplished using simpler algebraic operations. For example, convolution in the time domain can be accomplished by multiplication of the Laplace transforms of the input and the impulse response. Also, differentiation and integration in the time domain will be accomplished by multiplication and division when using Laplace transforms. Thus, with the help of Laplace transforms, as shown in Table 5.2, it is possible to solve differential equations using the much simpler tools of algebra.

The Laplace transform is computed from the time domain representation of a function using the following equation:

$$
f(s) = \int_{0}^{\infty} e^{-st} F(t)dt \tag{4}
$$

where s is a complex variable. Fortunately, as long as transform tables are available, it is not necessary to do this computation. As long as tables are available, the transformation is accomplished by table-look-up. If a good table containing most common transform pairs is available, then even the partial fraction expansion explained below will be unnecessary.

For example, suppose the problem was to integrate a unit step function. A step function is a function that is zero for times less than zero and is one for times equal to zero or greater. This function was introduced in chapter 4 in the context of Fitts' Law. The positioning task used in Fitts' experiments could be modeled as responding to a step input. The appearance of the target is considered time zero. The distance to the

TABLE 5.2
Laplace Transforms

Function or Operation	Time Domain	Laplace Domain
Function	$F(t) = 0$, for $t < 0$	
Impulse	$\delta(t) = 0$ for $t \neq 0$ $\int_{-\infty}^{\infty} \delta(t)dt = 1$	$f(s) = 1$
Step	$F(t) = 1$, for $t \geq 0$	$f(s) = \dfrac{1}{s}$
Ramp	$F(t) = t$, for $t \geq 0$	$f(s) = \dfrac{1}{s^2}$
Sine Wave	$F(t) = \dfrac{\sin(at)}{a}$, for $t \geq 0$	$f(s) = \dfrac{1}{s^2 + a^2}$
Exponential	$F(t) = e^{-at}$, for $t \geq 0$	$f(s) = \dfrac{1}{s + a}$
Operation		
Mult/Add	$aF_1(t) + bF_2(t)$	$af_1(s) + bf_2(s)$
Differentiation	$\dfrac{dF(t)}{dt}$ $\dfrac{d^2F(t)}{dt^2}$	$sf(s) - F(0)$ $s^2f(s) - sF(0) - \dfrac{dF}{dt}(0)$
Integration	$\int_0^t F(u)\,du$	$\dfrac{f(s)}{s}$
Convolution	$\int_0^t F(u)G(t-u)\,du$	$f(s)g(s)$
Time Delay	$\mu(t) = F(t - a)$ for $t \geq a$ $\mu(t) = 0$ for $t < a$	$e^{-as}f(s)$

target is the size of the step. Using the conventions of calculus, the problem of integrating a unit step would be represented as:

$$\int_0^t 1\,du \qquad (5)$$

This is a simple problem to solve using calculus. If individuals have not studied calculus, then this can still be a simple problem, if they use the Laplace transform. First, use Table 5.2 to look up the Laplace transform for a step input, which is:

$$\frac{1}{s} \qquad (6)$$

Then look up the operation in the Laplace domain that corresponds to the operation of integration in the time domain. (Again, remember logarithms—the operation of multiplication on real numbers corresponds to the operation of addition on the log transforms of those numbers). Table 5.2 shows that the operation of integration is accomplished by dividing by the Laplace operator, s. Thus,

$$\int_0^t 1du \Leftrightarrow \left(\frac{1}{s}\right)\left(\frac{1}{s}\right) = \left(\frac{1}{s^2}\right)$$

The final step is to look up the time function that corresponds to the Laplace function:

$$\left(\frac{1}{s^2}\right) \Leftrightarrow t, \text{ for } t \geq 0$$

Thus, the integral of a unit step function is a ramp with a slope of 1. If instead, a step of size 10 was integrated, then what would be the output? Using the tables of transforms, it is possible to deduce that it would be a ramp with slope of 10. Thus, the magnitude of the step input determines the slope of the integral response.

As a second example, the transform method can be used to solve for the step response of the first-order lag. To do this, look up the transform for the step input with magnitude A, $\left(\frac{A}{s}\right)$. Also, the transform for the impulse response (or transfer function) of the first-order lag must be known. This is:

$$\left(\frac{k}{s+k}\right) \tag{7}$$

where k is the value of the gain or $1/k$ is the time constant (as discussed in chap. 3, which introduced the first-order lag). The procedure for deriving the transfer function from the block diagram is discussed later. The step response of the first-order lag equals the convolution of the time domain functions: step input (A) and the impulse response of the first-order lag (ke^{-kt}).

$$\int_0^t Ake^{-k(t-u)}du \tag{8}$$

But using the Laplace transforms, convolution is accomplished by multiplication of the two functions:

$$\left(\frac{A}{s}\right)\left(\frac{k}{s+k}\right) \tag{9}$$

The difficulty at this point is to algebraically manipulate this product to a form that corresponds to the entries in Table 5.2. This can be accomplished using a technique called *partial fraction expansion*. The goal is to expand this multiplication into a sum of terms, each of which corresponds to an entry in the table. This is accomplished as follows:

$$\frac{n_1}{s} + \frac{n_2}{s+k} = \frac{Ak}{s(s+k)}$$

Solving this relation for n_1 and n_2 is equivalent to asking how the complex fraction on the right can be represented as the sum of simple fractions (where *simple* means in a form contained in the available Laplace table). The first step to solving this relation is to multiply both sides of the relation by $s(s + k)$. The result of this multiplication is

$$n_1 s + n_1 k + n_2 s = Ak$$

For this relation to hold, the following must be true:

$$n_1 k = Ak$$
$$n_1 + n_2 = 0$$

Which makes it possible to determine that $n_1 = A$ and $n_2 = -A$. Finally, the solution is

$$\frac{A}{s} - \frac{A}{s+k} = \frac{Ak}{s(s+k)} \tag{10}$$

Now it is a simple matter of substituting the time domain responses for their Laplace counterparts, which gives the same result that was presented at the beginning of this chapter. Because of the need to use the partial fraction expansion to find Laplace representations that correspond to the entries in the table, the process was not trivial, but using the Laplace transforms, it was possible to compute the convolution of two functions using nothing more than algebra. In most cases, using Laplace transforms makes it possible to do computations that would otherwise be difficult.

BLOCK DIAGRAMS

The use of Laplace transforms greatly facilitates the computation of the output response given knowledge of the input and knowledge of the system dynamics (i.e., impulse response). However, the equations and analytic functions do not always provide the best representation for understanding the nature of the dynamic constraints. For example, Fig. 5.1 clearly illustrates the closed-loop feature of the dynamic. It is a bit more difficult to "see" this structure in the impulse response or transfer function of the process:

$$\left(\frac{k}{s+k} \right)$$

Block diagrams will often provide a richer (from an intuitive perspective) representation of the dynamic system, whereas the transfer function will be the best representation for deriving analytic predictions.

Convolution

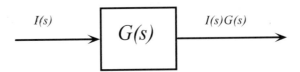

Addition/Subtraction

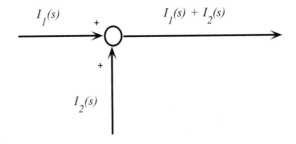

Integration

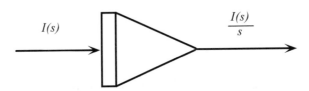

FIG. 5.4. Elements for constructing block diagrams. Arrows represent signals. Boxes represent the operation of convolution (multiplication in the Laplace domain). Circles represent addition (subtraction if negative sign is specified). Blocked triangles represent the special operation of integration (division by s in the Laplace domain).

There are three basic elements for building block diagrams (Fig. 5.4): blocks, arrows, and circles. Blocks generally represent the operation of convolution. It is standard practice to identify the impulse response for the box in terms of its Laplace transform. This is generally called the *transfer function*. Circles represent the operation of summation (subtraction if negative sign is indicated), and arrows represent signals (outputs and inputs). One additional symbol that is often used is a triangle with a narrow rectangle on the left edge, which is used to represent the operation of integration. These conventions are fairly standardized; however, variations will be encountered. Generally, the context will make any deviations from these conventions obvious.

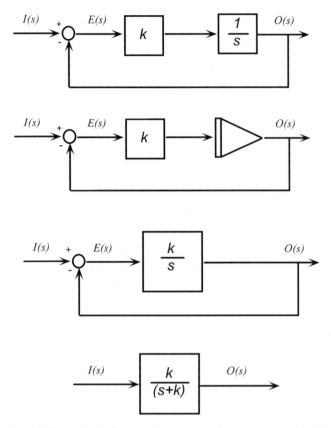

FIG. 5.5. Four different block diagrams for representing a first-order lag. These are equivalent representations for a single dynamic system.

Figure 5.5 shows various ways in which a first-order lag could be represented using these conventions. The top diagrams use the simplest elements and provide the most explicit representation of the dynamics. The bottom diagrams provide the more economical representations, but much of the structure is hidden.

To translate from the elementary block diagrams to analytical representations (transfer functions), the following procedure can be used. First, identify the inputs, the outputs, and for systems with negative feedback, the "error" signals. The error signal is the difference between the command input (e.g., target position in the Fitts' paradigm) and the output (e.g., moment-to-moment arm position). These signals are labeled in Fig. 5.5. From this representation, two relations can easily be seen. The first relation is between output, $O(s)$, and error, $E(s)$. This relation summarizes the constraints on the forward loop. The second relation is between output, input [$I(s)$], and error. This relation summarizes the constraints on the feedback loop. For the first-order lag, these relations are:

$$\text{Forward Loop:} \qquad E(s) \times k \times \frac{1}{s} = O(s) \qquad (11)$$

$$\text{Feedback Loop:} \qquad I(s) - O(s) = E(s) \qquad\qquad (12)$$

The second step is to reduce these two relations to a single relation between input and output. This is accomplished by substituting the expression for $E(s)$ in one equation into the other equation, thus eliminating the error term:

$$[I(s) - O(s)] \times k \times \frac{1}{s} = O(s) \qquad\qquad (13)$$

or

$$\left[I(s) \times \frac{k}{s} \right] - \left[O(s) \times \frac{k}{s} \right] = O(s) \qquad\qquad (14)$$

The next step is to rearrange terms so that all terms involving input are on one side of the equation and all terms involving output are on the other:

$$\left[I(s) \times \frac{k}{s} \right] = O(s) + \left[O(s) \times \frac{k}{s} \right]$$

Finally, the terms are arranged so that the output term is isolated on one side of the equation:

$$\left[I(s) \times \frac{k}{s} \right] = O(s) \left[1 + \frac{k}{s} \right]$$

$$I(s) \times \frac{k}{s} \times \frac{s}{s+k} = O(s)$$

$$I(s) \left[\frac{k}{s+k} \right] = O(s) \qquad\qquad (15)$$

Equation 15 shows the output as the product of the Laplace transform of the input and the Laplace transform of the impulse response (i.e., the transfer function). Remember that this multiplication in the Laplace domain corresponds to convolution in the time domain. These steps make it possible to derive the transfer function for a system from a block diagram. Later chapters illustrate this procedure for more complex systems.

CONCLUSIONS

This chapter introduced the Laplace transform and the block diagram as two ways for representing and analyzing linear dynamic systems. The Laplace transform permits using knowledge of algebra to solve differential and integral equations. The block diagram helps in visualizing the structural relations within a dynamic system. Later chapters provide ample opportunity to exercise and explore these new skills.

REFERENCE

Giffin, W. C. (1975). *Transform techniques for probability modeling.* New York: Academic Press.

6

The Step Response:
Second-Order System

The linear second-order system may be the most common and most intuitive model of physical systems.

—Bahill (1981)

The first-order lag described in chapter 4 provided a basis for predicting both total movement time and also the time history of the movement. The prediction for total movement time was consistent with Fitt's Law and also with empirical data on movement time. However, are the predictions for the movement time history consistent with empirical observations? Figure 6.1 compares the time histories predicted by the first-order lag with patterns for a second-order lag. The top curve shows position as a function of time. The bottom curve shows velocity as a function of time. The differences in response are most apparent for the velocity profile. The first-order lag predicts that maximum velocity is achieved early in the movement and that velocity declines exponentially from the initial peak value. Performance data obtained for humans show a gradual increase in velocity with a peak near the midpoint of the movement distance. This is qualitatively more similar to the response of the second-order system, although the velocity profile for humans is more nearly symmetrical (approximately bell shaped with peak velocity at the midpoint of the movement) than the data for the second-order system.

The mismatch between the velocity profile predicted by the first-order lag and the observed velocity profile might have been predicted on simple physical principles, in particular, Newton's Second Law of Motion. The arm or any other limb has a mass. Systems with mass cannot instantaneously achieve high velocity. The rate at which velocity builds up will depend on the amount of force applied to the limb and on the mass of the limb. Thus, it takes time for velocity to build up to its peak level. This constraint is represented by Newton's Second Law of Motion:

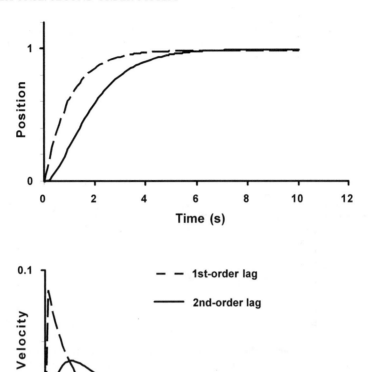

FIG. 6.1. A comparison of the step response for a first- and second-order lag. The top graph plots position versus time. The bottom shows velocity versus time. The first-order lag's instantaneous transition to peak velocity at time zero is drawn approximately.

$$Force = Mass \times Acceleration \tag{1}$$

Newton's second law can be represented in the form of a block diagram (top, Fig. 6.2). In this diagram, force is the input and position of the body with mass of m is the output. Position feedback will be insufficient to control a system that behaves according to the second law. For example, consider stopping a car at an intersection. At what distance from the intersection should the driver hit the brake—40 feet, 30 feet? Of course, the answer to this question is: "It depends on how fast you are going!" If a car is traveling at a high speed, then the driver must begin braking early. If a car is traveling at lower speeds, then a driver can wait until closer to the intersection before initiating braking. Thus, the driver must know both the distance from the intersection (position) and the speed and direction at which the car is traveling (velocity). This will be true for any system that behaves according to the second law of motion.

The bottom of Fig. 6.2 shows a negative feedback system, in which both position and velocity are fed back.

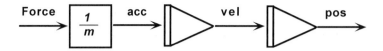

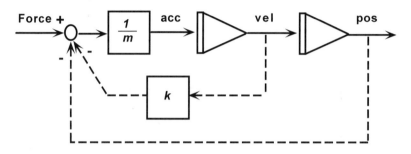

FIG. 6.2.   The top diagram illustrates Newton's Second Law of Motion in terms of a block diagram. The bottom diagram shows how continuous feedback of position and velocity might be used to control a second-order system (i.e., one governed by the second law).

In the feedback scheme shown in the bottom of Fig. 6.2, the new element is the parameter (k), which can be thought of as the weight given to the velocity feedback relative to the position feedback. Figure 6.3 shows the results of varying k over a range from .2 to 4.25 (the mass term, m, is set at 1.0). When very little weighting is given to the velocity feedback ($k = .2$), then the response shows a large overshoot and oscillations. As the weighting on velocity is increased, the oscillations become smaller (more damped) becoming negligible at $k = 1.4$. At k values of 2 and higher, there is no overshoot. At a k value of 4.25, the result is a very sluggish motion toward the target. The implications of this graph for controlling a second-order system is that if velocity is ignored, then the system will repeatedly overshoot the target. If too much weighting is given to velocity information, then the response will be slow. However, the right balance between velocity and position feedback can result in a fast response that stops at the target. Although the position profiles shown for $k = 1.8$ and 2.0 appear similar to the time history for the first-order lag, they are different. Figure 6.1 illustrates the differences. The time histories shown in Fig. 6.1 use a value of $k = 2.0$. Again, the differences are most distinct in terms of the velocity profile.

SPRINGS

The mass, spring, and dashpot, as shown in Fig. 6.4, are often presented as physical prototypes of a second-order system. Imagine the effect of adding a mass to the end of a spring plus dashpot and releasing it. It is easy to imagine different springs that

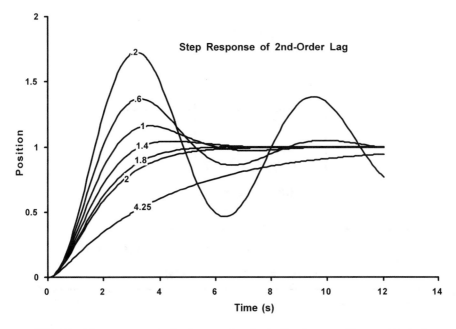

FIG. 6.3.   The step response for the second-order feedback system illustrated in the bottom of Fig. 6.2. The mass term is set at 1.0 and the weight on velocity feedback (*k*) is varied over a range from .2 to 4.25.

show the full range of outputs shown in Fig. 6.3. Some springs will slowly stretch to a new steady state length; others will oscillate before settling to a final steady state length.

The motion of the mass at the end of the spring can be described in terms of the following equation:

$$F(t) - k_2 \dot{x}(t) - k_3 x(t) = \frac{1}{k_1} \ddot{x}(t) \tag{2}$$

With "dot" notation, each dot signifies differentiation. Thus, one dot above *x* represents velocity, and two dots represent acceleration. Equivalently, in Laplace notation:

$$F(s) - k_2 sx(s) - k_3 x(s) = \frac{1}{k_1} s^2 x(s) \tag{3}$$

The right side of this equation represents three forces that combine to determine the motion of the spring. The first force, $F(s)$, is the input force. This is the pull on the end of the spring (as a result of some extrinsic source). The second force, $-k_2 sx(s)$, is the resistance due to friction or drag. The effect due to friction is in the opposite direction from that of the velocity (hence, the negative sign) and the magnitude of this force is proportional to the velocity. In general, for slow movements of a ball or anything moving through a viscous medium (e.g., oil), frictional drag is proportional to veloc-

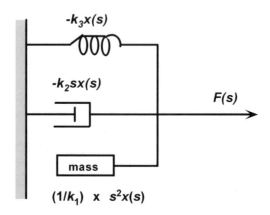

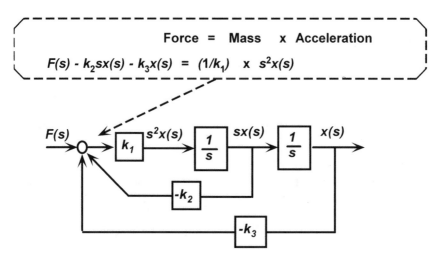

FIG. 6.4.   A mass, spring, and dashpot system. Three forces, the input [F(s)], the damping due to the dashpot [$-k_2s\,x(s)$], and the restorative spring force [$-k_3\,x(s)$] determine the
acceleration of the mass [$(1/k_1) \times s^2\,x(s)$].

ity. For faster motions (e.g., a flying airplane), which produce a degree of turbulence,
the drag becomes more nearly proportional to the square of velocity. Thus, in general
the greater the speed, the greater will be the friction. Finally, the third force, $-k_3x(s)$, is
the resistance due to the elastic properties of the spring. This force is proportional
to the displacement of the spring. This last force reflects Hooke's Law, which says
that the elastic force that tries to restore a spring to its original state when it is distorted is proportional to the distortion. Thus, the farther the spring is stretched from
its resting length, the greater will be the force to oppose further distortion.

The block diagram at the bottom of Fig. 6.4 illustrates the structural relations
among the three forces that determine the motion of the spring. The transfer function
for this system is:

$$\frac{k_1}{s^2 + k_1 k_2 s + k_1 k_3} = \frac{x(s)}{F(s)} \tag{4}$$

A standard convention is to replace the constants (k's) in this equation with two parameters ζ and ω_n, where

$$\zeta = \frac{k_2}{2} \sqrt{\frac{k_1}{k_3}} \qquad (5)$$

and

$$\omega_n = \sqrt{k_1 k_3} \qquad (6)$$

The parameters ω_n and ζ have physical significance. ω_n is the undamped natural frequency. It is the frequency at which the system would oscillate if the damping, k_2, were zero. The qualitative changes from oscillatory to nonoscillatory approach to a resting position, illustrated in Fig. 6.3, depend on the level of the second parameter, ζ, which is the damping ratio. When the damping ratio is equal to zero (i.e., there is no drag on the spring), then the system is said to be undamped and the response to a step input would be a continuous sinusoidal oscillation. When the damping ratio is greater than zero but less than one, then the system is said to be underdamped and the response to a step input will be a damped sinusoidal response (i.e., the amplitude of the oscillation would decrease gradually until the spring came to rest at its new length). When the damping ratio is one (critically damped) or greater (overdamped), then there will be no oscillations (e.g., a spring in thick oil). The transfer function using these alternative parameters is:

$$\frac{1}{k_3} \times \frac{\omega_n^2}{s^2 + 2\zeta\omega_n s + \omega_n^2} = \frac{x(s)}{F(s)} \qquad (7)$$

Table 6.1 (see Bahill's, 1981, Table 3-2, p. 106) shows both the Laplace and time domain forms for the step response of the second-order system (with $k_3 = 1$) as a function of the damping ratio. The table shows that there is a sinusoidal component for $\zeta < 1$, but no sinusoidal components for $\zeta \geq 1$.

FITTS' LAW AND THE STEP RESPONSE

Does the second-order lag provide a good model of human movements? Does it predict movement time? As shown by Langolf (1973; see also Langolf, Chaffin, & Foulke, 1976), it is possible to analytically predict movement time as a function of distance to the target or amplitude (A) and target width (W) in the same way as for the first-order system in chapter 4. The response of the underdamped second-order system is:

$$\text{Output} = 1 - \frac{e^{-\zeta\omega_n t}}{\sqrt{1-\zeta^2}} \sin\left(\omega_n t \sqrt{1-\zeta^2} + \tan^{-1}\left(\frac{\sqrt{1-\zeta^2}}{\zeta}\right)\right) \qquad (8)$$

TABLE 6.1
Step Response of a Second-Order System

Damping Ratio	Laplace Domain	Time Domain
$\zeta = 0$	$\dfrac{\omega_n^2}{s(s^2 + \omega_n^2)}$	$1 - \cos\omega_n t$
$0 < \zeta < 1$	$\dfrac{\omega_n^2}{s(s^2 + 2\zeta\omega_n s + \omega_n^2)}$	$1 - \dfrac{e^{-\zeta\omega_n t}}{\sqrt{1-\zeta^2}}\sin\left(\omega_n t\sqrt{1-\zeta^2} + \tan^{-1}\left(\dfrac{\sqrt{1-\zeta^2}}{\zeta}\right)\right)$
$\zeta = 1$	$\dfrac{\omega_n^2}{s(s + \omega_n)^2}$	$1 - (1 + \omega_n t)e^{-\omega_n t}$
$\zeta > 1$	$\dfrac{\omega_n^2}{s\left(s + \dfrac{\omega_n}{\alpha}\right)(s + \alpha\omega_n)}$	$1 + \dfrac{1}{\alpha^2 - 1}\left(e^{-\alpha\omega_n t} - \alpha^2 e^{\frac{-\omega_n t}{\alpha}}\right)$
	$\zeta \equiv \dfrac{1 + \alpha^2}{2\alpha}$	

Adapted from *Bioengineering: Biomedical, Medical, and Clinical Engineering* (p. 106) by A. T. Bahill, 1981, Englewood Cliffs, NJ: Prentice Hall. © Reprinted by permission of Pearson Education, Inc. Upper Saddle River, NJ.

If it can be assumed that target capture occurs when the points of maximum and minimum oscillation are within the target, then, as an approximation, the sinusoidal component of this equation can be ignored. Thus, the time to come within one half of a width (W) of the step input with amplitude (A) can be approximated in the following way:

$$A - \frac{1}{2}W = A - A\left(\frac{e^{-\zeta\omega_n t}}{\sqrt{1-\zeta^2}}\right) \tag{9}$$

$$\frac{1}{2}W = A\left(\frac{e^{-\zeta\omega_n t}}{\sqrt{1-\zeta^2}}\right)$$

$$\frac{W}{2A} = \left(\frac{e^{-\zeta\omega_n t}}{\sqrt{1-\zeta^2}}\right)$$

$$\left(\sqrt{1-\zeta^2}\right)\frac{W}{2A} = e^{-\zeta\omega_n t}$$

$$\ln\left(\sqrt{1-\zeta^2}\right) + \ln\frac{W}{2A} = -\zeta\omega_n t$$

$$\frac{1}{-\zeta\omega_n}\ln\left(\sqrt{1-\zeta^2}\right) + \frac{1}{\zeta\omega_n}\ln\left(\frac{2A}{W}\right) = t$$

$$\frac{1}{-\zeta\omega_n}\ln\left(\sqrt{1-\zeta^2}\right) + \frac{\ln 2}{\zeta\omega_n}\log_2\left(\frac{2A}{W}\right) = t \tag{10}$$

The analysis shows that second-order systems (e.g., damped springs) behave consistently with Fitts' Law. That is, the movement time is a linear function of the index of difficulty measured in bits, $\log_2\left(\dfrac{2A}{W}\right)$.

$\dfrac{1}{\zeta\omega_n}$ is the predominant time constant for the underdamped second-order system. Its oscillations will be within an exponential envelope that constrains over- and undershoots over time. Thus, the time constant determines the time for the second-order system to settle into a particular width of the target. This is consistent with the analysis of the first-order lag. The movement time for a dynamic system is predicted by the time constant for that system. However, although the second-order equation can be used to predict Fitts' Law, it fails as a model for human movement for the same reason that the first-order lag failed. The second-order system with a step input provides a closer match to the position and velocity time histories of human movements. However, as indicated at the beginning of the chapter, the velocity profile for human movements is more symmetrical than the profile for the second-order system that is shown in Figure 6.1 (e.g., see Carlton, 1981). The next chapter discusses a model that produces a velocity profile that shows the symmetry found for human movements.

THE EQUILIBRIUM POINT HYPOTHESIS AND HIERARCHICAL CONTROL

An important difference between a physical system like a spring (Fig. 6.4) and a biological control system involves the comparator problem. The comparator is where the input and feedback are "compared" in order to produce an "error" signal. For a physical system, like a spring, the comparator reflects fixed physical relations among opposing forces (e.g., the pull of an external load, $F(s)$, vs. the restorative spring force, $-k_3x(s)$). However, for a biological control system, the comparator may involve the comparison of a mental intention (e.g., to move to a target location) with perceptual feedback (e.g., visual and kinesthetic information about limb position and velocity relative to the target) to produce a difference that must be translated into a signal to the motor system (e.g., firing of motor neurons resulting in contractions of muscles). How are the different sources of information encoded within the nervous system so that perceived differences can be translated into the appropriate motor command?

Figure 6.5 illustrates a hierarchical control system. The outer-loop reflects the perception of a difference between the position of a limb and a target. The inner-loop reflects the springlike properties of the muscle assembly associated with arm/hand movements. An important question for theories of motor control involves the nature of the interactions between the peripheral dynamics of the limb and the central nervous system (CNS) in "controlling" the limb motion.

One possibility is that the burden of control falls completely on the CNS. This would mean that the signal from the CNS to the muscle system would be continuously modulated based on feedback in the outer loop. For example, the command from the CNS to the muscle system might be a continuous function of the visually observed error and error rates between the limb and the target. From this perspective, computations within the CNS are the dominant constraint determining the time course (i.e., position, velocity, acceleration) of the action.

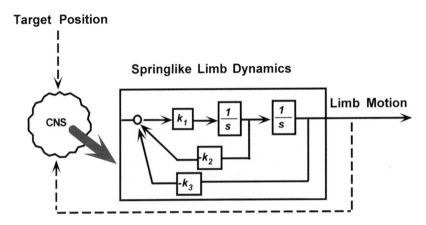

FIG. 6.5. A hierarchical control system. The outer loop perceives the need to move the arm to a new target position (e.g., based on visual feedback). The inner loop represents the mass-springlike dynamics of the limb.

An alternative style of control generally associated with Bernstein (1967) distributes the burden of control over the CNS and the peripheral muscle systems. This style of control reduces the dependence on outer loop feedback by taking advantage of natural constraints of the muscle system (e.g., the springlike properties). One example of this style of control is the equilibrium-point hypothesis (e.g., see Feldman, 1986, or Bizzi, Hogan, Mussa-Ivaldi, & Giszter, 1992, for reviews). The equilibrium-point hypothesis suggests that the commands from the CNS to the muscle system are adjustments to the parameters of the virtual mass-spring system. For example, a change in the stiffness or spring constant will change the resting length for the spring. This adjustment results in movement of the limb to the new resting length, but the time history of the movement is largely determined by the springlike dynamics of the muscle system. With this type of control, the CNS specifies a goal (or series of subgoals) for the movement, but the path between goals is largely shaped by the dynamics of the peripheral muscle system. As Bizzi et al. (1992) noted, "The important point is that according to the [equilibrium point] theory, neither the forces generated by the muscles nor the actual motions of the limbs are explicitly computed; they arise from the interplay between the virtual trajectory and the neuromuscular mechanics. Hence, neither the forces nor the motions need be explicitly represented in the brain" (p. 613).

These two different styles of control have important implications for the role of the outer feedback loop and for the nature of computations in the CNS. Bizzi, Accorneo, Chapple, and Hogan (1984) compared single-joint target acquisition movements of limited accuracy in deafferented and intact monkeys with their arms shielded from view. The deafferentation essentially cuts the outer feedback loop and alters the springlike properties as well. In addition to deafferentation, Bizzi et al. perturbed the movement following the initiation signal (e.g., restricted the motion for a brief time). If the outer loop played an important role in regulating these movements, then the deafferented monkeys should have had difficulty controlling their arm movements (adjusting to the perturbances). The results showed some quantitative differences, but no qualitative differences between intact and deafferented monkeys. This result supports the general notion of distributed control. That is, trajectory shape in this simple task did not appear to depend on continuous outer loop feedback.

Later chapters show that time delays can make control systems vulnerable to instability. One of the difficulties of placing the burden of control on the CNS is that the outer loop can have relatively long time delays (due both to the transport delays associated with sending signals up through the CNS and then back down to the peripheral muscles and to the effective time delays associated with the limb dynamics). Thus, a system that heavily depends on the outer loop for control might be prone to instability. Bizzi et al. (1992) suggested that some of the instabilities that occur in robotic control systems might result from use of this style of control. Conversely, the general stability of animal movement control suggests that these biological systems utilize more distributed styles of control that reduce the demands on the CNS and take advantage of the natural stability of the peripheral dynamics.

Debate continues about the details of the interactions between the CNS and peripheral system in controlling movement. It seems that it is possible to find evidence for a wide continuum of control styles in biological systems—from styles that rely heavily on the CNS to more distributed styles of control that take advantage of natural system constraints. The style of control is likely to vary as a function of the task demands (e.g., the demands for precision in space-time or the nature of external disturbances) and as a function of the experience and skill of the actor involved.

Rosenbaum and Krist (1996) used the task of throwing a ball for maximal distance to illustrate how natural physical constraints and central control signals can combine to produce optimal results. To meet the goal of a far throw, the system should release the ball with a maximum angular velocity. This might be achieved by taking advantage of the whiplike characteristics of the arm. According to Rosenbaum and Krist (1996),

> The external torque applied at the handle of the whip accelerates the system as a whole and gives it angular momentum. Because of the elastic properties of the whip, it does not rotate all at once; rather, the more distal the segment the later it is affected by the motion of the handle. The angular momentum of the system travels from the handle to the tip, expressing itself in ever smaller parts of the system. Therefore, the rotational inertia of the system decreases as well. According to the law of conservation of momentum, the angular velocity increases at the same rate as the rotational inertia decreases. This effect outstrips the concurrent reduction of the radius of rotation, so the tip of the whip eventually acquires a speed that can be supersonic; in fact, the "crack" of the whip is literally a sonic boom. What occurs in throwing, kicking, and whipping, then, is the transfer of kinetic energy and momentum from proximal to distal body segments and finally to the ball or whip itself. That a whip, by virtue of its physical structure, can allow for the transfer of kinetic energy and momentum from the source to its tip indicates that the transfer need not be deliberately planned or programmed. Throwing and related skills may similarly exploit the whip-like characteristics of the extremities. (p. 59)

Rosenbaum and Krist (1996) continued by observing that despite the natural whiplike characteristics of the arm "active muscular torque production, especially in the base segments, plays an important role in generating maximum velocities in the end-effector" (p. 59). They cited research by Alexander (1991) showing that the proper sequencing of muscular contractions is essential to perfecting throwing. It is likely that animals are capable of a wide range of solutions to the motor control problem and that "skilled" performance depends on an optimal balance between central (outer loop) control and natural constraints associated with the physical structure of the limbs. Rosenbaum and Krist (1996) provide a good introduction to research designed to uncover the role of the CNS in "controlling" motor actions. Also, Jordan (1996) provides a nice tutorial on different styles of control.

The optimal balance between central and peripheral constraints on skilled motor control is an important question for coaches and for designers of robotic control systems. In sports, there seems to be a distinction between Western and Eastern beliefs about skill. In the West, coaches tend to encourage their players to concentrate and to focus their minds on the game. This approach seems to reflect a belief in a strong centralized style of control. In the East, philosophies like Zen promote a "no mind" approach to skill. This approach seems to reflect belief in a more distributed style of control, where the mind "lets" the peripheral constraints take over much of the responsibility for action. It seems that the current trend in robotic design is moving from control systems that relied heavily on centralized control to systems that take better advantage of peripheral constraints.

The main point for this chapter is that the second-order, springlike characteristics seen in movement control tasks may reflect properties of a cognitive process (e.g., adjusting movements based on continuous information about error and error velocity) or may simply reflect the structural properties of the limb being controlled. Research to answer this question must be guided by sound intuitions about the nature of control systems.

CONCLUSIONS

This chapter raises a number of important issues discussed further in later chapters. First, the consideration of Newton's second law emphasizes the fact that movements are constrained by physical laws, as well as laws of information processing. The equilibrium point hypothesis raises the question of what is being controlled. Or, perhaps more precisely, how are intentions communicated within the motor control system? Are the intentions communicated directly in terms of the relevant output variable (e.g., force or muscle length) or are they communicated indirectly in terms of parametric constraints (e.g., the spring constant or equilibrium point) that determine the output variable? Are the commands from higher cognitive systems discrete or continuous functions of the error perception? Later chapters explain that hierarchical control strategies may prove to be very important for dealing with stability constraints that arise due to time delays in many control systems.

REFERENCES

Alexander, R. M. (1991). Optimal timing of muscle activation for simple models of throwing. *Journal of Theoretical Biology, 150,* 349–372.

Bahill, A. T. (1981). *Bioengineering: Biomedical, medical, and clinical engineering.* Englewood Cliffs, NJ: Prentice-Hall.

Bernstein, N. A. (1967). *The control and regulation of movements.* London: Pergamon Press.

Bizzi, E., Accorneo, N., Chapple, W., & Hogan, W. (1984). Posture control and trajectory formation during arm movement. *Journal of Neuroscience, 4,* 2738–2744.

Bizzi, E., Hogan, N., Mussa-Ivaldi, F. A., & Giszter, S. (1992). Does the nervous system use equilibrium-point control to guide single and multiple joint movements. *Behavior and Brain Sciences, 15,* 603–613.

Carlton, L. G. (1981). Movement control characteristics of aiming responses. *Ergonomics, 23,* 1019–1032.

Feldman, A. G. (1986). Once more on the equilibrium-point hypothesis (λ Model) for motor control. *Journal of Motor Behavior, 18,* 17–54.

Jordon, M. I. (1996). Computational aspects of motor control and motor learning. In H. Heuer & S. W. Keele (Eds.), *Handbook of perception and action: Vol. 2. Motor skills* (pp. 71–120). New York: Academic Press.

Langolf, G. D. (1973). *Human motor performance in precise microscopic work – Development of standard data for microscopic assembly work.* Unpublished doctoral dissertation, University of Michigan, Ann Arbor.

Langolf, G. D., Chaffin, D. B., & Foulke, J. A. (1976). An investigation of Fitts' Law using a wide range of movement amplitudes. *Journal of Motor Behavior, 8,* 113–128.

Rosenbaum, D. A., & Krist, H. (1996). Antecedents of action. In H. Heuer & S. W. Keele (Eds.), *Handbook of perception and action: Vol. 2. Motor skills* (pp. 3–69). New York: Academic Press.

Nonproportional Control

> Dynamical modeling is the art of modeling phenomena that change over time. The normal procedure we use for creating a model is as follows: first we identify a real world situation that we wish to study and make assumptions about this situation. Second, we translate our assumptions into a mathematical relationship. Third, we use our knowledge of mathematics to analyze or "solve" this relationship. Fourth, we translate our solution back into the real world situation to learn more about our original model. There are two warnings. First, the mathematical relationship is not the solution. . . . The second warning is to make sure that the solution makes sense in the situation being considered.
>
> —Sandefur (1990, p. 2)

For the first-order lag (chap. 4) and the second-order lag (chap. 6), control is accomplished by proportional adjustments to continuously monitored "error" signals. These are examples of continuous, proportional control systems. However, some control systems do not have a continuous error signal available. These systems must discretely sample and respond to the error signal. In other situations, particularly when there are large time delays inherent in the control problem (e.g., the familiar example of adjusting the water temperature of your shower) continuous, proportional control to error can lead to instability. To avoid being scalded by the shower, it is prudent to make small discrete adjustments to the water temperature and then to wait to see the effects of those adjustments. As individuals become familiar with a particular plumbing system, they may be able to fine-tune their control inputs. That is, they will learn how large to make the adjustments and how long to wait between adjustments so that they can reach the desired water temperature relatively efficiently. This chapter discusses a discrete strategy for controlling a second-order system.

For a discrete or intermittent control system, each command is executed in a ballistic fashion. That is, the command is executed as a package, without any adjustments due to feedback. Initiating the action is like firing a bullet. Once the bullet has been fired, no corrections are possible. After the command has been executed, then the output can be observed and further commands can be executed to compensate for any errors that are observed. Feedback has no effect during the execution of the pre-packaged command. However, future command packages may be tuned to reflect feedback obtained during the execution of previous commands — just as a marksman can adjust aim based on the outcome of previous shots.

What would be the simplest command signal to move an inertial system (i.e., governed by Newton's second law) from one position at rest to another position at rest? First, it would be necessary to begin acceleration in the appropriate direction (to effectively step on the gas), then at approach to the goal position it would be necessary to decelerate (brake) so that zero velocity is reached at the goal position. This simple control might be characterized as a *bang-bang control*. The first bang determines the acceleration. The second bang determines the deceleration. If acceleration and deceleration balance each other out, then the system will come to rest at a new position. Figure 7.1 illustrates the response of a simple second-order system to a bang-bang input.

Figure 7.2 shows how the bang-bang response maps to the movement kinematics. The height of the first bang determines the degree of acceleration. The height and duration of the first bang will determine the peak velocity of the movement. The velocity will peak at the switch between bangs that initiates the deceleration phase of the motion. The system will continue to decelerate until the area under the second bang is equal to the area under the initial bang. If the second bang is terminated at this point of equivalence, then the motion will stop (i.e., have zero velocity) at that point. If the second bang continues, then the system will begin accelerating in the opposite direction. A symmetric bang-bang will result in a movement that has peak velocity midway between the start and end position. The symmetry of the velocity profile that results is similar to profiles observed for human movements to fixed targets. However, human velocity profiles tend to look more like normal curves (bell shaped) than like the triangular curves shown in Fig. 7.2. (e.g., Carlton, 1981).

It is interesting to note that if there are symmetric control limits on the maximum height for the opposing pulse commands, then a symmetric bang-bang response whose heights are equal to the control limits will result in the minimum time movement from one position to another. This can be proven using optimal control theory (e.g., Athans & Falb, 1966). However, it should be intuitively obvious that maximum acceleration followed by maximum deceleration will result in the highest velocities and, thus, the quickest movement between two points. If humans are responding to discrete target acquisition tasks using a bang-bang control (as might be implied by the symmetric velocity profile), then they are behaving as minimum time optimal

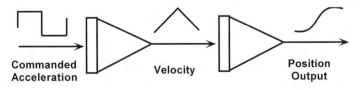

FIG. 7.1. The response of a simple second-order system to a bang-bang command.

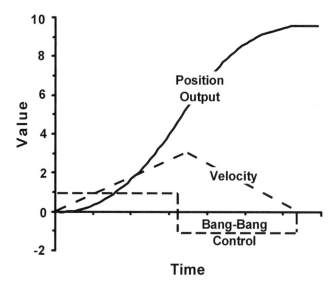

FIG. 7.2. The kinematics of the response of a simple second-order system to a bang-bang command (lower dashed line). The solid line shows the position. The triangular dashed line shows the velocity. The scaling of position, velocity, and time are arbitrary.

controllers. Such controllers will reach the target faster than proportional feedback controllers (i.e., first- or second-order lags). The accuracy of such controllers would depend on the timing of the switch from full acceleration to full deceleration. If the switch is made too early, then the movement will fall short of the target. If the switch is made too late, then the movement will overshoot the target.

The bang-bang profile might be thought of as a discrete control package. In this case, a bang-bang command is executed when a target appears. If the timing of the switch is correct, then the movement is successful. If the timing is in error, then a second discrete package is executed in response to the error resulting from the first movement. This process continues iteratively until the target is reached. Such discrete control packages have classically been referred to as motor programs. Kantowitz and Knight (1978) defined a motor program as a "central representation of a series of sequential actions or commands performed as a molar unit without benefit of feedback; that is, later segments of the program [e.g., decelerating bang] are performed with no reference to peripheral feedback that may arise from execution of earlier program segments [e.g., accelerating bang]" (p. 207). Kantowitz and Howard (reported in Kantowitz & Knight, 1978, Fig. 4) showed that Fitts' (1954) data could be fit quite well using a bang-bang control model. It is important to note that in current usage, the term *motor program* can sometimes be used more generally to refer to any modular action system and may include feedback loops within modules.

Alternatively, the bang-bang response might be composed of two discrete packages. The first package is the initial acceleration pulse. The second package is the deceleration pulse. This second package may be initiated based on observation of the response to the first command. This second strategy would be possible if, while the commands were discrete, there was continuous information about the position and

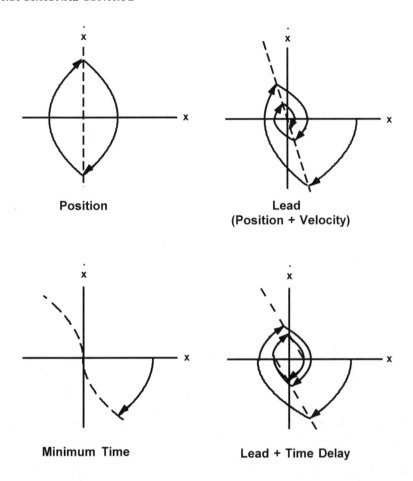

Position

Lead
(Position + Velocity)

Minimum Time

Lead + Time Delay

FIG. 7.3.   The dashed lines in the state space diagrams are the "switching curves." The solid lines show accelerations and decelerations from various initial distances from the target (the origin). From "A model of human controller performance in a relay control system." Paper presented at the Fifth National Symposium on Human Factors in Electronics by R. W. Pew, 1964. Adapted by permission.

velocity of the system. This can best be thought of in terms of a state space diagram that plots position (x) versus velocity ($\dot{x}$) and is called a *phase plane* (Fig. 7.3).

The trajectories for the minimum time criterion shown in the lower left quadrant of Fig. 7.3 can be derived in the following manner. First, for convenience, the values of 1 and –1 are assigned to the maximum acceleration (or deceleration) values. Thus, with the bang-bang command, acceleration equals either 1 or –1:

$$\ddot{x} = 1 \quad \text{or} \quad \ddot{x} = -1$$

Integration of both sides of these equations with respect to time (t) gives the velocity function for each of these inputs:

$$\dot{x} = t + c_1 \quad \text{or} \quad \dot{x} = -t + c_1$$

Assuming that the system is at rest (i.e., velocity = 0) initially, then the constants (c_1) are equal to zero and the equations reduce to:

$$\dot{x} = t \quad \text{or} \quad \dot{x} = -t$$

Integrating these equations with respect to time gives the position function for each of the inputs:

$$x = \frac{1}{2}t^2 + c_2 \quad \text{or} \quad x = -\frac{1}{2}t^2 + c_2 \tag{1}$$

The constants (c_2) in the aforementioned equation are the initial positions. Substituting for t yields an equation for position in terms of velocity:

$$x = \frac{1}{2}\dot{x}^2 + c_2 \quad \text{or} \quad x = -\frac{1}{2}\dot{x}^2 + c_2 \tag{2}$$

The trajectories shown in the bottom left of Fig. 7.3 correspond to these equations. The switching curves are the decelerating half of the parabolas with position intercepts at the target (zero position).

Thus, Fig. 7.3 shows the position and velocity of the response of the second-order system. The dashed parabolic curves in the lower left quadrant of the figure represent minimum time "switching curves." These switching curves show trajectories through state space that correspond to a maximum deceleration command that terminates at the target (position and velocity equal to zero). The solid curve represents maximal acceleration toward the target from a specific initial position. To capture the target, the switch from acceleration to deceleration must happen at the intersection of the solid acceleration curve with the dashed deceleration curve.

A controller might be designed that continuously monitors its state relative to boundaries in state space (e.g., the optimal switching curve). Upon reaching a critical boundary, a discrete control action (full braking) could be initiated. Such a controller might be characterized as a "motor program," or production system. That is, it is a finite collection of condition–action rules. The conditions are represented as regions or boundaries in state space. The actions are discrete responses to specific conditions (see Jagacinski, Plamondon, & Miller, 1987, for a review). This type of controller is sometimes referred to as a *finite state controller*. For example, a set of productions might be:

Step 1. If target position is different than arm position, then initiate maximum acceleration in the direction of the target (pulse).

Step 2. If position and velocity correspond with a point on the switching curve, then initiate deceleration (brake).

Step 3. If velocity equals zero, then cease deceleration.

Step 4. If position equals target, end.

Else, repeat Step 1.

If the control limits for the acceleration and deceleration pulses are symmetric and the arm is initially at rest, then Step 2 can be simplified. The deceleration response can be initiated on reaching the halfway point between the initial position and the target position. Thus, the controller would only need to monitor position relative to the target.

Humans may not switch at the time optimal criterion shown in the bottom left quadrant of Fig. 7.3. Three other possible criteria were described by Pew (1964; see also Graham & McRuer, 1961, regarding these and other examples of discrete control). The top left quadrant in Fig. 7.3 shows a switching criterion based only on position. This system does not attend to velocity. The result of using the zero position criterion for discrete control of a simple second-order system would be an equilibrium oscillation (limit cycle) whose amplitude depended on the initial condition. That is, this controller will repeatedly overshoot the target, first in one direction, then in the other. It will never converge (come to rest) on the target. The upper right quadrant of Fig. 7.3 shows a criterion based on a linear combination of position and velocity (also called "lead"). Using this criterion, the system will converge to zero error, but it will not accomplish this in minimum time as would be true if the parabolic criterion shown in the lower left quadrant were used to control the "switch." Depending on the ratio of position to velocity (slope of the switching line), this criterion will lead to an early or late switch. Thus, this results in a series of under- or over-shoots that get successively smaller on each iteration. This is a satisfactory, but not optimal strategy. The switching criterion shown in the lower right of Fig. 7.3 is a combination of lead and a time delay. As a result of this switching criterion, the system would converge to a limit cycle whose amplitude was a function of the intersection of the velocity axis with the switching line. The system would oscillate with small amplitude in the target region (e.g., think about balancing an inverted broom or stick—it is not atypical to see the steady state control to be a continuous series of small oscillations centered around the "target" vertical orientation).

Pew (1964) measured human performance in which the control was accomplished using a discrete bi-stable relay controller. With this controller, the "operator controlled the position of the target along the horizontal dimension of an oscilloscope by alternately switching between two response keys. . . . The left and right keys applied a constant force toward the left and right respectively" (p. 241). This paradigm forces the operator to behave as a discrete controller. A switching function description of the performance data was closest to the pattern for a lead plus time delay as shown in the bottom right of Fig. 7.3. The data showed limit cycles and the amplitudes of the limit cycles seemed to depend more on operator characteristics (an internal time delay), than on task constraints (e.g., initial position or control gain).

Evidence for the discrete nature of control can be found in the time history of movements in target acquisition tasks. Crossman and Goodeve (1983) found velocity peaks and short periods of zero velocity in the time histories of participants in a wrist-rotation target acquisition task. These patterns suggest a sequence of discrete control pulses. This led Crossman and Goodeve (1983) to propose a discrete model alternative to Fitts' Law (see also Keele, 1968). This model assumed a discrete series of movements of constant relative accuracy and constant duration. The constant relative accuracy constraint means that the ratio of the distance from the target center (error) after a correction to the distance before a correction is constant:

$$X_{n+1}/X_n = k$$

where $0 < k < 1$. Equivalently,

$$X_{n+1} = k \ X_n$$

If X_0 is the amplitude (A), and movements continue until the error remaining is less than or equal to one half of the target width (W), then the number of iterative control actions (N) can be estimated from the following relation:

$$W/2 = k^N A$$

or

$$-\log_2(2A/W) = N\log_2(k)$$

or

$$N = -\log_2(2A/W)/\log_2(k)$$

The constant duration constraint means that each correction will take a fixed time (T). Total movement time (MT) will be:

$$MT = NT$$

Noting that $\log_2(k)$ is negative because $0 < k < 1$, and substituting for N:

$$MT = [-T/\log_2(k)] \times \log_2(2A/W)$$

Thus, the iterative correction model is consistent with Fitts' Law, which states that movement time will be a log function of $2A/W$. The information-processing rate for the iterative correction model is a function of the precision of each correction (k) and the time for each correction (T). The assumptions of constant time and accuracy have proven not to be consistent with human performance. The first submovement tends to be slower and more accurate than subsequent episodes and these measures vary considerably with amplitude (A) (e.g., Jagacinski, Repperger, Moran, Ward, & Glass, 1980). However, as long as the increase in time is proportional to the increase in log accuracy among first submovements, and this proportionality matches the ratio of time to log accuracy for subsequent submovements, then Fitts' Law will hold (Jagacinski et al., 1980). More recent iterative control models are discussed in chapter 8.

NONPROPORTIONAL CONTROL

A critical difference between the bang-bang style of control and the first-order lag or second-order lag model is the relation between instantaneous magnitude of error and the magnitude of control action. For the first- and second-order lags described in the previous chapters, there is a proportional relation between error and control. Thus, there is effectively an infinite number of different possible control actions corresponding to the infinite number of points in state space. For the bang-bang style of control, there is a degree of independence between the magnitude of error and the magnitude of control. That is, there will be a single action associated with a region of state space. With a bang-bang control system, the control magnitude will be constant for extended periods of time, while error is varying. This is why this control is sometimes called "discrete," because there are periods where the command actions seem to be independent from the changing error signal being fed back. Thus, corrections appear to be made at discrete intervals in time. If the intervals are constant, then the control is said to be *synchronous*. If the intervals are not constant in time, then the control is said to be *asynchronous*.

It is important to note that the apparent independence of control and feedback can reflect different underlying processes. One possibility is that the error is sampled discretely. For example, drivers in automobiles may sometimes take their eyes off the road (e.g., to consult a map or to adjust the radio). Thus, the control may be fixed to reflect the last sample of the error. The control does not vary with the error signal because the error signal is temporarily not available. Some have argued that, at a fundamental level, the human information-processing system is a discrete sampling system.

There are many control situations where the human operator must sample multiple sources of information. For example, a driver must periodically sample the rear-view mirror to keep in tune with the evolving traffic situation. Also, the driver's attention may be required to search for landmarks or street names when navigating in a new part of town. Although drivers are continuously steering while monitoring traffic and looking for the next landmark, they will be responding to intermittent information about steering error. Research on human information sampling shows that humans are able to adjust their sampling rates to reflect the importance and likelihood of events on different "channels" of information. However, the sampling behavior does show some consistent biases that reflect limitations in human memory and decision making. See Moray (1986) or Wickens (1992) for a summary of research on human sampling behavior.

A synchronous discrete control system is one that operates on discrete samples of data measured at fixed intervals (a constant sampling frequency). Bekey (1962) described two types of synchronous control strategies. One strategy uses a zero-order hold. This control system observes the position of a continuous function and extrapolates a constant position based on this observation. A first-order hold extrapolates a constant rate based on discrete observations of position and velocity. Figure 7.4 illustrates these two strategies for signal reconstruction. Note that the first-order discrete reconstruction produces continuous adjustments reflecting the velocity of the last sample. When these reconstructions are further transformed into a response and

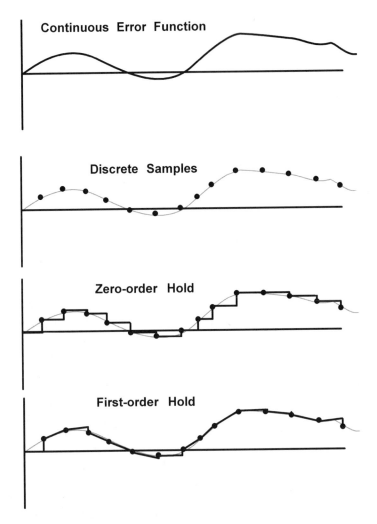

Continuous Error Function

Discrete Samples

Zero-order Hold

First-order Hold

FIG. 7.4. Two methods for reconstructing a signal from sampled data. The zero-order hold makes estimates based on the current position. The first-order hold makes estimates based on the current position and velocity. From "The human operator as a sampled-data system" by G. A. Bekey, *IRE Transactions on Human Factors in Electronics, HFE-3*, 43–51, copyright by IEEE. Adapted by permission.

smoothed by dynamics of the motor system, the output can resemble continuous control. Thus, even though the control system is discrete, the time history of the output can appear to be continuous. An important constraint of discrete sampling of position is that the signals that can be accurately reconstructed have bandwidths not exceeding one half of the sampling frequency. If both position and velocity are sampled, then the limit on signal bandwidth can be twice as large, i.e., equal to the sampling frequency (Fogel, 1955). Also, a nearly constant sampling rate generates harmonics in the output that extend over the entire frequency spectrum. These har-

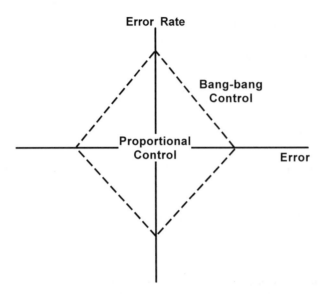

FIG. 7.5. This control system uses a proportional style of control when error and error velocity are small, but uses a discrete time-optimal style of control to correct large errors. From "The surge model of the well-trained human operator in simple manual control" by R. G. Costello, *IEEE Transactions on Man-Machine Systems, MMS-9*, 2–9, copyright by IEEE. Adapted by permission.

monics can contribute "remnant" or noisiness in the frequency response for these systems. The frequency response for human tracking is discussed in much greater detail in chapters 14 and 19.

A second reason why there may be some independence between control and feedback is that the fixed control response reflects control limits. For example, drivers might slam on the brakes to avoid a collision. Even though they are attending to the road and have continuous information about the changing error signal, the control signal cannot increase beyond its limits. Such a system may be proportional for some range of error signal, but if the error exceeds some level, then the control response may hit a limit beyond which a proportional response is not possible (e.g., the brakes are slammed on). An example of this style of control is Costello's (1968) surge model (Fig. 7.5). With the surge model, small errors (shown as the central region within the diamond) are responded to proportionally. However, large errors are responded to with a stereotypical response (e.g., bang-bang control). The bang-bang would have pulse durations that could bring the system back within the "ballpark" of the proportional control zone in a time optimal fashion.

This combination of two control styles is consistent with Woodworth's (1899) classical analysis of human movement. He described two phases in the movement to a target: the *initial adjustment* and the *current control*. The initial adjustment seemed to be relatively independent of visual feedback; in Woodworth's terms, the initial movement is specified as a "whole." This phase of the movement might reflect a control that is calibrated to get the arm within the general vicinity (i.e., ballpark) of the target. Visual feedback becomes important in the last phase of the movement, current control. In this phase, the human utilizes visual feedback to make any fine adjustments required in order to capture the target. The type of control used in Costello's and Woodworth's models is called hierarchical. That is, there appear to be not just multiple control actions, but multiple control laws, or styles of control, that correspond to different regions of the state space. In one region of state space the control system may be relatively independent of visual feedback, and in another region the control system may be tightly coupled to visual feedback.

A third possibility why there may be independence between control action and feedback is that the fixed control responses reflect discrete commands. Thus, even though continuous feedback is available, the control system may be designed to implement a set of discrete control responses. The production system described earlier in this chapter (see p. 62) is an example of this style of control. Bekey and his colleagues (e.g., Angel & Bekey, 1968) suggested that humans may use such asynchronous discrete control. It is called asynchronous because action is not based on constant sampling rates. Rather, action is contingent on the status of the system (e.g., magnitude and rate of error). This style of control is an example of finite state control. This is because the state space is divided into a finite set of regions (states), and each region is associated with a particular control strategy. This is illustrated in Fig. 7.6, which shows five regions with a different discrete response associated with each region. Angel and Bekey (1968) argued that the finite state style of control provides a good match to human tracking performance with a second-order control system. Burnham and Bekey (1976) described a finite state model for car following in which different acceleration and braking strategies are implemented in different regions of the state space.

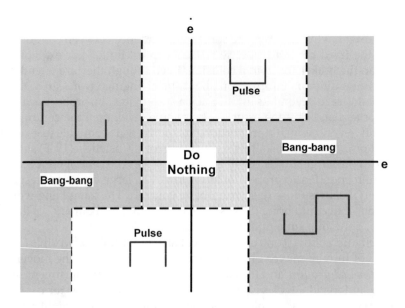

FIG. 7.6.   A finite state control strategy. Bang-bang control is used to correct position errors and a pulse control is used to correct velocity errors. The central region represents an acceptable level of error (e) and error rate ($\dot{e}$) (after Bekey & Angel, 1966).

Finite state styles of control, such as those suggested by Bekey, are particularly likely to occur in situations where people are controlling systems with long inherent time lags. Plumbing systems often have very long lags between the initial adjustment and the ultimate change of water temperature. Thus, it is unlikely that an individual will continuously adjust the temperature control until the water reaches the desired temperature. Rather, the individual probably makes a discrete adjustment and then waits for some period for the effects of the adjustment to have an impact on the water temperature. If the water temperature is not at the right temperature after this waiting period, then a second adjustment is made, followed by another period of waiting. This process continues iteratively until a satisfactory temperature is reached.

Discrete, finite state control has been observed with such slow systems as waterbaths (Crossman & Cooke, 1974), large ships (Veldhuyzen & Stassen, 1976), and tele-operated robots (Sheridan, 1992). It has also been observed in faster systems that are at the margins of stability such as high performance aircraft (e.g., Young & Meiry, 1965). Figure 7.7 illustrates a hypothetical example of a discrete style of control that might be used to dock a spacecraft. The switching boundaries are set at constant ratios of position to velocity (diagonal lines in state space). From an initial position a thrust command would cause the craft to accelerate toward the docking station. On encountering the first diagonal boundary, the thrust command would be terminated (control would be set at a neutral position) allowing the craft to coast toward the docking station at a constant velocity. For many control situations (e.g., stopping an automobile at an intersection) the neutral control (i.e., coast) region may show some deceleration due to friction or drag. Upon encountering the second diagonal boundary, a reverse thrust would be initiated causing the craft to decelerate as it

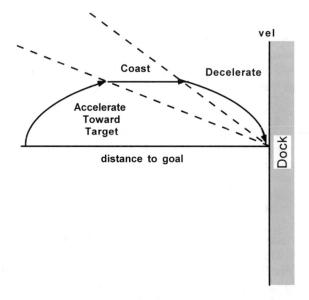

FIG. 7.7. A finite state controller. The switching criteria are diagonal lines (constant time-to-contact). Three controls are an acceleration toward the goal (bang), a coast (zero control input) resulting in a constant velocity, and a deceleration (bang) into the target.

approaches contact with the dock. Note that because the diagonal boundary is not an ideal switching curve, a step thrust command will typically not result in a contact with zero velocity. Thus, the final phase of control may involve a more proportional style of control in which fine adjustments of thrust are input in order to make a "soft" contact with the dock.

Control strategies that include a *coast* phase tend to reduce fuel consumption (e.g., Athans & Falb, 1966). Such strategies may reflect a concern for fuel economy; however they may also reflect limitations of the perceptual system or the internal model of the spacecraft (Veldhuyzen & Stassen, 1976). For example, a controller may have only a fuzzy notion of an ideal switching curve (see Fig. 7.3) and the coast region may be a reflection of this uncertainty. The first diagonal might reflect a conservative control strategy that allows ample opportunity to brake in time to avoid a hard collision with the dock. An error of stopping too short would normally be less costly (and easier to correct) than colliding with the dock at a high velocity. Also, note that the inverse slope of each diagonal line in state space reflects a constant time-to-contact with the goal position (i.e., meters/meters/s = s) for a constant velocity motion. There is some evidence that biological systems have neurons that are sensitive to this ratio (e.g., Sun & Frost, 1998; Wagner, 1982). Thus, a "coast" region might reflect constraints within the feedback loop (i.e., perceptual system). Perceptual limitations and implications for finite state control will be considered in more detail in chapter 22.

PROPORTIONAL OR DISCRETE CONTROL?

Is human movement control discrete or proportional? This question has often been posed as a dichotomy and some researchers have expressed strong opinions about the true nature of human perceptual motor skills. For example, Craik (1947) made the general claim that human motor skill reflects an intermittent (discrete) servo system. Pew (1974) presented an interesting discussion of the difference between *discrete* and *continuous* models of human tracking. It is interesting to read the conclusion of his analysis:

> After working for several years to try to decide whether a discrete or a continuous representation was more appropriate, I have found no prediction that unambiguously distinguishes the two possibilities and have concluded that while the discrete representation is more intuitively compelling, both kinds of analyses are useful and provide different perspectives and insights into the nature of performance at the level of the simple corrective feedback system. (p. 12)

Pew's comments serve as an important reminder that models provide ways of looking at the world. This is part of the inductive process of science. A model is a hypothesis. It may be proven wrong, but it cannot be proven absolutely true. In fact, a stronger claim might be made that due to constraints of the modeling process (e.g., the assumptions required by the particular mathematical formalisms involved), models are never completely accurate. Their value depends on the insights that they provide into the phenomenon of interest and on the empirical tests that they suggest.

Continuous and discrete models provide different perspectives on perceptual-motor skill. Thus, it may be counterproductive to frame the question of continuous control versus discrete control as if there is a right answer. Rather, it is a question of the utility of the perspective that the alternative types of models provide. This will typically depend on the goals of the analysis and the precision of the measurements available. It also depends very much on the time scale used in observations. At a very fine time scale action may involve a sequence of discrete or ballistic responses. For example, the firing of neurons is a set of discrete, all-or-none responses. Yet, the movement of an arm that depends on integration over many neurons may best be described at a coarser time scale as continuous. At this point, it might be best to treat proportional and discrete models as complementary perspectives that can both contribute to a deeper understanding of skilled performance.

It is possible that most natural movements (e.g., a tennis swing) will have a blend of open-loop and closed-loop control as suggested by the Successive Organization of Perception Model (Krendel & McRuer 1960; McRuer, Allen, Weir & Klein, 1977) illustrated in Figure 7.8. The outer of the two open-loop paths reflects an ability to respond directly to the input (e.g., a shot from the opponent) with a pre-packaged "motor program" that reflects knowledge about both the input (e.g., the ball trajectory) and the control system obtained over years of training. The inner of the two open-loop paths reflects an ability to respond directly to the input (e.g., the oncoming ball) in a continuous fashion that anticipates the movement dynamics without reference to an error signal. This feedforward path, reflected by the $O(s)$ operator, is called pursuit control. The closed-loop path, $C(s)$, reflects an ability to refine either form of open-loop response based on moment-to-moment monitoring of error feedback. This refinement will allow the actor to compensate for both unexpected disturbances and aspects of the movement that may be incompletely specified in the open-loop path. For example, the open-loop path of a skilled tennis player may be capable of getting the tennis racket within the "ballpark" of the desired outcome, and the closed-loop path may contribute the fine adjustments needed to perfect the shot. When playing a highly practiced piece, a musician's fingers may move in an open-loop manner to the

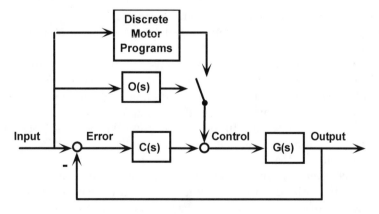

FIG. 7.8.   A multi-path control system. O(s) is the pursuit control path. C(s) is the closed loop control path. G(s) is the system dynamics (after Krendel & McRuer, 1960).

appropriate keys, strings, or frets on the instrument (taking advantage of the consistent or predictable features of the instrument). Simultaneously, a closed-loop path may adjust the phasing of the actions to keep in synchrony with the other musicians. A system where the two paths are designed to complement each other could provide control that is both more precise (i.e., more effective in achieving the goal) and more robust (i.e., stable over a broader range of conditions) than could be achieved by either path alone.

The appropriate balance between the open-loop and closed-loop paths is likely to depend on the consistency of the task situations (e.g., practice in consistent task situations may allow the open-loop path to dominate) and on the skill level of the actor. Novices may need to rely heavily on the closed-loop path. Experts might be better tuned to the task dynamics. The open-loop path of an expert should be tuned to take advantage of the natural consistencies or invariants within the task environment (e.g., positions of the keys on the instrument). And the closed-loop path should be tuned to provide stable adjustments demanded by the natural sources of variability (e.g., the behavior of other musicians). Note that the tuning of the expert can lead to performance errors when a normally consistent aspect of the task environment is unexpectedly changed. This was illustrated when the vault was set at the wrong height during the 2000 Summer Olympics in Sydney. As a result, highly skilled gymnasts were losing control, suggesting that there were open-loop components to this skill that were out of tune with the unexpected height of the vault.

OVERVIEW

This chapter introduced the concept of discrete control systems. These systems respond in a ballistic fashion. A discrete correction is executed in an all-or-none manner without being influenced by feedback. With discrete control systems, responses are not continuously linked to error as occurs in continuous proportional adjustments. Rather, the connection is an iterative one — with adjustments being made discretely. This style of control can result from intermittent sampling of error or from constraining the control actions to prepackaged programs (i.e., motor programs) that operate ballistically.

Also, it is important to realize that a continuous, smooth arm movement can be the product of discrete control inputs. Thus, an output function that is well-described mathematically using continuous differential equations does not necessarily indicate that the control processes contributing to the output are continuous. Further, a discrete control may be the optimal solution to some control problems. Not only might a discrete control get to the goal quicker, but it may be less likely to become unstable due to delays in the feedback path.

Finally, this chapter introduced the state space as a way that can help researchers to visualize the dynamic constraints operating on control systems. This representation will appear again in later chapters. The state space is an important tool for studying dynamic systems and it becomes particularly important for visualizing nonlinear dynamics (e.g., see Abraham & Shaw, 1985).

REFERENCES

Abraham, R. H., & Shaw, C. D. (1982). *Dynamics: The geometry of behavior*. Santa Cruz, CA: Aerial Press.

Angel, E. S., & Bekey, G. A. (1968). Adaptive finite state models of manual control systems. *IEEE Transactions on Man–Machine Systems, MMS-9*, 15–20.

Athans, M., & Falb, P. L. (1966). *Optimal control*. New York: McGraw-Hill.

Bekey, G. A. (1962). The human operator as a sampled data system. *IRE Transactions on Human Factors in Electronics, HFE-3*, 43–51.

Bekey, G. A., & Angel, E. S. (1966). Asynchronous finite state models of manual control systems. In *Proceedings of the Second Annual Conference on Manual Control* (NASA-SP-128, pp. 25–37). MIT, Cambridge, MA.

Burnham, G. O., & Bekey, G. A. (1976). A heuristic finite-state model of the human driver in a car-following situation. *IEEE Transactions on Systems, Man, and Cybernetics, SMC-6*, 554–562.

Carlton, L. G. (1981). Movement control characteristics of aiming responses. *Ergonomics, 23*, 1019–1032.

Costello, R. G. (1968). The surge model of the well-trained operator in simple manual control. *IEEE Transactions on Man–Machine Systems, MMS-9*, 2–9.

Craik, K.J.W. (1947). Theory of human operator in control systems. I. The operator as an engineering system. *British Journal of Psychology, 38*, 56–61.

Crossman, E.R.F.W., & Cooke, J. E. (1974). Manual control of slow response systems. In E. Edwards & F. P. Lees (Eds.), *The human operator in process control* (pp. 51–66). London: Taylor & Francis.

Crossman, E.R.F.W., & Goodeve, P. J. (1983). Feedback control of hand-movement and Fitts' Law. *Quarterly Journal of Experimental Psychology, 35A*, 251–278.

Fitts, P. M. (1954). The information capacity of the human motor system in controlling the amplitude of movement. *Journal of Experimental Psychology, 47*, 381–391.

Fogel, L. J. (1955). A note on the sampling theorem. *IRE Transactions on Information Theory, IT-1*, 47–48.

Graham, D., & McRuer, D. (1961). *Analysis of nonlinear control systems*. New York: Dover.

Jagacinski, R. J., Plamondon, B. D., & Miller, R. A. (1987). Describing movement control at two levels of abstraction. In P. A. Hancock, (Ed.), *Human factors psychology* (pp. 199–247). Amsterdam: North Holland.

Jagacinski, R. J., Repperger, D. W., Moran, M. S., Ward, S. L., & Glass, B. (1980). Fitts' Law and the microstructure of rapid discrete movements. *Journal of Experimental Psychology: Human Perception and Performance, 6*, 309–320.

Kantowitz, B. H., & Knight, J. L. (1978). Testing tapping time-sharing: Attention demands of movement amplitude and target width (pp. 205–227). In G. E. Stelmach (Ed.), *Information processing in motor control and learning*. New York: Academic Press.

Keele, S. W. (1968). Movement control in skilled motor performance. *Psychological Bulletin, 70*, 387–403.

Krendel, E. S., & McRuer, D. T. (1960). A servomechanisms approach to skill development. *Journal of the Franklin Institute, 268*, 24–42.

McRuer, D. T., Allen, R. W., Weir, D. H., & Klein, R. H. (1977). New results in driver steering control models. *Human Factors, 19*, 381–397.

Moray, N. (1986). Monitoring behavior and supervisory control. In K. R. Boff, L. Kaufman, & J. P. Thomas (Eds.), *Handbook of perception and human performance* (pp. 40.1–40.51). New York: Wiley.

Pew, R. W. (1964, May). *A model of human controller performance in a relay control system*. Paper presented at the Fifth National Symposium on Human Factors in Electronics, San Diego, CA.

Pew, R. W. (1974). Human perceptual-motor performance. In B. H. Kantowitz (Ed.), *Human information processing: Tutorials in performance and cognition* (pp. 1–39). New York: Wiley.

Sandefur, J. T. (1990). *Discrete dynamical systems: Theory and application*. Oxford, England: Clarendon.

Sheridan, T. B. (1992). *Telerobotics, automation and supervisory control*. Cambridge, MA: MIT Press.

Sun, H., & Frost, B. J. (1998). Computation of different optical variables of looming objects in pigeon nucleus rotundus neurons. *Nature Neuroscience, 1*, 296–303.

Veldhuyzen, W., & Stassen, H. G. (1976). The internal model. In T. B. Sheridan & G. Johannsen (Eds.), *Monitoring behavior and supervisory control* (pp. 157–170). New York: Plenum.

Wagner, H. (1982). Flow-field variables trigger landing in flies. *Nature, 297*, 147–148.

Wickens, C. D. (1992). *Engineering psychology and human performance* (2nd ed.). New York: Harper Collins.

Woodworth, R. S. (1899). The accuracy of voluntary movement. *Psychological Review, 3*(3), 1–114.

Young, L. R., & Meiry, J. L. (1965). Bang-bang aspects of manual control in higher-order systems. *IEEE Transactions on Automatic Control, 6*, 336–340.

Interactions Between Information and Dynamic Constraints

> *Any physical system will be imbedded in a "heat bath," producing random microscopic variations of the variables describing the system. If the deterministic approximation to the system dynamics has a positive entropy, these perturbations will be systematically amplified. The entropy describes the rate at which information flows from the microscopic variables up to the macroscopic. From this point of view, "chaotic" motion of macroscopic variables is not surprising, as it reflects directly the chaotic motion of the heat bath.*
> —Shaw (1984, p. 34)

Fitts' model of movement time used an information metric. This metric is very useful for characterizing the noise properties of a communication channel. The arm is viewed as a channel through which an intention to move to a specific location is communicated. The output variability reflects the signal-to-noise properties of the communication channel. The problem with this metaphor, however, is that it ignores dynamics. The arm is a physical system; it has mass and is governed by the physical laws of motion. The first-order and second-order lags presented in the previous chapters address the issue of dynamics. The second-order lag is more consistent with the physical laws of motion. However, neither feedback model of movement addresses the issue of noise or motor variability. A complete theory of movement must address both noise or information and dynamics or the mechanics of motion.

SCHMIDT'S LAW

Schmidt and his colleagues (Schmidt, Zelaznik, & Frank, 1978; Schmidt, Zelaznik, Hawkins, Frank, & Quinn, 1979) addressed the issue of dynamics and variability head on. First, they noted that the target width may not reflect the actual movement variability as assumed by Fitts' Law. That is, the distribution of movement endpoints can be narrower (or wider for very small targets) than the target width. Fitts treated movement amplitude and target width as independent variables and movement time (MT) as a dependent variable. Schmidt and his colleagues, however, chose to manipulate amplitude and movement time and measure the movement variability or effective target width (W_e). They argued that

> with the goal of coming to an understanding of movement control in mind, we think that it makes more sense to control experimentally both A and MT (the latter via metronome-paced movements or through previous practice with knowledge of results about movement time), using W_e as the single dependent variable; subjectively it seems to us that A and MT are determined in advance by the subject, and W_e "results" from these decisions, making W_e a logical choice for a single dependent variable. (p. 185)

Empirical studies that have constrained distance (A) and movement time (MT) and that have directly measured W_e have found a linear relation between W_e and movement speed (A/MT) for brief movements on the order of 200 ms or less. This relation:

$$W_e = K \times A/MT \qquad (1)$$

is sometimes called Schmidt's Law (Jagacinski, 1989; Keele, 1986; Schmidt et al., 1978; Schmidt et al., 1979). The logic behind this model comes from simple physical principles. The argument is that the human controls movement via discrete impulses of force, and the variability in these impulses is due to variation in the magnitude of the force and variation in the duration over which it is applied. The impulse force determines the speed and distance covered by the movement. To cover the same distance at a greater speed or a longer distance at the same speed would require a proportionally larger impulse. If variability in force and duration are each proportional to their respective magnitudes, and if the impulse has a certain shape (Meyer, Smith, & Wright, 1982; see chap. 19), then the spatial variability of the movement endpoint will be proportional to (A/MT).

MEYER'S OPTIMIZATION MODEL

It is not immediately obvious why the relation between amplitude, target width (movement variability), and movement time should be linear under experiments where time is constrained (Schmidt's Law), but log-linear under experiments where accuracy is constrained (Fitts' Law). Meyer and colleagues (Meyer, Abrams, Kornblum, Wright, & Smith, 1988; Meyer, Smith, Kornblum, Abrams, & Wright, 1990) developed the "optimized dual-submovement model" that provides a framework consistent with both the predictions of Schmidt's and Fitts' Laws. The idea of optimization introduces another constraint on the control problem. An optimal control solution will be a control solution that minimizes (or maximizes) some performance criterion. In the context of the positioning task, subjects are typically instructed to move to the target "as quickly as possible." Thus, they are explicitly asked to choose a control solution that minimizes movement time.

Consistent with Schmidt's Law, Meyer et al.'s optimization model assumes that the variability for each submovement is proportional to its average velocity. Because of this variability, acquiring a target may require more than one submovement (this model assumes at most two submovements). Thus, total movement time in acquiring a specified target will depend on the sum of the movement times for the submovements. Here is where the assumption of optimality comes in. Meyer et al. (1990) provided an intuitive description:

> Finally, another key assumption is that the average velocities of the primary and secondary submovements are programmed to minimize the average total movement duration (T). This assumption stems from the demands of the typical time-minimization task. Confronted with these demands, subjects presumably try to reach the target region as quickly as possible while attaining some set high proportion of target hits. To achieve their aim, they must adopt an appropriate strategy for coping with the effects of motor noise. Such a strategy requires making an optimal compromise between the mean duration (T_1) of primary submovements and the mean duration (T_2) of secondary submovements, whose sum determines the average total movement duration (i.e., $T_1 + T_2 = T$).
>
> In particular, the primary submovements should not take too much time. If they are very slow, then this would allow the noise in the motor system to be low, yielding greater spatial accuracy . . . without a need for secondary submovements. However, it would also tend to over inflate the average total movement duration because of an excessive increase in the mean primary-submovement duration (T_1).
>
> On the other hand, the primary submovements should not take too little time either. If they are very fast, then this would generate lots of noise, causing them to miss the target frequently. As a result, many secondary corrective submovements would then have to be made, and the average total movement duration would again tend to be over inflated because of an excessive increase in the mean secondary-submovement duration (T_2).
>
> So under the optimized dual-submovement model, there is a putative ideal intermediate duration for the primary submovements, and associated with this ideal, there is also an ideal intermediate duration for the secondary submovements. (pp. 205–206)

To summarize, the speed for the first movement is chosen so that many, but not all, of the initial movements terminate in the target. In essence, the first movement is designed to reach the general ballpark of the target. For those cases where the initial submovement terminates outside of the target, rapid corrections can make up the difference. This strategy is chosen to minimize the average time to reach the target.

Meyer et al.'s optimization model predicts that the average total movement time will be closely approximated by the following equation:

$$MT = k_1 + k_2 \, (A/W)^{1/2} \qquad (2)$$

where k_1 and k_2 are constants, A is the distance to the target (amplitude), and W is the target width. The power function is similar to the log function and provides a very good fit to data from spatially constrained movements (e.g., Ferrell, 1965; Gan & Hoffmann, 1988b; Hancock, Langolf, & Clark, 1973; Kvålseth, 1980; Sheridan &

Ferrell, 1963). The exponent, $1/2$, reflects the constraint on the maximum number of submovements. A more generalized model has been proposed (Meyer et al., 1990) in which the exponent in Equation 2 is $1/n$, where n reflects the maximum number of submovements and is a strategic parameter chosen by the performer. As n approaches infinity, the power function approaches a log function, that is, Fitts' Law. Meyer et al. (1990) reanalyzed their data and found a better fit with n equal to 4 or more. They also reanalyzed movement time data from Fitts (1954) and estimated n to be 3 for those data.

The optimal control metaphor, like the servomechanism metaphor, views movement as a control problem. However, the optimal control metaphor puts additional layers on the control system. First, each individual submovement is considered to be ballistic or open-loop. At another level, the system is an intermittent feedback control system in which a series of submovements are generated until the target is acquired. This control system operates in the short run (i.e., this reflects control adjustments within a trial). At a higher level, adjustments are made in terms of the parameters by which the individual submovements are generated. Settling in on the appropriate control strategy that minimizes movement time will happen over many trials of a particular movement task. Thus, the optimization model assumes learning in which the system tunes its own dynamic properties to satisfy some cost functional or higher order constraint. Thus, the control system is adaptive. The topic of adaptive control is addressed more thoroughly in later chapters.

Key parameters of the optimization model are the maximum number of submovements and the velocity of individual submovements. A possible, but not necessary, implication of this model is that the human makes some decision about the trade-off between the speed of individual submovements and the maximum number of submovements that is acceptable. Presumably, the appropriate balance between these two parameters might be discovered in the course of iterative adjustments during practice. Thus, this optimal control model for movement control is a more sophisticated version of the feedback control metaphor. An important question, suggested by this metaphor, is the need for a theory of the learning process that guides the evolution of the optimal performance seen with well-practiced subjects. Peter Hancock (1991, personal communication) suggested that there might be a boundary seeking process where speed is gradually increased until an error threshold is exceeded. This would be followed by a regression to a slower speed. This process would continue in an iterative fashion throughout practice as the actor hunts for the optimal balance between speed and accuracy. Evaluation of this process requires analysis of sequential dependencies across trials and requires a look at higher moments of the response distributions than means and standard deviations (Newell & Hancock, 1984).

DRIPPING FAUCET

A common complaint of insomniacs is a leaking faucet. No matter how severely the tap is wrenched shut, water squeezes through, the steady, clock like sound of the falling drops often seems just loud enough to preclude sleep. If the leak happens to be worse, the pattern of the drops can be more rapid, and irregular. A dripping faucet is an example of a system capable of a chaotic transition, the same system can change from a periodic and predictable to an aperiodic, quasi-random pattern of behavior, as a single parameter (in this case, the flow rate) is varied. Such a transition can readily be seen by eye in many faucets, and is an experiment well worth performing in the privacy of one's own kitchen. If you slowly turn up the flow rate, you can often find a regime where the drops, while still separate and distinct, fall in an irregular, never-repeating pattern. The pipe is fixed, the pressure is fixed; what is the source of the irregularity? (Shaw, 1984, pp. 1–2).

The aforementioned quote and the quote at the beginning of this chapter are from Shaw's (1984) thesis on the dripping faucet, which was very important in the emerging field on nonlinear dynamics. In order to characterize the path to chaos taken by the dripping faucet, Shaw found it useful to consider both a dynamic model and information statistics to characterize the data. This combination of metaphors led Flach, Guisinger, and Robison (1996) to speculate that there may be parallels between the arm as a dynamic system and the dripping faucet, although there is no evidence to indicate that the arm exhibits chaotic dynamics in these target acquisition tasks. At slow rates, each drop interval is well predicted by the preceding interval and each arm movement is precise, but at faster rates the noise increases (i.e., intervals become less regular and the accuracy of movement decreases). Why? What is the source of the irregularity? How can a deterministic system create "entropy"?

The arm and the droplet are dynamic systems that should be governed in a deterministic fashion by the laws of mechanics. Thus, there should be a lawful or functional relation between the forces input to the system (I) and the motion that results as output (O). Suppose, however, there is some variability or uncertainty associated with the input (δI). How does this variability scale to the output? In other words, what is the variability in the output (δO) with respect to the input (δI) or what is $\delta O / \delta I$. Suppose the functional relation between I and O is:

$$O = I^2 \tag{3}$$

as shown in Fig. 8.1. Then the uncertainty around the input will be magnified by the deterministic dynamics so that the uncertainty at the output will be:

$$\delta O / \delta I = 2I \tag{4}$$

Thus, the uncertainty at the output will increase as a function of the magnitude of the input. The variability of the output will be larger in proportion to the size of the input. This is an example of a deterministic system with positive entropy. In Shaw's words, "the 'purely deterministic' map acts as an information source" (p. 33). Thus, the question is, how does the deterministic map magnify the noise? Figure 8.1 shows a nonlinear map in which the noise is amplified through a second-order function. The variability of the input is constant, but the variability of the output scales with the magnitude of the input.

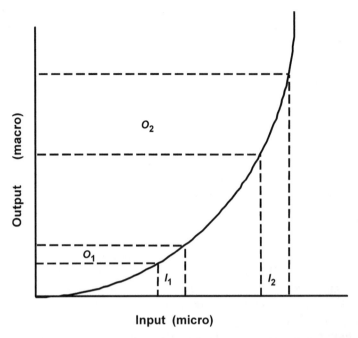

FIG. 8.1. A nonproportional scaling of noise at the microlevel at which an intention is specified (INPUT) to variability at the macrolevel in terms of arm movements. Although the variability for the two inputs is equivalent, the output variability increases with increased magnitude of the input.

It is important to appreciate that "information" is used synonymously with "entropy," or the "uncertainty" or "variability" associated with the outcome. As Shannon and Weaver (1963) noted:

In the physical sciences, the entropy associated with a situation is a measure of the degree of randomness, or of "shuffled-ness" if you will in the situation. . . . That information be measured by entropy is, after all, natural when we remember that information, in communication theory, is associated with the amount of freedom of choice we have in constructing messages. Thus for a communication source one can say, just as he would also say it of a thermodynamic ensemble, "This situation is highly organized, it is not characterized by a large degree of randomness or choice — that is to say the information (or entropy) is low." (pp. 12–13)

The association with information and entropy can sometimes create confusion. When the information, variability, or uncertainty increases due to "noise" in a communication channel, the increase in information is "bad." Again, it is instructive to read Shannon and Weaver's (1963) comments about noise:

If noise is introduced, then the received message contains certain distortions, certain errors, certain extraneous material, that would certainly lead one to say that the received message exhibits, because of the effects of noise, an increased uncertainty. But if the uncertainty is increased, the information is increased, and this sounds as though the noise were beneficial!

It is generally true that when there is noise, the received signal exhibits greater information — or better, the received signal is selected out of a more varied set than is the
transmitted signal. This is a situation which beautifully illustrates the semantic trap into
which one can fall if he does not remember that "information" is used here with a special
meaning that measures freedom of choice and hence uncertainty as to what choice has
been made. It is therefore possible for the word information to have either good or bad
connotations. Uncertainty which arises by virtue of freedom of choice on the part of the
sender is desirable uncertainty. Uncertainty which arises because of errors or because of
the influence of noise is undesirable uncertainty. (p. 19)

For the water droplet, the uncertainty on the input side may arise from the random motion of molecules within the water drop and the air surrounding the water
drop, as suggested by the quote at the beginning of the chapter. For the arm, there is
also a "heat bath" of neurons, neurochemicals, and tissue between the intention to
move and the resulting movement. This "heat bath" is a potential source of noise.
Variability in communicating this intention through the heat bath may also be amplified by the dynamics of the arm. A key issue in modeling the arm as a dynamic system will be to understand how the microscopic sources of variability (i.e., communication noise) that arise in the communication of the intention are amplified by the
dynamics of the arm. In other words, it is important to understand how microscopic
variability is scaled to the macroscopic directed arm movement variability.

As a first hypothesis, Flach et al. (1996) offered a very simple linear model for the
scaling of variability. They propose that the variability will scale proportionally with
the forces associated with the movement:

$$W_e = k_1 + k \times F \tag{5}$$

Assuming a symmetric bang-bang command and no variability in the timing of the
command (see Schmidt, Zelaznik, & Frank, 1978), then the force can be approximated
from the kinematics of the movement using the following relation:

$$F \propto M \times A/MT^2 \tag{6}$$

where M is the mass of the arm, A is the amplitude or distance moved, and MT is the
duration of the movement. The (A/MT^2) term is proportional to acceleration for an
ideal force pattern, but does not necessarily correspond directly to an actual kinematic of the movement (e.g., average or peak acceleration). Substituting for the F in
Eq. 5 results in the prediction that variability of the output will scale directly with
movement distance (A) and inversely with the square of movement duration (MT):

$$W_e = k_1 + k \times M \times A/MT^2 \tag{7}$$

or

$$W_e = k_1 + k_2 (A/MT^2) \tag{8}$$

where W_e is the macroscopic variability of the movement, k_1 is a constant that reflects
minimum variability, k_2 is a constant that reflects the mass of the arm (M) and the rate
at which entropy scales with force (k).

This model was tested against the data collected by Schmidt et al. (1978; data were estimated from Figure 8). The value estimated for k_1 was 2.82; and the value for k_2 was 4093.07 and the correlation between W_e and (A/MT^2) was .96. Schmidt et al. found a correlation of .97 between W_e and A/MT. So the model in which variability scales with force fits the data nearly as well as Schmidt's model in which the variability was assumed to scale with speed.

In an accuracy-constrained movement task, as employed by Fitts, ultimate success depends on the relation between W_e and the specified target width W. If W_e is much larger than W, then initial movements will often fall outside the target and additional submovements will be required. If W_e is much smaller than W, then the force used and the resulting speeds may be less than optimal. The movements will be slower than necessary. This rationale is essentially identical to the rationale motivating Meyer's model. The task then is to discover how the actor balances speed/force and variability in a way that accomplishes the task of acquiring the target in minimum time. As a first approximation, the target width, W, can be substituted for W_e and then it is possible to solve for MT and to make a prediction with regard to the data from Fitts' experiments.

$$W = 2.82 + 4093.07(A/MT^2) \tag{9}$$

$$MT = [(4093.07 \times A)/(W - 2.82)]^{1/2} \tag{10}$$

This prediction assumes only one submovement. Note that there are no free parameters in this equation. The constants used are based on Schmidt et al.'s data obtained in a timing constrained task. W and A are parameters of the task (set by the experimenter). Note, also, that the result is a power function with exponent of $1/2$. This is similar to Meyer's optimization model. However, in Meyer's model, the $1/2$ was based on limiting the maximum number of submovements to two. In the Flach et al. model, the $1/2$ results from the assumption that variability scales with acceleration (Meyer and Schmidt assumed that variability scaled with average velocity). Figure 8.2 shows a comparison of the predictions of this equation with the data obtained by Fitts. The correlation was .935. In evaluating this correlation, keep in mind again that no free parameters were adjusted to get a "best fit."

Note that whereas the shape of the function shown in Fig. 8.2 is in close agreement with the data, there is a constant error between the predictions and the data obtained by Fitts. Of course, if W_e is set equal to W as in the previous equation, then some movements will terminate outside the target (because W_e is the standard deviation of the movement variability). Thus, additional submovements will be required. To see how these additional movements contributed to the overall movement times, Flach et al. conducted Monte Carlo simulations. For these simulations, the logic outlined was used to determine the movement duration for each submovement. The duration of these submovements varies because A (distance from the target) will be different for each submovement. The endpoint of the movement was treated as a random variable with mean at the target center and variability about the mean determined by the previous equations (W_e). The Monte Carlo simulations generated submovements until a movement terminated inside the target. Figure 8.3 shows the movement times

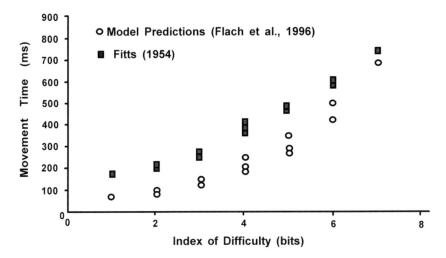

FIG. 8.2. A comparison of movement time data (msec) obtained by Fitts (1954) (filled squares) and the predictions of Flach et al.'s (1996) model assuming a single submovement (open circles).

that result as a function of averaging 100 simulated trials. The simulation produced both movement times and the number of submovements as dependent variables. The target was acquired on the first submovement on approximately 40% of the trials and was acquired on the second submovement for about 25% of the trials. It is interesting to note that the mean number of submovements ranged from slightly greater than 2 for the lowest *ID*s to slightly greater than 3 for the higher *ID*s. This is in close agreement with the assumptions of Meyer's model. As already noted, Meyer et al. (1990)

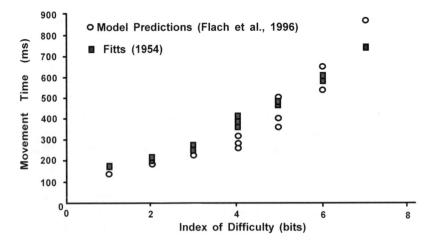

FIG. 8.3. Movement time as a function of the index of difficulty for a discrete movement task. The filled squares are empirical results obtained by Fitts (1954). The open circles are results from a Monte Carlo simulation using the dynamic scaling model with multiple submovements (Flach et al., 1996).

argued for $n = 3$ for their reanalysis of the Fitts (1954) data and $n = 4$ or more for their own 1988 data.

ADDITIONAL TESTS OF SUBMOVEMENT STRUCTURE

There are striking similarities between the Flach et al. (1995) model and Meyer at al.'s (1988, 1990) optimization model. Both models assume that the movement times obtained in the Fitts' paradigm reflect a trade-off between speed and accuracy in which the speed/accuracy of individual submovements is traded off against the speed costs associated with additional submovements. However, there are also important differences between the two models. In Meyer et al.'s model, spatial variability is assumed to scale with average submovement velocity. For the Flach et al. model, spatial variability is assumed to scale with force (acceleration). For Meyer's model, the number of submovements is a parameter (independent variable). For the Flach et al. model, the number of submovements is predicted (dependent variable).

Liao, Jagacinski, and Greenberg (1997) tested the relation between the spatial variability of submovement endpoints and average submovement velocity in a target acquisition task that used a wide range of movement amplitudes. Contrary to the assumption of Meyer et al. (1988, 1990), this relation was curvilinear rather than linear. The relation between spatial variability of submovement endpoints and movement amplitude/(submovement duration)^2 was also found to be curvilinear for younger adults, although nearly linear for older adults (Liao, personal communication, January 1999). The data for the younger adults are contrary to the assumption of the Flach et al. (1996) model. These analyses call into question an important aspect of each of the previous models.

As an alternative way of describing submovement variability, Liao et al. (1997) noted that mean submovement duration was linearly proportional to $\log_2(1/\text{mean}$ absolute relative error) (Fig. 8.4). In the tradition of Crossman and Goodeve (1983), Keele (1968), and Langolf (1973), relative error was defined as the ratio of distance from the target center after a submovement to distance from the target center before a submovement. Relative error was assumed to be normally distributed and centered over the target. Mean absolute relative error is then proportional to the standard deviation of relative error, and is a measure of spatial variability. The inverse of relative error can be considered a measure of relative accuracy. The speed–accuracy relation for individual submovements depicted in Fig. 8.4 is thus similar to Fitts' Law for overall movements. However, it leaves open the question of how this speed–accuracy relation is related to the dynamics of the underlying movement generation process.

In the style of the Meyer et al. (1988, 1990) model, Liao et al. (1997) assumed there was a maximum number of submovements that constrained the performer's choice of submovements from the speed–accuracy trade-off function to compose overall movements. Comparison of the data with simulations suggests that: (a) the maximum number of submovements was 3 for both older and younger adults in this experiment, and (b) rather than choosing submovements to minimize the total move-

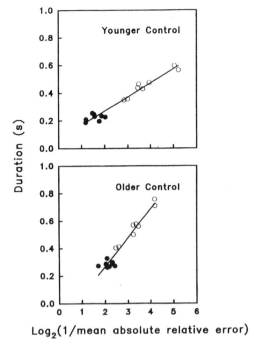

FIG. 8.4. Plots of mean submovement duration versus the logarithm of 1/(mean absolute relative error) for first (open circles) and second (dark circles) submovements for each of eight targets for older and younger adults. From "Quantifying the performance limitations of older and younger adults in a target acquisition task" by M. Liao, R. J. Jagacinski, & N. Greenberg, 1997, *Journal of Experimental Psychology: Human Perception and Performance, 23*, p. 1657. Copyright 1997 APA. Reprinted by permission.

ment time, performers used a heuristic choice of submovements that approximates optimal performance. The heuristic appears to be that the first submovement accuracy (or duration) is a function of movement amplitude, A (see Gan & Hoffmann, 1988), and the relative accuracy is sufficiently high that subsequent submovements can have a constant, low level of relative accuracy (and constant short duration) (Fig. 8.4) and still meet the overall constraint on maximum number of submovements. In other words, only the first submovement is strategically varied across different targets, and subsequent submovements are all relatively inaccurate, but sufficient to capture the target without using too many of them. Use of such a heuristic would limit the number of parameters that a performer would have to adaptively adjust in order to achieve good performance.

CONCLUSIONS

Control systems must deal with both dynamic and information constraints. Fitts' Law emphasizes the information constraints on movement. Control theoretic models (e.g., first- and second-order lags) emphasize the dynamic constraints on movement. Meyer et al.'s (1988, 1990), and Flach et al.'s (1996) models and Shaw's dripping faucet analysis consider the conjunction of stochastic (information/entropy) and dynamic constraints. A critical question is how input or command uncertainty scales through the dynamics to determine output variability? Or, in other words, how do the two sources of constraint interact to bound performance? Liao et al. (1997) suggested that modification of the models reviewed here may still be necessary to account for this relation.

The dripping faucet metaphor was adapted from the literature on nonlinear dynamics. Shaw used a nonlinear spring model for the water drip. Flach et al., however, used a linear 2nd order model to describe the amplification of the microlevel information source to the resulting output variability. Thus, this model does not predict chaotic behavior in Fitts' Law tasks. The metaphor was chosen to illustrate the coupling between constraints on perception/communication (information theory) and constraints on action (dynamics) within a complex system. Complex systems must deal with uncertainty and with dynamic limits. Different languages (information statistics, differential equations) offer different perspectives and unique insights into the complex system. When exploring these systems, it is important not to assume constraints of a particular language always reflect properties of the system being studied. And perhaps some of the more interesting features of complex systems can only be appreciated by taking multiple distinct perspectives on the phenomenon.

Nonlinearity and variability are considered to be essential to the creative and adaptive development of biological systems. However, accepting nonlinearity as important does not mean that linear assumptions and models should be discarded altogether (Graham & McRuer, 1961). Linear assumptions have great power for helping us to understand the behavior of control systems. Many treatments throughout this book depend on assumptions of linearity. The views afforded by these assumptions of linearity provide an important context for appreciating the value of nonlinear dynamics. So, a healthy respect for the power of linear assumptions is recommended, but readers should be alert to the costs of these assumptions. Researchers should be aware that the answers obtained from nature sometimes have more to do with the assumptions behind the questions asked, than with the fundamental properties of nature. Hopefully, those who study this book will learn how to use the assumptions of linearity effectively, but cautiously. Readers should not be trapped by the assumptions of linearity. The goal is to provide a broad appreciation for the power of the language of control theory that spans both linear and nonlinear assumptions. In fact, it is probably a mistake to strictly categorize phenomena as being either "linear" or "nonlinear." These adjectives refer to the modeling tools, not the phenomena. For some phenomena, linear models will provide important insights. However, for other phenomena, nonlinear models will offer insights not possible with linear models. Good carpenters know when and when not to use a hammer. Likewise, good carpenters do not throw away one tool, when they find occasion to use another.

REFERENCES

Crossman, E.R.F.W., & Goodeve, P. J. (1983). Feedback control of hand-movement and Fitts' Law. *Quarterly Journal of Experimental Psychology, 35A*, 251–278. (Original work presented at the meeting of the Experimental Psychology Society, Oxford, England, July 1963.)

Ferrell, W. R. (1965). Remote manipulation with transmission delay. *IEEE Transactions on Human Factors in Electronics, 6*, 24–32.

Fitts, P. M. (1954). The information capacity of the human motor system in controlling the amplitude of movement. *Journal of Experimental Psychology, 47,* 381–391.

Flach, J. M., Guisinger, M. A., & Robison, A. G. (1996). Fitts' Law: Nonlinear dynamics and positive entropy. *Ecological Psychology, 8,* 281–325.

Gan, K., & Hoffmann, E. R. (1988). Geometrical conditions for ballistic and visually controlled movements. *Ergonomics, 31,* 829–839.

Graham, D. & McRuer, D. (1961). *Analysis of nonlinear control systems.* New York: Dover.

Hancock, W. M., Langoff, G., & Clark, D. O. (1973). Development of standard data for stereoscopic microscopic work. *AIII Transactions, 5,* 113–118.

Jagacinski, R. J. (1989). Target acquisition: Performance measures, process models, and design implications. In G. R. McMillan, D. Beevis, E. Salas, M. H. Strub, R. Sutton, & L. Van Breda (Eds.), *Applications of human performance models to system design* (pp. 135–149). New York: Plenum.

Keele, S. W. (1968). Movement control in skilled motor performance. *Psychological Bulletin, 70,* 387–403.

Keele, S. W. (1986). Motor control. In K. Boff, L. Kaufman, & J. Thomas (Eds.), *Handbook of perception and human performance* (Vol. 2, pp. 30.1–30.60). New York: Wiley-Interscience.

Kvålseth, T. O. (1980). An alternative to Fitts' Law. *Bulletin of the Psychonomic Society, 16,* 371–373.

Langolf, G. D. (1973). *Human motor performance in precise microscopic work – Development of standard data for microscopic assembly work.* Unpublished doctoral dissertaion, University of Michigan, Ann Arbor.

Liao, M., Jagacinski, R. J., & Greenberg, N. (1997). Quantifying the performance limitations of older and younger adults in a target acquisition task. *Journal of Experimental Psychology: Human Perception and Performance, 23,* 1644–1664.

Meyer, D. E., Abrams, R. A. Kornblum, S., Wright, C. E., & Smith, J.E.K. (1988). Optimality in human motor performance: Ideal control of rapid aimed movements. *Psychological Review, 95,* 340–370.

Meyer, D. E., Smith, J. E., Kornblum, S., Abrams, R. A., & Wright, C. E. (1990). Speed–accuracy tradeoffs in aimed movements: Toward a theory of rapid voluntary action. In M. Jeannerod (Ed.), *Attention and performance* (Vol. 13, pp. 173–226). Hillsdale, NJ: Lawrence Erlbaum Associates.

Meyer, D. E., Smith, J.E.K., & Wright, C. E. (1982). Models for the speed and accuracy of aimed limb movements. *Psychological Review, 89,* 449–482.

Newell, K. M., & Hancock, P. A. (1984). Forgotten moments: Skewness and kurtosis as influential factors in inferences extrapolated from response distributions. *Journal of Motor Behavior, 16,* 320–355.

Schmidt, R. A., Zelaznik, H. N., & Frank, J. S. (1978). Sources of inaccuracy in rapid movement. In G. E. Stelmach (Ed.), *Information processing in motor control and learning* (pp. 183–203). New York: Academic Press.

Schmidt, R. A., Zelaznik, H. N., Hawkins, B., Frank, J. S., & Quinn, J. T., Jr. (1979). Motor output variability: A theory for the accuracy of rapid motor acts. *Psychological Review, 86,* 415–451.

Shannon, C. E., & Weaver, W. (1963). *The mathematical theory of communication.* Chicago: University of Illinois Press.

Shaw, R. (1984). *The dripping faucet as a model chaotic system.* Santa Cruz, CA: Aerial.

Sheridan, T. B., & Ferrell, W. R. (1963). Remote manipulator control with transmission delay. *IEEE Transactions on Human Factors in Electronics, 4,* 25–29.

Order of Control

> *In performing manual skills we often guide our hands through a coordinated time-space trajectory. Yet at other times we use our hands to guide the position of some other analog system or quantity. At the simplest level, the hand may merely guide a pointer on a blackboard or a light pen on a video display. The hand may also be used to control the steering wheel and thereby guide a vehicle on the highway, or it may be used to adjust the temperature of a heater or the closure of a valve to move the parameters of a chemical process through a predefined trajectory of values over time.*
>
> —Wickens (1992, p. 452)

The previous chapters have focused on arm movements, discussing how these movements can be approximated as first- or second-order control systems. This chapter, however, considers order of control as a property of the controlled system or plant (e.g., a computer input device or a vehicle). In this case, the *order of control* (or control order) refers to the dynamic relation between displacement of a control device (e.g., a joystick or steering wheel) and the behavior of the system being controlled. The order of control specifies the number of integrations between the human's control movement and the output of the system (i.e., the plant) being controlled. This usage of the term *order* is consistent with more technical usage of the term where it refers to the order of the highest derivative in a differential equation or to the number of linked first-order differential equations used to model a system. In addition to the number of integrations between control input and plant output, the time delay between input and output is also a very important feature of the plant dynamics.

ORDER OF CONTROL

As mentioned previously, the control order refers to the number of integrations between the control input to a plant and the output of the plant. Figures 9.1, 9.2, and 9.3 graphically illustrate both the step response for zero-, first-, and second-order sys-

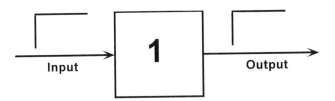

FIG. 9.1. A zero-order system has no integrations between input and output. A step input results in a step output. The proportional relation between input and output is determined by the gain. Here the gain is equal to one.

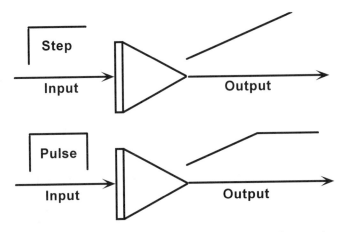

FIG. 9.2. A first-order system has one integrator between input and output. A step input results in a ramp output. The slope of the ramp is determined by the system gain. To get an approximate step output, a pulse input is required. The output stops only when the input is in the null (zero commanded velocity) position.

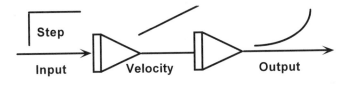

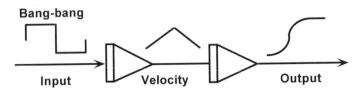

FIG. 9.3. A second-order system has two integrators between input and output. A step input results in a quadratic response with velocity increasing at a constant rate. In order to get an approximate step output response, a bang-bang input is required. The first bang determines the acceleration. The second bang determines the deceleration. When the two bangs are equal and opposite, an approximate step output is produced.

tems and the input required to produce a "steplike" output (displace the system from one fixed position to another fixed position). In discrete positioning tasks such as those modeled by Fitts' Law, the task of the human is to produce a steplike output from the plant. That is, the task is to move the plant from one position to another in minimum time and with the accuracy specified by the target width. Positioning the computer cursor on an item in a menu or on a word within a manuscript (e.g., when word processing) are examples of positioning tasks where the cursor represents the response of the "plant" and the control input is generated by the human using a mouse or other input device. In this case, the order of control would refer to the number of integrations between the mouse motion and the cursor motion.

Zero Order

A plant with zero integrations between control input and output is a position control system. That is, there is a proportional relation between displacement of the control and the output of the plant, as illustrated in Fig. 9.1. Mouse controls typically employ a zero order of control—there is a proportional relation between position of the cursor and position of the mouse on the mouse pad. When the mouse moves, the cursor also moves. When the mouse stops, the cursor stops. The scaling of mouse movement (e.g., mics) to cursor movement (e.g., pixels) refers to the gain of the system. Most computers allow users to adjust this scaling. When the gain is high, a small displacement of the mouse on the pad results in a relatively large displacement of the cursor on the screen. When the gain is low, a large displacement of the mouse on the pad results in a relatively small displacement of the cursor on the screen.

First Order

A plant with one integration between control input and output is a velocity control system. That is, there is a proportional relation between displacement of the control and the velocity or rate of the plant output. As can be seen in Fig. 9.2, in order to get a step output with a velocity control system, the operator must first displace the control in the direction of the target. The size of this displacement determines the speed of approach to the target. A small displacement will result in a slow approach. Larger displacements will result in proportionally faster approach speeds. The exact proportional relation is determined by the gain of this system. That is, the gain of the velocity control system determines the proportionality between position of input and velocity of output. Note that when the control is stopped in any position other than the null (zero input), the output continues in motion at a velocity proportional to the displacement from the null position. In order to stop the cursor at the target, the control must be returned to the null position. Thus, the commanded velocity is zero and the plant stops at whatever position it is in when the control input reaches zero.

A velocity control works best for control devices that have a well-defined null or zero position. For example, velocity control works well for a spring-centered joystick. A great advantage of a velocity control is that the extent of motion in the output space is not limited by the range of motion of the input device. For example, if an individual wanted to scroll over an effectively limitless workspace (e.g., a large field of data), then a position control would require an effectively limitless range of motion for the input device. However, a velocity control allows a limitless range of motion on the output side, even when the extent of control motion is limited. For the first-order control system, the limits on the input device constrain the velocity of output, but not the range of motion of the output.

Second Order

A plant with two integrations between control input and plant output is called an *acceleration control*. For the acceleration control, there is a proportional relation between displacement of the input and the acceleration of the plant output. As can be seen in Fig. 9.3, a displacement of the input results in an acceleration proportional to the displacement. A small displacement results in a small acceleration. Larger displacements result in proportionally larger accelerations with the proportional relation determined by the gain. If the input is then returned to the null position, zero acceleration is commanded, but the output continues to increase at a constant velocity. In order to get the output to stop at a specified target, acceleration must be commanded in the opposite direction from the initial command (in effect deceleration). When the output comes to zero velocity, then the control input must return to the zero position. Thus, in order to get a step output with the second-order system, a bang-bang input is required. The first bang determines the peak velocity. Assuming an equal displacement and duration for the second bang (the area under the two "bangs" is equal), these two parameters will determine where the system will come to rest.

This control system is somewhat more difficult than either the zero- or first-order control systems, because the reversal of the input must be made in anticipation of the final stopping position. This dynamic is typical of many real-world control tasks (e.g., vehicular control, remote manipulation, etc.). For example, in order to move a car from a stop at one corner to a stop at the second corner, drivers first must initiate acceleration in the direction of the second corner. Then they must initiate deceleration. This deceleration must be initiated in anticipation of the coming intersection. If drivers wait until they reach the intersection before braking, then they will slide into the intersection.

Video games, particularly those that simulate vehicles (e.g., cars, planes, or spacecraft), often employ second-order controls. The problem of controlling second-order systems is more difficult and more interesting than controlling zero- or first-order systems. Also, the characteristics of second-order systems are more representative of the problem of controlling systems that have inertia. Acceleration control is difficult, but most people can become skilled at using these systems with practice.

Third and Higher Order Control

Higher order control systems can also be found in domains such as aviation, as Roscoe, Eisele, and Bergman (1980) described:

Short of the submarine, the helicopter, and the hot-air balloon, the fixed-wing airplane is among the most contrary vehicles to control. When flying a specific course at constant altitude, the pilot is operating a machine that requires fourth-order lateral and third-order longitudinal control. . . . The lateral, or crosscourse, aircraft dynamics constitute a fourth-order system wherein the response to a control deflection creates a roll acceleration (ϕ''), roll rate (ϕ'), bank angle (ϕ), heading (ψ), and displacement (D). In the third-order longitudinal (vertical) mode, a control deflection initially creates a pitch acceleration (θ''), and its integrals are, successively, pitch rate (θ'), which is roughly proportional to vertical acceleration (h''); pitch(θ), which is roughly proportional to vertical speed (h'); and altitude (h). (pp. 36–37).

It is up to the reader to work out what type of input would be required to produce a step output with a third- or fourth-order control system. Think about the control actions required to move an aircraft from one altitude to another or from one course to another parallel course. People typically have great difficulty controlling third-order and higher order systems. However, as in the case of skilled pilots, people can become quite proficient with proper training and with proper feedback displays.

HUMAN PERFORMANCE

As the figures in the previous sections illustrate, the input movements required to produce a step output are strikingly different, depending on the dynamics of the plant or vehicle being controlled. Because of these differences, it might be expected that the log-linear relation of Fitts' Law that applies to arm movements might not work for modeling performance with first- or second-order control systems. The arm movements required are quite different than for the position control system. However, a number of studies have found that Fitts' Law holds well, even with higher order control devices.

Jagacinski, Hartzell, Ward, and Bishop (1978) evaluated performance in a one degree of freedom target capture task similar to that used by Fitts. The task was to position a cursor on targets that appeared on a CRT screen using a joystick control. Two control dynamics were studied. In one condition, cursor position was proportional to stick displacement (position or zero-order control). In another condition, cursor velocity was proportional to stick displacement (velocity or first-order control). Fitts' Law was found to hold in both cases. However, the slope was steeper for the velocity control system (200 ms/bit) than for the position control system (113 ms/bit).

One explanation for the increased slope for the velocity control system is based on the concept of stimulus–response (SR) compatibility. SR compatibility refers to the degree of correspondence between the response pattern and the initiating stimulus (e.g., the target). For example, Fitts and Seeger (1953) found that reaction time depended on the relation between the spatial layout of the stimuli and the spatial arrangement of the responses. The fastest responses were obtained when there was a simple spatial correspondence between the two layouts. A related construct is *population stereotype*. This refers to the degree to which stimulus–response mappings reflect cultural expectations (Loveless, 1963). For example, in the United States people expect the upward position of a switch to correspond to "lights on." The opposite convention (up = off) is common in Europe. Responses are fastest when the stimulus–response arrangement is consistent with stereotypes (i.e., expectations).

It is possible that the shape of the input movement for the position control system is more compatible and/or is more consistent with peoples' expectations given the objective of a step output. Thus, the position control system results in less processing time for each corrective iteration (i.e., submovement). The spatial correspondence between input and output in the case of the velocity control is more complex and less consistent with expectations, therefore requiring more processing time per iteration. The more corrective iterations that are required, the greater the difference in response times that might be expected between two dynamics (Jagacinski, 1989; Jagacinski, Repperger, Moran, Ward, & Glass, 1980).

Hartzell, Dunbar, Beveridge, and Cortilla (1982) evaluated performance for step changes in altitude and airspeed in a simulated helicopter. Altitude was regulated through a lagged velocity control system by manipulating a collective (joystick). Air speed was regulated through an approximate lagged acceleration control by manipulating a second joystick (cyclic). The movement times for the individual control axes were consistent with Fitts' Law. The slope of the movement time function was more than twice as large for the acceleration control than for the velocity control (1,387 vs. 498 ms/bit). This is consistent with the aforementioned arguments. The more complicated the control input relative to the output, the greater the costs associated with precise control (i.e., high indexes of difficulties).

It appears that Fitts' relation does not simply characterize limb movements; it represents a general constraint on performance of a target acquisition task. This general constraint reflects the "step" response of the human–machine control system, regardless of the particular arm movements required by the plant dynamics. The information-processing rate does vary with the plant dynamics. In general, the information-processing rate will decrease (i.e., MT slopes will increase) with increasing order of control.

Quickening and Prediction

The challenge of controlling second-order systems may not only reflect the geometry of the motion relations (SR compatibility), but it may also reflect the requirement that the controller take into account both the position and velocity of the system. For example, in order to stop a car at the threshold of an intersection, at what distance (position) should drivers initiate braking? The correct response to this question is—it depends! It depends on how fast the car is going. If traveling at a high speed, it will be necessary to initiate braking at a relatively larger distance, than if traveling at a slower speed. The driver must consider both speed and distance in order to respond appropriately. This fact is discussed in more detail in the context of frequency representations and in terms of state variables in later chapters. In other words, a driver cannot wait until the car reaches the desired position before initiating a control action. The driver must act in anticipation of reaching the intersection. This anticipation must take into account both speed and position of the vehicle.

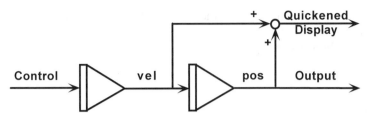

FIG. 9.4. A block diagram for a second-order system with a quickened display. The output to the quickened display is the sum of position and velocity. Effectively, the quickened display projects the output into the future based on the current velocity.

Figure 9.4 shows a block diagram for a "quickened" second-order control system. Quickening is a method for reducing the difficulty of controlling second-order and higher order systems that was proposed by Birmingham and Taylor (1954). A quickened display for an acceleration control system shows the operator a weighted combination of output position and velocity. This weighted summation effectively anticipates the future position of the plant. Thus, when operators respond to the position of the quickened display element, they are responding to position and velocity of the vehicle or plant. Quickening can greatly improve human performance in controlling second-order and higher order systems. For higher order systems, higher derivatives of the output (e.g., acceleration) are also combined in the quickened display. More generally, quickening is a prediction of the future position of the vehicle based on the current position, velocity, acceleration, and so on. The number of derivatives required depends on the order of the control system. In some sense, the addition of display quickening can effectively reduce the control task to a zero- or first-order system — although the dynamic response of the vehicle is not changed, only the dynamics of the display. The difficulty of the control task is greatly reduced with the quickened display.

The terms *quickened* and *predictive* are sometimes used as synonyms. However, prediction is a more general term than quickening. A predictive display makes some guess about the future state of a system. This guess can be based on a direct or indirect measure of the system derivatives. However, it might also be based on some other computational model of the system performance (e.g., a "fast" time simulation might be used to extrapolate t seconds into the future). In addition to derivative information, the model might include assumptions about future control inputs or about future disturbances. With simple quickening, predictions are made only on the basis of the currently available derivatives. Kelley (1968) showed that predictive displays based on fast time simulation significantly improved performance in controlling the depth of a submarine. This is a third-order control task.

There are also differences in the information that is displayed to a human operator. In simple quickened displays, only the predicted position is shown. However, predictive displays can combine predictions with other information. For example, an aircraft might be represented on an air traffic control display as a vector. The base of the vector might represent the current position of the craft and the head of the vector might represent a prediction of the future position of the craft, estimated from a combination of position with higher derivatives. In this case, the length of the vector would reflect the contribution of the higher derivative input. This type of display allows a human operator to see both the current status and a prediction of the future status of a vehicle. Wickens (1992) presented a nice discussion of prediction and quickening in displays for manual control tasks.

TIME DELAYS

Another significant attribute of the plant, in addition to its integral properties, is the presence of time delays. Whereas the integral properties reflected in the control order alter the shape of the output, relative to the input, time delays have no effect on the shape of their immediate output. Time delays affect only the temporal relation between input and output. Sometimes the term *pure time delay* is used to emphasize that the effect is only temporal; that there is no effect on the shape of the response. With a pure time delay, the shape of the response will be identical to the shape of the input. For example, a step input to a plant, characterized by a pure time delay of 100 ms, would produce a step output. If the input occurred at time T, then the response would be a step initiated at time $T + 100$ ms.

The term *pure time delay* is also used to differentiate the response of a time delay from that of a *lag*. First- and second-order lags were discussed in previous chapters. The response of these systems to a step input is a rounded step output. Thus, there is a lag or delay, from the onset of the step input until these systems approximately reach their steady state output levels. The lag's initial, gradual response starts at the time of the step input. In contrast, for a time delay, the initiation of the output is delayed relative to the input. Figure 9.5 shows the unit step responses for a first-order

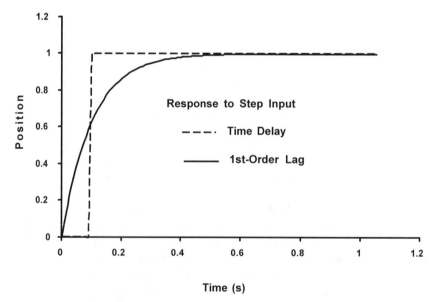

FIG. 9.5.   A graph of the step response of a pure time delay of 100 ms (solid line) and of a lag with a 100 ms time constant (dotted line).

lag with time constant of 100 ms and for a pure time delay of 100 ms. As can be seen in Fig. 9.5, the lag and time delay are not identical. Despite the technical differences, it is not unusual for lag, time lag, or time delay to be used as synonyms in common language. Usually, the meaning will be clear from the context. However, communications might be improved if researchers were more careful in the way they used these various terms.

Most natural systems have both time delays (due to the time to transmit information through some medium—transmission time) and lags (due to the inertial dynamic properties). For example, the time to communicate an intention through the nervous system (from eye to hand) might be modeled as a pure time delay. However, the dynamic response of the muscles once the nerve impulses arrive might be modeled as a second-order lag (reflecting the inertial dynamics of the body).

A situation where human operators must deal with significant time delays is in remote control of manipulators and vehicles in space. This control task can involve telemetry time delays wherein video feedback is delayed by several seconds. Note this delay is due to the transmission of information to and from the remote location in space and that it is in addition to any lags due to the dynamics of the controlled system. Noyes and Sheridan (1984) showed that a predictor display could speed up the time for accomplishing simple manipulation tasks by 70%. This predictor display was achieved by sending the control signals in parallel to a computer model of the system being controlled. The output of the computer model was then used to drive a corresponding graphic model on the feedback video. The graphic model led the video picture to indicate what the actual video would show several seconds hence.

DESIGN CONSIDERATIONS

In a target acquisition task, the goal is to move a control system output into alignment with a fixed or moving target. In this task, the movement path is generally not so important as the final position of the system output (e.g., the cursor). Examples of target acquisition tasks include: moving a cursor to the correct position in a line of text when editing an electronic document; positioning a laser pointer on a projected image to direct the audience's attention; moving a cursor (piper) onto an image of an enemy aircraft so that the automatic targeting system can lock-on; directing a telescope or binoculars to a target (e.g., a bird or planet). The step response, or Fitts' Law paradigm, discussed in the previous chapters is an example of a target acquisition task. The various chapters show many different ways to model performance in target acquisition tasks. Note that the differences among these models may have important implications for understanding the basic information processes involved in target acquisition, but there is little difference in terms of variance accounted for in target acquisition time across the models. Most of the models can account for at least 90% of the variance in controlled laboratory tasks. Thus, Fitts' Law (that response time will be linearly related to the Index of Difficulty, $ID = 2A/W$) provides an important metric of performance. In comparing alternative control devices (e.g., mouse vs. trackball) for target acquisition tasks, it will be important to choose a range of IDs that is representative of the application domain. Knowing that target acquisition time is a well-defined linear function of the ratio A/W constitutes a simple theory of target acquisition that is very useful to system designers (e.g., see Card, 1989). The slope of the Fitts' Law function can provide a useful metric for evaluating alternative controls. Even if such functions were nonlinear, they would still provide a useful basis for comparing competing designs for target acquisition systems. A typical pattern of results might show that the performance differences across different designs are small for easy targets, but large for difficult targets (e.g., Epps, 1986).

In many applications, the target to be acquired is stationary. However, there are situations that require acquisition of a moving target. Skeet shooting is a simple example. Another example is the automatic targeting systems in some advanced fighter aircraft. In these systems, once the pilot acquires the moving target (locks-on), automatic targeting systems will track the target. A modified Index of Difficulty that includes the target velocity (Hoffmann, 1991; Jagacinski, Repperger, Ward, & Moran, 1980) can be used to compare designs for acquiring moving targets in a manner paralleling the evaluation of stationary target acquisition systems.

In general, the difficulty in learning and using a control system will increase with increasing orders of control. Position (zero-order) and velocity (first-order) controls are relatively easy to use. Acceleration (second-order) controls are more difficult. Third-order and higher order controls are very difficult and can lead to unstable performance. For most human computer interfaces, a position or velocity control will result in the best target acquisition performance.

In some systems, the order of control is a design option. For example, in many newer cars the electronic device for opening and closing the side windows is a velocity control system. Depressing a switch determines the rate of change of window position. In contrast, the older manual cranks for window opening and closing were part of a position control system. Rotating the window control lever a certain number of degrees resulted in a proportional change in the vertical window position. The newer velocity control system is convenient to use if a passenger wants to open a window fully so that the mechanics of the system automatically stop it. It is also easy to use if someone only has a very rough criterion for how much the window should be open (e.g., about half). However, if there is a more precise criterion (e.g., open the window just a half centimeter to avoid heat build up in the parking lot on a sunny day), then the velocity control system is more difficult to use than the traditional position control. That is, there may be a series of over- and undershoots, before reaching the desired target.

Another common situation where order of control is a design option is in the design of pointing devices and cursors for computer systems. Common design options include a zero-order (position) or a first-order (velocity) control system, although each can be augmented with additional dynamics (e.g., chap. 26). In general, precision of placement will be best for position controls, but they require space (like a mouse pad) to achieve a comfortable scaling (i.e., gain) between cursor movement on the screen and movement of the control. An advantage of velocity controls is that the range of output is not artificially constrained by the range of the control device. As in the example of the car window, the range of movement for the switch with the velocity control is very small relative to the range of movement required for the manual crank. For target acquisition tasks, velocity control systems will generally work better when the null position (i.e., zero velocity command) is well defined. For example, this is the case for a spring-centered control stick or for a force stick. With both systems, the control automatically returns to the null position when released. In summary, if precise control is required for target acquisition (high *ID*s), then a position control will generally be the best choice. If relative range of movement is an issue, then a velocity control should be considered.

A common example that at first seems contrary to this generalization is the scroll bar typically found to the side of a computer text editor. This controller includes both a position and a velocity control. A typical scroll bar is a narrow column that has an upward pointing arrow, a downward pointing arrow, and a movable square that indicates the relative position in the manuscript of the text that is presently being displayed on the computer screen (e.g., halfway down the column corresponds to the middle of the manuscript). Clicking on one of the arrows with the mouse cursor moves the text up or down, typically at a constant rate, and thus is a velocity control system. However, a second mode of control is to place the mouse cursor directly over the movable square, drag it to the desired position in the manuscript, and then release it. In some text editors, the text display is only then advanced to the indicated position. This mode is typically a position control, but with a delayed display of the corresponding text. Due to this lack of concurrent text updating during the dragging motion, this position control permits only approximate positioning, and more precise positioning is performed with the rate control, which does have concurrent text updating. In other word processors, the displayed text is updated concurrently when the movable square is dragged to a new position. However, for lengthy manuscripts, this position control can become very sensitive, because many pages map into a limited range of control movement. In this case as well, the movable square position control is commonly used for approximate positioning, and the less sensitive velocity control arrows are subsequently used for fine positioning.

The integral properties of velocity and higher order control systems can sometimes be desirable because of their filtering properties. An example where the input space is large relative to the output space is microsurgery. In this case, the low pass filtering properties of the integration (discussed more thoroughly in later chapters) can dampen or effectively filter out the effects of arm and hand tremors on the movement of the microsurgical instruments. In effect, the higher order dynamics can have a smoothing effect on performance.

A velocity control can also be better than position control when the goal is to acquire a moving target. For example, Jagacinski, Repperger, Ward, and Moran (1980) found that for capturing small fast targets, performance with a velocity control was superior to performance with a position control.

Maximizing human performance is not always the goal of design. Sometimes the goal of design is to make the task challenging or interesting (as in video games), and sometimes the goal is to simulate another system (as in driving or flight simulators). In both these cases, second-order control systems might be the most appropriate choice. Second-order control (acceleration) systems are challenging, but with practice most people can achieve stable control with this order of plant. For higher orders of control (third or fourth order), stable performance is very difficult and would only be achieved with very extended levels of practice if at all.

There are many situations where the order of control is not a design option. In these systems, the order for the controlled process is determined as a result of the physics of the system. Prominent examples include vehicular control (chap. 16) and industrial process control. In these situations, laws of motion or laws that govern the physical processes that are being controlled determine the order of the dynamics. In these cases, it is important for the designer to be able to identify the order of the controlled process. If the order of the natural process is second-order or higher, then it may be desirable or even necessary to provide systematic support to the human controller.

Most physical systems have dynamics that are at least second order. A second-order system requires that a controller have information about both position and velocity. If the order of control is greater than second order, then the controller will be required to have information about acceleration and higher derivatives. In general, humans are very poor at perceiving these higher derivatives. Thus, the human will have great difficulty achieving stable control without support. Two ways in which designers can support humans in controlling second-order and higher order systems are quickening and aiding.

As noted earlier, quickened displays show the effects of the anticipated system output rather than the actual system output (Birmingham & Taylor, 1954). The anticipation typically consists of a sum of the output position, velocity, acceleration, and so on. This technique works well for stationary target acquisition (Poulton, 1974). Such a display reduces excessive overshooting by making the eventual effects of acceleration and higher temporal derivatives in a sluggish system more apparent early in the course of a movement. Also, as the system output converges to the stationary target value, the output velocity, acceleration, and so forth, diminish toward zero. The quickened display then approximates the actual unquickened output, giving the person an accurate assessment of the effects of their movements. However, this display technique is problematic for tracking constant velocity targets (Poulton, 1974). For example, if a quickened display were to be used with an acceleration control system for tracking a constant velocity target, the quickened display would consist of an additive combination of the system output position and velocity. If the system were tracking with a constant offset error from the target, then the display might show zero error because a displacement proportional to the system output velocity is added to the displayed output position. To overcome such problems, a better anticipatory display would indicate both the actual system output as well as the anticipated output. Also, instead of simply providing a single point estimate of anticipated output, it may be beneficial to display an anticipated trajectory extending over time (e.g., Kelley, 1968).

Aiding provides support by changing the effective dynamics between the control input and the system output. For example, rate aiding is a simple example of a system that responds like a position control and a velocity control working in parallel (Birmingham & Taylor, 1954; Poulton, 1974; chap. 26). The parallel position control overcomes some of the sluggishness of the rate control. This principle can be extrapolated to higher order systems. For example, the integrations required by the mechanical links and aerodynamics of aircraft can be effectively bypassed in electronic fly-by-wire systems. Thus, a control axis (aircraft heading) that would naturally be third order can be made to respond as if it were a zero- or first-order system. This is possible because automated control systems manage the integrations. Thus, the aircraft can be made to respond proportionally (zero order) to the pilots commands. Even though the pilot may be providing continuous control inputs, these inputs are effectively instructions to autopilot systems that are directly controlling the flight surfaces (e.g., Billings, 1997).

In sum, as the order of control of a system increases, the information-processing requirements on the controller also increase. Each additional integration increases the phase lag between input and system response and adds another state variable that must be taken into account by the controller. If the loop is closed through a human operator, then the designers are responsible to make sure that the state variables are perceivable and they are configured so that the operator can make accurate predictions about the system response. For systems that have dynamics of greater than second order, this will generally require either supplementary display information (e.g., quickening) or control augmentation (e.g., aiding).

Time delay is another important consideration in the design of control systems. In many systems, there is a transport delay between the input of a control and the response of the system. A familiar example is plumbing systems where there is a relatively long time delay between when the temperature control is adjusted and when the water actually changes temperature. Significant time delays can also be present in remote teleoperated control systems (e.g., uninhabited air vehicles, UAVs) or remote systems for space exploration.

In some respects, increasing time delay is like increasing order of control, in that both make a control system less immediately responsive and often less stable. That is, the response is "lagged" relative to the input. This lagging of the response shows up in the frequency domain as a phase lag (chap. 13). However, they do this in different ways that can be characterized in the frequency domain (chap. 13). Each integrator introduces a constant (frequency independent) phase lag of 90 degrees, as well as an amplitude change that is inversely related to frequency (i.e., amplification at low frequencies and attenuation at high frequencies). In contrast, a time delay introduces no amplitude change, but has a phase lag that increases with frequency. The impact of this increasing phase lag is that it sets a stability limit on the forward loop gain. The lower the forward loop gain, the less responsive the system is to error. The larger the time delay, the narrower the range of gains that will yield stable control. With large time delays, there may be no stable gain available that results in a sufficiently rapid response. That is, proportional control strategies will not work well.

For systems with long time delays, a discrete strategy in which prepackaged (ballistic) controls are input, followed by a waiting period, can be very effective (e.g., Ferrell, 1965; Sheridan & Ferrell, 1963). These types of strategies were discussed in chapter 7. For example, consider, again, the problem of adjusting the shower. The person might begin with a discrete adjustment of the temperature. Then the person waits until the water reaches a stable temperature. If this temperature is too hot or cold, then a second discrete adjustment is made and the person waits again. This process is iterated until a satisfactory water temperature is reached. This is the kind of performance that Crossman and Cooke (1974) observed when they had humans adjust the temperature of an experimental waterbath. Angel and Bekey (1968) provided a good example of a discrete control strategy in their illustration of finite state control for tracking with a second-order system. Figure 7.6 illustrates the logic of Angel and Bekey's discrete control system. This system had five different discrete responses (two bang-bang responses to position errors, two pulse responses to velocity errors, and a "do nothing" response). The regions in the state space represent the state of the processes to which each response is appropriate.

A discrete controller such as Angel and Bekey's suggests the existence of an underlying model of the process being controlled that is a bit more sophisticated than the internal models required for proportional control (i.e., gains on the state variables). The model must specify the associations between states of the process and specific discrete response forms (e.g., where to set the controls for the initial setting of the shower temperature). The model must also specify the criteria for shifting control—for example, the length of time to wait in order to assess the impact of the previous response (How long does it take the water temperature to stabilize in response to the initial adjustment?). Thus, an important consideration for designers is to provide support to the operator in developing a model of the process. Training can help the operator to develop an internal model of the process. Also, representations that show the time evolution of the process (e.g., time histories and state space displays; Miller, 1969; Pew, 1966; Veldhuysen & Stassen, 1977), the logic of the underlying processes (engineering diagrams), or predictions (fast time simulations) might provide valuable support to the operator.

CONTROL DEVICES

Another consideration when designing systems is the type of input or control device that is used. A wide variety of devices is available (e.g., mouse, spring-centered joystick, force-stick, space tablet, light pen, trackball, keys, etc.). What is the best device for the task? And, in particular, what properties of a control device contribute to the appropriateness of that device for a plant with a specific control order?

One important consideration is the prominence of the null position for the input device. For velocity and higher order control systems, it is useful to have a well-defined null (zero) position. This position is important, because the stick must be in this position for the output to stop on a target. For a position (zero-order) control system, this is not a problem because the output stops, whenever the control device stops moving. Thus, a mouse is a good device for controlling a position system, but not the best device for higher order systems. Spring-centered joysticks or force sticks are generally better suited for velocity and higher order control tasks. One way to make the null position more distinct is to include a *dead-band* in the zero region. The dead-band is a nonlinearity. The effect illustrated in Fig. 9.6 is that there is a distinct region in the vicinity of the null position where the output will be zero, independent of the input value. The boundaries of the dead-band can also be identified by a tactual "notch," which allows the human operator to feel the transition between the zero and proportional regions of the control space.

Another common nonlinear feature of control devices is *hysteresis*. This can be thought of as a kind of moving dead-band associated with the reversal of control. Hysteresis can be visualized as a wide cylinder (like an open juice can) sitting loosely on a thin control stick. When the juice can is in contact with the stick, control input is

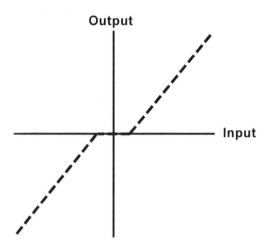

FIG. 9.6. A dead-band that can be used to insure that there is a well-defined null position for a control device. In the dead-band region, the output is zero for a range of input positions. Outside the dead-band region, output is proportional to input.

proportional to displacement of the can. However, when the action on the can is reversed, there is a brief time when the can is moved, but because there is empty space in the can, the stick remains fixed. Thus, immediately following a reversal in direction of the control action, there will be a period in which the output is constant, even though the control input is changing. Then when the opposite side of the juice can contacts the stick, the proportional relation between motion of the can and control is reinstated. A small amount of hysteresis is typically designed into keys (e.g., on your computer keypad). That is, the motion required to activate the key (close the switch) is smaller than the reverse motion required to reopen the switch. Thus, the key remains in the closed position, even if the pressure is partially released past the initial closing position. The benefit of this nonlinear asymmetry is that it helps to prevent accidently double clicking on the same letter. Sanders and McCormick (1993) reported that "too little hysteresis can produce key bounce and cause inadvertent insertions of extra characters; too much hysteresis can interfere with high-speed typing" (p. 363).

In interpreting the results of empirical comparisons, it is important to remember that the particular control device that people hold in their hand is only one part of the target acquisition system. Other dynamic aspects of the system may be designed independently. This point is not always obvious. For example, one well-respected human factors handbook summarized data that showed an advantage for a mouse over a control stick for target acquisition tasks in text editing. The handbook did not mention that the evaluation had used position dynamics with the mouse and rate dynamics with the control stick. This confounding of dynamics with control device may have put the control stick at a disadvantage. These control devices should have been tested with the same dynamics or sets of dynamics.

CONCLUSION

This chapter introduces the topic that is typically referred to as *manual control*, that is, the study of humans as operators of dynamic systems. Early research focused on the human element in vehicular control (in particular, modern high performance air vehicles). Designers of these systems realized that the human was an important element in the system control loop. In order to predict the stability of the full system, they had to include mathematical descriptions of the human operators along with the descriptions of the vehicle dynamics. Most of this research used tracking tasks and frequency domain modeling tools, which are discussed in later chapters. However, today the computer is one of the most common "vehicles." Humans are routinely called on to position cursors in menus and to perform other positioning tasks. So it is appropriate to introduce the problem of controlling dynamical systems in the context of the discrete position task. These ideas are revisited in later discussions of the tracking task and the frequency domain.

REFERENCES

Angel, E. S., & Bekey, G. A. (1968). Adaptive finite state models of manual control systems. *IEEE Transactions on Man–Machine Systems, MMS-9*, 15–20.

Billings, C. E. (1997). *Aviation automation: The search for a human-centered approach*. Hillsdale, NJ: Lawrence Erlbaum Associates.

Birmingham, H. P., & Taylor, F. V. (1954). A design philosophy for man–machine control systems. *Proceedings of the IRE*, 1748–1758.

Card, S. K. (1989). Theory-driven design research. In G. R. McMillan, D. Beevis, E. Salas, M. H. Strub, R. Sutton, & L. V. Breda (Eds.), *Applications of human performance models to system design* (pp. 501–509). New York: Plenum.

Crossman, E.R.F.W., & Cooke, J. E. (1974). Manual control of slow response systems. In E. Edwards & F. P. Lees (Eds.), *The human operator in process control* (pp. 51–66). London: Taylor & Francis.

Epps, B. W. (1986). Comparison of six cursor control devices based on Fitts' Law models. *Proceedings of the Human Factors Society Thirtieth Annual Meeting* (pp. 327–331). Santa Monica, CA: The Human Factors & Ergonomics Society.

Ferrell, W. R. (1965). Remote manipulation with transmission delay. *IEEE Transactions on Human Factors in Electronics, HFE-6*, 24–32.

Fitts, P. M., & Seeger, C. M. (1953). S-R compatibility: Spatial characteristics of stimulus and response codes. *Journal of Experimental Psychology, 46*, 199–210.

Hartzell, E. J., Dunbar, S., Beveridge, R., & Cortilla, R. (1982). Helicopter pilot response latency as a function of the spatial arrangement of instruments and controls. In *Proceedings of the Eighteenth Annual Conference on Manual Control* (AFWAL-TR-83-3021). Wright-Patterson Air Force Base, OH.

Hoffmann, E. R. (1991). Capture of moving targets: A modification of Fitts' Law. *Ergonomics, 34*, 211–220.

Jagacinski, R. J. (1989). Target acquisition: Performance measures, process models, and design implications. In G. R. McMillan, D. Beris, E. Salas, M. H. Strub, R. Sutton, & L. van Breda (Eds.), *Applications of human performance models to system design* (pp. 135–149). New York: Plenum.

Jagacinski, R. J., Hartzell, E. J., Ward, S., & Bishop, K. (1978). Fitts' Law as a function of system dynamics and target uncertainty. *Journal of Motor Behavior, 10*, 123–131.

Jagacinski, R. J., Repperger, D. W., Moran, M. S., Ward, S. L., & Glass, B. (1980). Fitts' Law and the microstructure of rapid discrete movements. *Journal of Experimental Psychology: Human Perception and Performance, 6*, 309–320.

Jagacinski, R. J., Repperger, D. W., Ward, S. L., & Moran, M. S. (1980). A test of Fitts' Law with moving targets. *Human Factors, 22*, 225–233.

Kelley, C. R. (1968). *Manual and automatic control*. New York: Wiley.

Loveless, N. E. (1963). Direction of motion stereotypes: A review. *Ergonomics, 5*, 357–383.

Miller, D. C. (1969). *Behavioral sources of suboptimal human performance in discrete control tasks*. (Tech. Rep. No. DSR 70283-9). Cambridge, MA: MIT Engineering Projects Laboratory.

Noyes, M. V., & Sheridan, T. B. (1984). A novel predictor for telemanipulation through a time delay. In *Proceedings of the Twentieth Annual Conference on Manual Control* (NASA Conference Pub. 2341). NASA Ames Research Center, CA.

Pew, R. W. (1966). Performance of human operators in a three-state relay control system with velocity-augmented displays. *IEEE Transactions on Human Factors in Electronics, HFE-7,* 77.

Poulton, E. C. (1974). *Tracking skill and manual control.* New York: Academic Press.

Roscoe, S. N., Eisele, J. E., & Bergman, C. A. (1980). Information and control requirements. In S. N. Roscoe (Ed.), *Aviation psychology* (pp. 33–38). Ames, IA: The Iowa State University Press.

Sanders, M. S., & McCormick, E. J. (1993). *Human factors in engineering and design.* New York: McGraw-Hill.

Sheridan, T. B., & Ferrell, W. R. (1963). Remote manipulative control with transmission delay. *IEEE Transactions on Human Factors in Electronics, HFE-4,* 25–29.

Veldhuysen, W., & Stassen, H. G. (1977). The internal model concept: An application to modeling human control of large ships. *Human Factors, 19,* 367–380.

Wickens, C. D. (1992). *Engineering psychology and human performance* (2nd ed.). New York: Harper Collins.

Tracking

> *In contrast to the skills approach, the dynamic systems approach examines human abilities in controlling or tracking dynamic systems to make them conform with certain time-space trajectories in the face of environmental uncertainty (Kelley, 1968; Poulton, 1974, Wickens, 1986). Most forms of vehicle control fall into this category, and so increasingly do computer-based cursor positioning tasks. The research on tracking has been oriented primarily toward engineering, focusing on mathematical representations of the human's analog response when processing uncertainty. Unlike the skills approach, which focuses on learning and practice, the tracking approach generally addresses the behavior of the well-trained operator.*
>
> —Wickens (1992, pp. 445–446)

The previous chapters have focused on movements to a fixed target. The tasks generally required the actor to move from one position to another as quickly and accurately as possible. This task requirement was compared to a step input and the resulting movements were compared to step responses. The step input is one way of interrogating a dynamic system. An alternative is to use sine waves as the input. The resulting response is called the *sinusoidal response*, or the *frequency response*. The following sections show how the response of a dynamic system to sine waves of various frequencies can be used to make inferences about the properties of the dynamic system. The next chapter presents some of the properties of sine waves that make these inputs so valuable for interrogating dynamic systems. This chapter introduces the tracking task as an experimental paradigm that will allow the frequency response of human actors to be measured.

DISPLAY MODES

In a tracking task, the actor's job is typically to minimize error between the control object (e.g., a simulated vehicle) and the target track (e.g., a simulated roadway). In simple tracking tasks, the subject would use a joystick to control a cursor presented on a video display. This simple tracking task can be presented in a compensatory or a pursuit mode.

In a *compensatory mode*, a marker, generally at the center of the screen, indicates zero error between the controlled object and the target track. A second cursor moves relative to this fixed cursor in proportion to the size of tracking error. The actor "compensates" for the error by moving the joystick in the appropriate direction to null out the displayed error. To the extent that the actor is successful in nulling out error, the moving cursor will remain near the center of the screen. In a compensatory tracking task, only the relative error between the controlled object and the target track are displayed.

There are two moving cursors in a *pursuit* tracking task. One cursor represents the target path. The second cursor represents the controlled object. The actor "pursues" the moving target cursor by moving his control stick in such a way that the controlled object becomes aligned with the target. In a pursuit tracking task, the motion of the target, of the controlled object, and the relative error between the two are each visually represented. Figure 10.1 illustrates the distinctions between compensatory and pursuit tracking tasks.

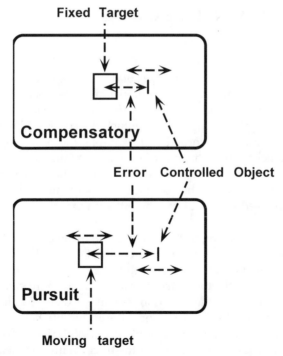

FIG. 10.1.   An illustration of two modes for presenting a tracking task. In the compensatory mode, the target is fixed (typically in the center of the display) and the position of the controlled object moves with the size and direction of the error signal. In the pursuit mode, the target moves independently to show the track, the controlled object reflects the effects of control actions only, and the error is shown as the relative position between the two moving cursors.

It is important to distinguish between the pursuit display mode for a tracking task and the pursuit strategy of control discussed in chapter 7 (see Fig. 7.8). Remember that the pursuit strategy includes a continuous open-loop pursuit response (directly to input based on the controller's internal model of the plant) and a closed-loop compensatory response to error. The common name "pursuit" reflects the idea that information about the track, the control dynamics, and the error are each explicitly represented in the pursuit display mode. Thus, the pursuit display mode may provide better support for the development of a pursuit control strategy. However, a pursuit display mode is neither necessary nor sufficient for development of a pursuit control strategy (Krendel & McRuer, 1968), as Wickens (1992) noted:

> Although it appears logical that pursuit behavior will occur with pursuit displays and compensatory behavior with compensatory displays, this association is not necessary in a pursuit display because the operator may focus attention either on error or input. . . . Conversely, pursuit behavior is possible even with a compensatory display when there are no disturbances, although this is more difficult. (p. 472)

GLOBAL PERFORMANCE MEASURES

With the positioning task, either time (e.g., Fitts' paradigm) or variability (e.g., Schmidt et al.'s task) could provide an index of the quality of performance. There are several conventions for characterizing the global quality of performance in tracking tasks. One measure is *time on target*, or *percent time on target*. Here the amount of time (or percentage of the total tracking time) in which the controlled object is within the bounds of the target track is recorded. For example, the percentage of time that a driver keeps the vehicle within the appropriate lane would be a measure of driving performance. Good drivers would be expected to be in the appropriate lane most of the time. This measure is useful when the target track has an appreciable width (e.g., a traffic lane), as opposed to situations where the track is defined by a line (e.g., the center of the lane).

A second measure of tracking performance is the mean error. This is, in effect, an integration or summation of error over the length of the track divided by the duration of the trial or number of samples in the sum. Zero mean error would result if the controlled object never left the target track or if the errors were symmetric about the target track (i.e., the errors in one direction, positive, exactly balanced out errors in the opposite direction, negative). To eliminate the ambiguity between these two situations, mean absolute error is more commonly used. With this measure, the absolute value of error is summed over the length of the track. With this measure, zero mean absolute error would indicate that the controlled object never left the target track. It is more typical to define the target track as a line with this measure (e.g., center of the lane). However, either mean error or mean absolute error could also be measured as the distance by which the controlled object exceeds the bounds of a lane with finite width.

$$\text{Mean Absolute Error} = \frac{\int_{t=0}^{T} |X_{Target} - X_{Cursor}| dt}{T} \approx \frac{\sum_{i=1}^{N} |X_{Target_i} - X_{Cursor_i}|}{N} \qquad (1)$$

where   T = the duration of tracking

N = the number of samples

i = index of discrete samples

X_{Target} = the position of the target track

X_{Cursor} = the position of the controlled object

The most common measures of tracking performance are *mean squared error* (MSE) or *root mean squared error* (RMSE or RMS). With these measures, each sample of error is squared first, before it is added with the other samples. The total is then divided by the number of samples to get a mean. Squaring the samples has two effects. First, squaring insures that there will be no negative components of the sum (squaring a negative number yields a positive number). Thus, as with the absolute error, there is no possibility of balancing out positive and negative errors. The greater the error, independent of the direction, the greater will be MSE or RMSE. The second implication of squaring is that larger errors will impact the total proportionally more than smaller errors. Note that the square of a number less than one is actually smaller than the absolute number. Thus, squaring gives a proportionally greater penalty for larger errors. Taking the square root of the MSE converts the result back to units comparable with the original units in which errors were measured (as opposed to squared units):

$$\text{MSE} = \frac{\int_{t=0}^{T} (X_{Target} - X_{Cursor})^2 dt}{T} \approx \frac{\sum_{i=1}^{N} (X_{Target_i} - X_{Cursor_i})^2}{N} \qquad (2)$$

$$\text{MSE} = \frac{\int_{t=0}^{T} e^2 dt}{T} \approx \frac{\sum_{i=1}^{N} e_i^2}{N} \qquad (3)$$

$$\text{RMSE} = \sqrt{MSE} \qquad (4)$$

Note that the MSE can also be computed as the sum of two components, the mean error squared and the variable error squared:

$$\text{MSE} = \text{Mean Error}^2 + \text{Variable Error}^2 \qquad (5)$$

$$\text{Mean Error} = \bar{e} = \frac{\sum_{i=1}^{N} e_i}{N} \qquad (6)$$

$$\text{Variable Error} = \sqrt{\frac{\sum\limits_{i=1}^{N}(\bar{e}-e_i)^2}{N}} \qquad (7)$$

$$\text{MSE} = \left(\frac{\sum\limits_{i=1}^{N}e_i}{N}\right)^2 + \frac{\sum\limits_{i=1}^{N}(\bar{e}-e_i)^2}{N} \qquad (8)$$

When comparing the different error measures it is useful to visualize the tracking error as a distribution of errors that could result from sampling the error magnitude and direction frequently over the course of a trial. This distribution may or may not be normally distributed. The distribution could be centered on the target or it could be biased to one side or the other. The value of the time-on-target measure will depend on the position and size of the target relative to the mean and variability (spread) of the distribution. The value of the time-on-target measure will vary nonlinearly with the size of the target. The mean error measure will reflect the distance between the mean of the error distribution and the target (i.e., zero error). The variable error measure will reflect the spread of the error distribution. MSE or RMS measures will be joint functions of these two components (mean and variance of the distribution). See Bahrick, Fitts, and Briggs (1959) or Schmidt and Lee (1999) for further discussion of various error measures.

Perhaps it would be useful to give a numerical example. Table 10.1 shows tracking error for three different subjects and the results of different error computations for each. Readers should try to carry out the computations to check their understanding and to verify the results in Table 10.1.

Note that using percent time on target, Subject C is the best tracker and Subjects A and B are equivalent. With mean error as the performance measure, Subject B is the

TABLE 10.1
Tracking Error for Three Subjects

	Subjects		
	A	*B*	*C*
Sample 1	3	5	0
Sample 2	–3	–1	0
Sample 3	3	–1	9
Sample 4	–3	5	0
Sample 5	3	–1	0
Sample 6	–3	5	–9
% Time on Target	0	0	66.66
Mean Error	0	2	0
Mean Absolute Error	3	3	3
MSE	9	13	27
RMSE	3	3.61	5.20
Variable Error	3	3	5.20

worst and Subjects A and C are equivalent. With mean absolute error, all three subjects are equivalent. With MSE and RMS error, Subject C is worst and Subject A is best. With Variable Error, Subject C is worst and Subjects A and B are equivalent. It is important to realize that comparisons are not independent of the choice of performance metrics.

Note also that for each subject, the mean squared error is equivalent to the mean error squared plus the variable error squared. Mean error and variable error are mathematically independent components of the MSE. However, it is not unusual for these measures to be correlated within a group of trackers, because as Poulton (1974) noted, "Poor trackers tend to have a large constant position error and also a large variable error. While good trackers tend to have a small constant position error and a small variable error" (p. 36). In practice, it is a good idea to examine both of these independent measures of tracking performance. This is particularly important in cases where the target is not explicitly or continuously presented to the subject. For example, if the task is to fly at a constant altitude using the natural optical cues seen through the windshield, then the pilot may lose track of the reference and stabilize the aircraft at some higher or lower altitude. In this case, the reference altitude would have to be remembered and recognized by the pilot. In some conditions, keeping track of the reference may be more difficult than regulating against any variable disturbance. This would result in a significant mean error component in the RMS score. If the mean errors are small and do not vary systematically within an experiment, then RMS error would be the recommended performance metric. If there are systematic differences with the mean errors, then it is good practice to report and discuss the two independent components of MSE.

It is typical to present the goal as minimizing RMS error in laboratory tracking tasks. RMS error scores are often presented at the end of a trial so subjects have an index of how well they are doing. As is pointed out later, the requirement to minimize RMS error will be consistent with computational requirements for identifying optimal control solutions (i.e., the quadratic cost functional). However, in natural control tasks, like driving cars or flying aircraft, people do not typically try to minimize error with regard to the center lane of a highway. Rather, there is generally an envelope, lane, or region within which small deviations are inconsequential in terms of the goal. Also, the active control required in order to minimize RMS error would not result in the most comfortable ride. Often, comfort is valued as well as reducing error, so slow gradual corrections are preferred to the fast corrections required to minimize RMS error. Finally, it is important to note that the global measures of performance shown in Table 10.1 allow simple comparisons among subjects, but they are very impoverished descriptions of the tracking data. The following chapters discuss richer ways to analyze the tracking data that will help to more fully characterize performance and to better model the human operator.

THE TRACK

There are various ways of challenging or "driving" the subject in a tracking task. One way is to create a winding road that the subject must follow (command input). Obviously, the more bends in this track and the tighter the bends, the more difficult it will be to follow the track. The winding road could also be the trail of a lead vehicle (e.g., in formation flying) that changes its course to create a variable path. In both these cases, the deviations in the path determine the complexity of the tracking task.

An alternative way to challenge the subject is to have a simple track (e.g., a straight highway), but to introduce a disturbance (e.g., wind gusts) that will cause the controlled object (e.g., vehicle) to leave the track unless the subject counters or compensates for the effects of this disturbance. Again, the more variable the patterns of disturbance, the more interesting (demanding) the tracking task.

A mathematically simple way to create an interesting track or disturbance (that is easy to program on either digital or analog computers) is to add together sine waves of different frequencies, amplitudes, and phases. The next chapters introduce the sine wave and discuss some of the distinct advantages of using a sum-of-sines signal as the driving function in tracking tasks. One important advantage of sine waves is that many dynamic systems will have distinctive effects on sine waves of different frequencies. These distinctive effects can be used to make inferences about the properties of the control system. For this reason, the use of sine waves as test signals is an important method for building quantitative descriptions of dynamic control systems. Also, even if a track is not constructed from sine waves (e.g., the track may represent a natural path or disturbance or might be generated from filtered "noise"), it may be valuable to utilize Fourier analysis to describe the track as a sum of sinusoidal components. Fourier analysis provides a mathematical procedure for approximating a time varying signal as a sum of sine waves.

DESIGN CONSIDERATIONS

Continuous tracking tasks involve situations where the human must attempt to match the output of a control system to a continuously varying signal (e.g., steering a car along a winding road), or where the human operator must attempt to nullify a continuous disturbance (e.g., maintain a constant altitude in the presence of turbulence). In these tasks, the goal is to make the output of the control system coincident with the path of some target. Chapters 11 through 19 are directly relevant to modeling performance in continuous tracking tasks. For this task, modeling in terms of sinusoidal components, i.e., in the frequency domain, can often lead to important insights. Landmarks in the frequency domain, such as the crossover frequency and the gain and phase margins, have important implications for system stability (as discussed in chap. 13). In evaluating designs for continuous tracking tasks, it will be important that the bandwidths of the signals and/or disturbances are representative of the signals and disturbances in the application domain. Models such as the McRuer crossover model (chap. 14) or the optimal control model (chap. 17) can be an important complement to global performance metrics such as RMS tracking error. These models can be used to fit performance in the time or frequency domain, and the parameters of these models can be valuable for comparing alternative design solutions. For example, utilization of the crossover model allows the human–machine system to be evaluated using the intuitions of classical servo theory. Thus, alternative control solutions can be compared in terms of the gain and/or phase margins that result. Much of the early work on the crossover and optimal control models was motivated by concerns about the design of high performance aircraft.

Whereas classical control theory focuses on system stability and tracking error, there are often additional criteria for good system performance that also demand explicit consideration. For example, smoothness of response might be an important additional consideration, so that the person might be characterized as minimizing an additive combination of mean squared tracking error, control movement, and/or rate of change of control movement. Drivers who accelerate and brake very rapidly seem intuitively to place a low weighting on smoothness. Mathematical models that can explicitly demonstrate the effect of smoothness considerations on performance have been developed under the general approach of optimal control theory (e.g., Levison, 1989). Smoothness might therefore be explicitly considered by the system designer using optimal control models.

Many of the design considerations discussed in the previous chapter in the context of target acquisition apply equally well to tracking. For example, the difficulty of tracking tends to increase with the order of control. Also, time delays can be very disruptive to tracking performance and can have important implications for whether a discrete or continuous style of control is adopted by the human operator.

REFERENCES

Bahrick, H. P., Fitts, P. M., & Briggs, G. E. (1959). Learning curves facts or artifacts? *Psychological Bulletin,* *54*, 256–268.

Kelley, C. R. (1968). *Manual and automatic control.* New York: Wiley.

Krendel, E. S., & McRuer, D. T. (1968). Psychological and physiological skill development: A control engineering model. In *Proceedings of the Fourth Annual NASA/University Conference on Manual Control* (NASA SP- 192). University of Michigan, Ann Arbor, MI.

Levison, W. H. (1989). The optimal control model for manually controlled systems. In G. R. McMillan, D. Beevis, E. Salas, M. H. Strub, R. Sutton, & L. Van Breda (Eds.), *Applications of human performance models to system design* (pp. 185–198). New York: Plenum.

Poulton, E. C. (1974). *Tracking skills and manual control.* New York: Academic Press.

Schmidt, R. A., & Lee, T. D. (1999). *Motor control and learning: A behavioral emphasis.* Champaign, IL: Human Kinetics.

Wickens, C. D. (1992). *Engineering psychology and human performance* (2nd ed.). New York: Harper Collins.

Wickens, C. D. (1986). The effects of control dynamics on performance. In K. Boff, L. Kaufman, & J. Thomas (Eds.), *Handbook of perception and performance* (Vol. 2, pp. 39.1–39.60). New York: Wiley Interscience.

There Must Be 50 Ways
to See a Sine Wave

> *Rhythmic phenomena abound in biology at all levels of analysis. . . . At the behavioral level, rhythms can be observed in which the entire body or parts of the body move in a cyclic, repetitive fashion. Running, swimming, flying, breathing, chewing, and grooming are the most cited examples, but rhythmic behaviors need not be so stereotypic, a fact attested to by the musical accomplishments of humans.*
>
> —Kugler and Turvey (1987, p. 2)

As Kugler and Turvey noted, rhythmic phenomena abound in nature. As a small step towards understanding such complex rhythmic phenomena, it is useful to consider one of the simpler rhythmic patterns. A *sine wave*, or a *sinusoid*, refers to the spatiotemporal pattern shown in Fig. 11.1. All sine waves have this stereotypic, wavelike, repetitive shape. However, particular sine waves can differ in terms of amplitude, frequency, and phase:

1. *Amplitude* refers to the magnitude of the maxima and minima of the sine wave. Sine waves *a* and *b* differ only in their amplitudes, which are respectively 1 and 2.

2. *Frequency* refers to how many times the sinusoidal pattern repeats itself in a given period of time. Sine waves *a* and *c* differ in their frequency. In one second, sine wave *a* completes a full cycle. Therefore, its frequency is one cycle per second, which is also known as 1 Hertz, which is abbreviated as 1 Hz. (Although this naming convention honors a famous scientist, it is less obvious in its meaning than "cycles per second.") In one second, sine wave *c* completes just one half a cycle. Therefore, its frequency is .5 Hz. In any given time period, sine wave *a* will go through twice as many cycles as sine wave *c* (i.e., the frequency of sine wave *a* is twice as large as that of sine wave *c*).

3. *Phase* refers to what aspect of the repeating cycle is occurring at any particular instant of time. Sine waves *a* and *d* differ only in their phase. Sine wave *d* is one quar-

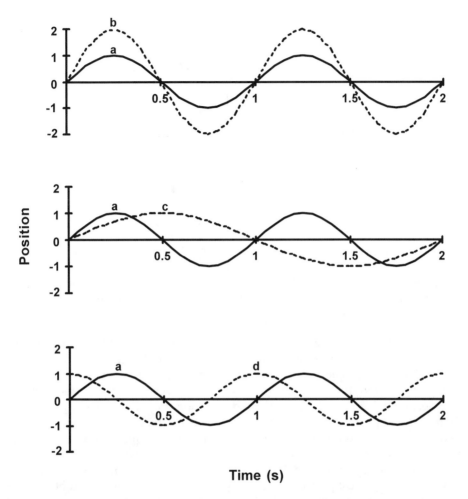

FIG. 11.1. The sine wave. The top graph shows two sine waves that differ only in amplitude. The middle graph shows two sine waves that differ in frequency. The bottom graph shows two sine waves that differ only in phase.

ter of a cycle ahead of sine wave *a*. Namely, when sine wave *a* is at zero and going toward its maximum, sine wave *d* is already at its maximum.

SINE WAVES, CIRCLES, AND TRIANGLES

One way of defining a sine wave is simply to define it to be the shape shown in Fig. 11.1. Another approach is to describe a procedure that will produce that shape. For example, consider a radius of length 1 rotating counterclockwise about the center of its circle at some constant angular rate such as once around the circle every second (Fig. 11.2). In other words, the radius sweeps along the circumference of the circle and completes one circle or one cycle every second. The projection of this radius onto

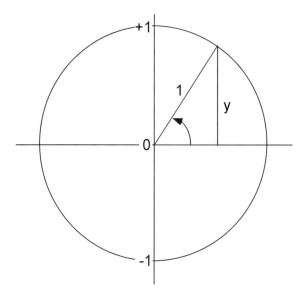

FIG. 11.2.   A sine wave can be generated by letting a radius rotate at a constant angular rate and recording its vertical projection, *y*, as a function of time.

the vertical axis can be determined by drawing a vertical line from the tip of the radius to the horizontal axis, thus forming a right triangle. This projection, *y*, will vary as the radius rotates. When the radius is pointing directly to the right, the projection will be equal to zero (i.e., *y* = 0). When the radius is vertical one quarter circle later, the projection will equal +1. When the radius is pointing directly leftward one quarter circle later, the projection will again be 0. One quarter circle later, the radius will be pointed downward and the projection will be –1. One quarter circle later, the radius will be pointed rightward, and the whole cycle of values will repeat again.

Note that starting with the initial rightward position of the radius, the succession of values for its vertical projection corresponds to the same succession of values exhibited by sine wave *a* in Fig. 11.1 (0, +1, 0, –1). In fact, sine wave *a* is precisely equal to the vertical projection of the radius in Fig. 11.2 over time. Now there is a procedure for generating a sine wave—let a radius rotate at constant angular rate and take its vertical projection as a function of time. This procedure demonstrates the close connection between the sinusoidal shape and the circle. The sinusoid is a function that translates circular motion into linear motion.

Note that each of the other sine waves in Fig. 11.1 could be generated in a similar manner. The radius would be twice as long to generate sine wave *b*. The angular rate of rotation would be half as fast to generate sine wave *c*. The initial position of the radius would be upward rather than rightward to generate sine wave *d*.

SYMBOLIC REPRESENTATION OF A SINE WAVE

Another way of referring to the shape of the sine wave is with the mathematical symbol "sin." Actually, when talking about a particular sine wave, its amplitude, *A*, frequency, f, and phase, θ, must be designated. These elements can be combined in the mathematical expression

$$A \sin[360°(f)(t) + \theta]$$

The symbol sin represents a function that translates location on the circle into linear location on the vertical axis (i.e., the sine function). The position of the radius on the circle in Fig. 11.2 can be measured by the angle it makes with the positive horizontal axis, demarcated by the curved arrow. At time zero, this angle is θ degrees, and it steadily increases as time passes. If t represents time in seconds, and f represents frequency in cycles per second, then in t seconds the radius will have rotated through $(f)(t)$ cycles (cycles/second × seconds = cycles). The number of cycles can be translated into degrees by multiplying by 360 degrees per cycle (degrees/cycle × cycles = degrees). The sine function translates the number of degrees the radius has rotated into its vertical projection. Thus:

$$\sin(0°) = 0$$
$$\sin(90°) = +1$$
$$\sin(180°) = 0$$
$$\sin(270°) = -1$$
$$\sin(360°) = \sin(0°) = 0$$

This is the same succession of values for sine wave a in Fig. 11.1 that was discussed previously, only now there is a symbolic representation. Because the amplitude, A, is 1, the frequency, f, is one cycle per second, and the initial phase, θ, is zero for sine wave a, it corresponds to $A\sin[360°(f)(t) + \theta] = (1)\sin[360°(1)(t) + 0°] = \sin(360°t)$. Referring again to Fig. 11.2, $y = \sin(360°t)$. Similarly, sine wave b is $2\sin(360°t)$, because its amplitude is twice as large. Sine wave c is equivalent to $\sin[360°(0.5)(t)] = \sin(180°t)$, because it has one half the frequency of sine wave a. Finally, sine wave d is $\sin(360°t + 90°)$, because it is one quarter cycle ahead of sine wave a. Equivalently, sine wave d has a 90° phase lead relative to sine wave a. These examples provide a symbolic representation of the three parameters of the sinusoidal shape, amplitude, frequency, and phase.

The sine of the angle between the radius and the horizontal axis (demarcated by the curved arrow in Fig. 11.2) can be defined more generally as the ratio of y to the length of the hypotenuse of the right triangle formed by the radius, the vertical projection, and the horizontal axis. In Fig. 11.2, the ratio is $y/1 = y$. However, more generally, the radius, which is also the hypotenuse, could have a length different from 1. Note that when dealing with triangles, the lengths of the sides are always positive (there cannot be a negative length), but the vertical projection, y, can be negative or positive, and the sine of the corresponding angle can be positive or negative.

Although the angle that the sine function acts on has been expressed in degrees such that 360° equals a full circle, the angle can also be expressed in other units. For example, a "radian" is another unit of angular measurement that is equal to 57.3°. This is the number of degrees covered by the radius of the circle in Fig. 11.2 if it is laid out along the circumference of the circle. The circumference is equal to 2π radii or

6.28 radii, so one radian is just a bit less than one sixth of 360° or just less than 60°. If sine wave *a* is written in terms of radians, it becomes sin(2π*t*). Every one second this sine wave goes through a full cycle that corresponds to 2π radians. And 2π radians per second is equivalent to 1 cycle per second, or 360° per second. So sin[360°(*f*)(*t*)] is equivalent to sin[2π(*f*)(*t*)], where the latter angular measure is in radians.

SINE WAVES AND INTEGRATORS

Another very interesting property of a sine wave is that if the mathematical operation of integration is performed on it, the resulting shape is still sinusoidal. For example, consider a generalized version of sine wave *d* in Fig. 11.1, $A\sin[360°(f)(t) + 90°]$, represented by the top graph in Fig. 11.3. This sine wave has amplitude, *A*, and frequency, *f*. Given that this sine wave goes through *f* cycles in one second, it must go through a full cycle in 1/*f* seconds (*f* cycles/second × 1/*f* seconds = 1 cycle). One quarter of a cycle therefore corresponds to 1/4*f* seconds, one half cycle corresponds to 1/2*f* seconds, and so on (Fig. 11.3).

Now integrate this function over time. The mathematical operation of integration is equivalent to summing up the area under the curve. Therefore, the cumulative area under this sine wave becomes increasingly positive for the first one quarter cycle from 0 to 1/4*f*, and reaches a maximum at time 1/4*f*. How large is this maximum? Looking at the top curve in Fig. 11.3, it is possible to see that the area under the sine

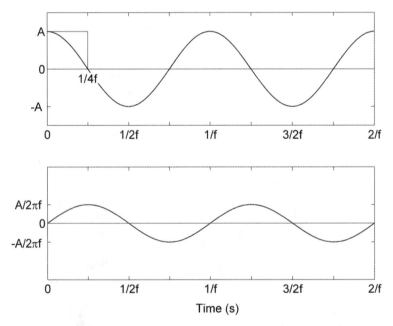

FIG. 11.3. This figure illustrates the process of integration. The bottom curve, $(A/2\pi f)\sin[360°(f)(t)]$, represents the integral of the top curve, $A\sin[360°(f)(t) + 90°]$. The integral is the accumulated area under the top curve.

function over this quarter cycle is less than the area of the rectangle, which is $A \times 1/4f$. On the other hand, because of the curvature of the sine wave, the area must be greater than half the area of this rectangle (i.e., greater than $1/2 \times A \times 1/4f = A/8f$). Using the rules of calculus, the exact answer turns out to be $A/2\pi f$, which is between $A/4f$ and $A/8f$, as expected. Given the close connection between the sine wave and the circle, the factor of 2π is perhaps not totally surprising. The bottom curve in Fig. 11.3 represents the integral of the top curve. The maximum of the integral is $A/2\pi f$.

Continuing on to summate area under the sine wave (top curve) beyond time $1/4f$, the integral (bottom curve) then goes to zero one quarter cycle later at $1/2f$, because the summation of the positive area from 0 to $1/4f$ in the top curve is just canceled by the negative area from $1/4f$ to $1/2f$ (Fig. 11.3). One-quarter cycle later, the integral achieves a minimum due to the summation of the negative area between $1/2f$ and $3/4f$. The minimum is $-A/2\pi f$. One quarter cycle after that the addition of the positive area from $3/4f$ to $1/f$ in the top curve brings the total summation of area or integral back to zero and thereby completes the cycle.

The integral of the sine wave has a sinusoidal shape and it also has the same frequency as the sine wave itself. However, the amplitude and phase of the integral are different. The amplitude of the integral is equal to $1/2\pi f$ of the amplitude of the original sine wave. The phase of the integral is one quarter cycle, or 90°, behind the phase of the original sine wave.

Note that if $2\pi f$ is equal to 1, then the integral of the sine wave would have the same amplitude as the sine wave itself. To achieve this condition, f must equal $1/2\pi$ cycles per second. As mentioned previously, $1/2\pi$ of a circle is exactly one radian. Therefore, if the frequency of the sine wave is one radian per second, its integral will have the same amplitude as the sine wave itself. Sine waves with frequencies slower than one radian per second have integrals with amplitudes larger than that of the sine wave itself. The base of the rectangle in Fig. 11.3 will be large and will permit large amplitudes for the integral. Sine waves with frequencies faster than one radian per second have integrals with amplitudes smaller than the sine wave itself, because the base of the rectangle will be small.

Take a sine wave with a frequency of one radian per second and integrate it twice. Each integration does not change the amplitude of the sine wave, but adds a 90° phase lag. Therefore, the sum of the two phase lags would be 180° or one half of a cycle. This effect is the same as multiplying the initial sine wave by –1, because a half-cycle phase shift simply reverses the temporal occurrences of the maxima and minima. If a sine wave is integrated twice and then is multiplied by –1, this brings the phase back to that of the original sine wave.

The previous observation leads to another way of generating a sine wave. Suppose there are two integrators and they are connected in series (Fig. 11.4). Next take the output of the second integrator, multiply it by –1, and reconnect it to the input of the first integrator. The operation of making that connection from the output of the –1 multiplier to the input of the first integrator forces a constraint on the system. Namely, it forces the input to the first integrator to be equal to the output of the multiplier. The mathematical function that satisfies this constraint is:

$$A \sin(t + \theta)$$

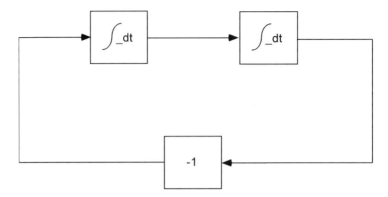

FIG. 11.4. An alternative way of generating a sine wave. Two integrators are connected in series and the output of the second integrator is multiplied by –1 and fed back as input into the first integrator. This circuit produces a sine wave with a frequency of one radian per second (1 cycle = 2π seconds).

namely, a sine wave with frequency one radian per second. Therefore, if an electrical circuit or a computer simulation is built with the structure shown in Fig. 11.4, it will exhibit an oscillation with a frequency of one radian per second. (The particular values of A and θ will depend on the initial values of the integrator outputs when the circuit is started; these initial values are not shown in Fig. 11.4.)

Suppose that instead of multiplying the output from the second integrator by –1, the output was multiplied by some other negative number such as –9. Would the result still be sinusoidal? How would it be different? In general, you will find that the response will be sinusoidal, and the frequency of the resulting sinusoid will equal the square root of the absolute magnitude of the multiplier. With a multiplier of –9, the frequency would be 3 radians per second.

These are surprising results in some ways. Each of the integrators does not inherently look sinusoidal or exhibit sinusoidal properties by themselves. However, when connected together in a loop structure, a sine wave emerges because the physical constraints embodied in the electrical circuit are analogous to the mathematical constraints associated with double integration and multiplication (as represented in Fig. 11.4). This constitutes still another way to generate a sine wave, and it demonstrates the special relation between the sinusoidal shape and the operation of integration. The circuit in Fig. 11.4 also provides a way of producing exactly 2π seconds, because that is the period of an oscillation with frequency one radian per second.

This chapter has discussed some of the basic properties of the sine wave. Similar discussions of these ideas can be found in more advanced texts on control theory (e.g., Milsum, 1966), physics texts (e.g., Feynman, Leighton, & Sands, 1963), and general engineering texts. For an introductory discussion of oscillations with different shapes, see Glass and Mackey (1988). Chapter 12 discusses how sets of sine waves can be used to approximate more complex temporal patterns via the mathematical technique of Fourier analysis. Chapter 21 discusses various ways that people may generate sinusoidal motion patterns.

REFERENCES

Feynman, R. P., Leighton, R. B., & Sands, M. (1963). *The Feynman lectures on physics* (Vol. 1, chaps. 9 & 21). Reading, MA: Addison-Wesley.

Glass, L., & Mackey, M. C. (1988). *From clocks to chaos: The rhythms of life*. Princeton, NJ: Princeton University Press.

Kugler, P. N., & Turvey, M. T. (1987). *Information, natural law, and the self-assembly of rhythmic movement*. Hillsdale, NJ: Lawrence Erlbaum Associates.

Milsum, J. H. (1966). *Biological control systems analysis* (chap. 7). New York: McGraw-Hill.

A Qualitative Look at Fourier Analysis

He: So, how long have you been undergoing Fourier analysis?
She: Don't ask how long. Ask how frequently.
He: Didn't you ever get a permanent wave?

— Anonymous

Jean-Baptiste-Joseph Fourier's Theorie analytique de la chaleur (The Mathematical Theory of Heat) inaugurated simple methods for the solution of boundary-value problems occurring in the conduction of heat. But, this "great mathematical poem," as Fourier analysis was called by Lord Kelvin, has extended far beyond the physical applications for which it was originally intended. In fact, it has become "an indispensable instrument in the treatment of nearly every recondite question in modern physics," communication theory, linear systems, etc.

—Hsu (1970, preface)

Fourier analysis provides a mathematical technique for decomposing signals that extend over time and/or space into a sum of sinusoidal components. As illustrated in Fig. 12.1, a periodic signal such as a square wave can be approximated by a sum of sinusoidal components. The top of Fig. 12.1 shows the sum of three sinusoids (a sinusoid with the same frequency and $4/\pi$ times the amplitude of the square wave plus a sinusoid with $4/3\pi$ times the amplitude and three times the frequency of the square wave plus a sinusoid with $4/5\pi$ times the amplitude and five times the frequency of the square wave). The bottom of Fig. 12.1 shows the sum of seven sinusoids, the same three as in the top graph plus four more sine waves with higher frequencies. Note that with the addition of the higher frequency components, the sum better approximates the square wave.

As noted in chapter 11, a sine wave has three parameters: a frequency, an amplitude, and a phase. Therefore, if n sine waves have to be added together to approxi-

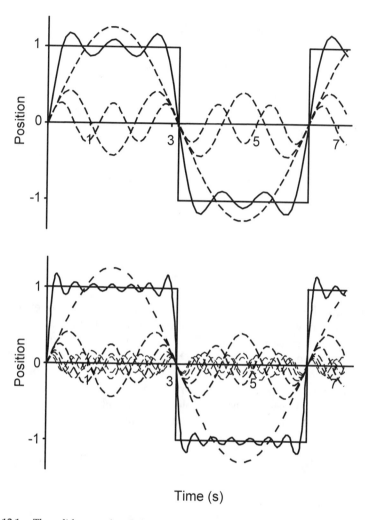

Time (s)

FIG. 12.1. The solid curve shows the sum of three sinusoids in the top diagram and the sum of seven sinusoids in the bottom diagram. With the addition of the higher frequency components, the sum better approximates the periodic square wave.

mate a given signal, there are a total of $3n$ parameters that need to be specified. If n is large, then this technique could be a very cumbersome way to describe the original signal. However, in practice, when amplitude and phase are plotted as a function of frequency, they often form simple patterns that can be described with just a few parameters. It is the occurrence of simple patterns of amplitude and phase as a function of frequency that makes Fourier analysis a useful way of describing signals. This chapter introduces the mathematical basis of Fourier analysis and describes some qualitative patterns of amplitude and phase. Subsequent chapters use this approach to describe the behavior of some of the elementary dynamic systems that have already been introduced and consider the response patterns in terms of amplitude ratios and phase shifts.

CALCULATING A FOURIER SPECTRUM

The technique of approximating a signal with a set of sinusoidal components begins with a choice of n frequencies. For example, suppose a one-dimensional temporal signal, $y(t)$, has a duration of T seconds and a mean value u_y. Then choose a set of n frequencies, ω_k, that have an integer number of cycles over this duration. In other words, ω_k (radians per second) multiplied by T (seconds) yields an integer multiple (k) of 2π radians. 2π radians correspond to a complete cycle of a sine wave:

$$\omega_k T = k2\pi, \quad \text{where } k = 1, 2, \ldots, n$$
$$\omega_k = k2\pi/T$$

Using sine waves with these frequencies to approximate $y(t)$:

$$y(t) = u_y + \sum_{k=1}^{n} [A_k \sin(\omega_k t + \theta_k)] + e(t) \tag{1}$$

Note that because the sine waves each have a mean value of zero over the interval T, their sum will also have a mean value of zero over this interval. In order to approximate $y(t)$, it is necessary to add an additional constant term, u_y, equal to the mean of $y(t)$. $e(t)$ represents a residual aspect of $y(t)$ that is not captured by the sine waves and u_y. The remaining problem is to choose the values of the amplitudes, A_k, and the phases, θ_k, so as to come up with a best approximation to $y(t)$ [i.e., to make the mean squared valued of $e(t)$ as small as possible]. Given that $\sin(u + v) = \cos(v)\sin(u) + \sin(v)\cos(u)$, Equation 1 can be transformed as follows:

$$y(t) = u_y + \sum_{k=1}^{n} [A_k \cos(\theta_k) \sin(\omega_k t) + A_k \sin(\theta_k) \cos(\omega_k t)] + e(t) \tag{2}$$

Or equivalently,

$$y(t) = u_y + \sum_{k=1}^{n} [B_{1k} \sin(\omega_k t) + B_{2k} \cos(\omega_k t)] + e(t) \tag{3}$$

where $B_{1k} = A_k \cos(\theta_k)$
$\quad\quad\quad B_{2k} = A_k \sin(\theta_k)$

In other words, comparing Equations 1 and 3 shows that a sine wave with a phase shift θ_k is equivalent to the sum of a sine wave and cosine wave with zero phase shift if the latter two are appropriately weighted by constants B_{1k} and B_{2k}. Although at first it looks as though this transformation has simply made the approximation to $y(t)$ more complicated, in fact the latter form is easier to deal with mathematically. If it is possible to calculate the B_{1k}'s and B_{2k}'s, then it is always possible to solve for the A_k's and the θ_k's and return to the $\sin(\omega_k t + \theta_k)$ form of the decomposition by using trigonometric transformations as follows:

$$B_{2k}/B_{1k} = [A_k\sin(\theta_k)]/[A_k\cos(\theta_k)] = \tan(\theta_k)$$

$$\text{Therefore,} \quad \theta_k = \tan^{-1}(B_{2k}/B_{1k})$$

$$B_{1k}^2 + B_{2k}^2 = [A_k\cos(\theta_k)]^2 + [A_k\sin(\theta_k)]^2 = A_k^2[\cos^2(\theta_k) + \sin^2(\theta_k)] = A_k^2$$

$$\text{because} \quad \cos^2(\theta_k) + \sin^2(\theta_k) = 1$$

$$\text{Therefore,} \quad A_k = (B_{1k}^2 + B_{2k}^2)^{1/2} \tag{4}$$

Determining the values of the B_{1k}'s and B_{2k}'s basically consists of determining how strongly each of the $\sin(\omega_k t)$'s and $\cos(\omega_k t)$'s is correlated with $y(t)$. Sines and cosines that are strongly correlated with $y(t)$ will be given large weightings, and sines and cosines that are weakly correlated with the signal will be given small weightings.[1] "To be correlated with $y(t)$" means that $y(t)$ rises and falls about its mean value in a manner that is systematically related to the way each sine or cosine rises and falls. This trend can be captured mathematically by multiplying $[y(t) - u_y]$ by each of the sines and cosines, integrating over the duration of the signal, and dividing the result by T to get an average. Technically, the result is called the *covariance* of $y(t)$ and each of the sine waves or cosine waves. If the two temporal patterns covary in a systematic fashion, then this integral will have a large positive or negative magnitude. However, if the two temporal patterns are generally unrelated, then the integral will be near zero.[2] For example, to calculate the covariance of $y(t)$ and $\sin(\omega_1 t)$:

$$\left(\frac{1}{T}\right)\int_0^T [y(t) - u_y]\sin(\omega_1 t)dt$$

$$= \left(\frac{1}{T}\right)\int_0^T \left\{\sum_{k=1}^n [B_{1k}\sin(\omega_k t) + B_{2k}\cos(\omega_k t)] + e(t)\right\}\sin(\omega_1 t)dt$$

$$= \left(\frac{1}{T}\right)\int_0^T B_{11}\sin^2(\omega_1 t)dt = 0.5B_{11} \tag{5}$$

Consider why the previous integral gives such a simple result. By choosing the frequencies ω_k so that they each complete an integral number of cycles over the signal duration T, the integral of the product of any of these two sine waves or cosine waves will be zero unless the two frequencies are identical. In other words, these sine waves and cosine waves of different frequencies are uncorrelated with each other over the interval T. This lack of a systematic relation between two sine waves or cosine waves of different frequencies can be visualized by crossplotting their values over time. For example, Fig. 12.2 crossplots $\sin[10(2\pi)t]$ versus $\sin[11(2\pi)t]$. The first sine wave completes 10 cycles per second (the phase goes from 0 to 20π in the first second), and the second sine wave completes 11 cycles per second (the phase goes from 0 to 22π in the first second). The top graph shows the two sine waves over a .25 s period. The motion

---

[1]The reader familiar with statistics can see this problem to be an example of multiple regression.
[2]For more detailed discussions of this issue see Papoulis (1965).

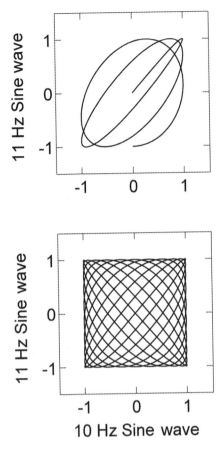

FIG. 12.2. Crossplot of two sine waves with frequencies of 10 and 11 cycles per second. The top graph shows the co-variation over 0.25 s, and the bottom graph shows the covariation over 1.00 s.

of the 10 cycle per second sine wave is plotted on the horizontal axis. It starts at the center (0) and completes 2.5 cycles of back and forth horizontal motion in .25 s. The 11 cycle per second sine wave is plotted on the vertical axis. It is somewhat faster, and although it also starts at 0, it gradually gets farther and farther ahead of the 10 cycle per second sine wave. The bottom graph in Fig. 12.2 shows a full second of motion, which crossplots 11 cycles of the faster vertical motion versus only 10 cycles of the horizontal motion. Looking at the overall pattern, the two motions are uncorrelated. When the slower horizontal sine wave is positive, the vertical sine wave might be either positive or negative. Similarly, when the horizontal sine wave is negative, the vertical sine wave might be either positive or negative. On average, there is no simple linear relation between their values over the 1-second period. They are uncorrelated, and the integral of their product over the 1-second period will be zero.

In a similar manner, a sine wave and cosine wave of the same frequency are also uncorrelated over an integral number of cycles. For example, Fig. 12.3 (left) shows a crossplot of $\sin[10(2\pi)t]$ on the horizontal axis versus $\cos[10(2\pi)t]$ on the vertical axis over one full cycle. The pattern is circular, indicating they are uncorrelated (i.e., there is no systematic linear relation). When the sine is positive, the cosine can be positive or negative; when the sine is negative, the cosine can be positive or negative. Therefore, the integral of their product over an integral number of cycles will be zero. Similarly, $e(t)$, which may contain frequencies higher than ω_n, is uncorrelated with any of the sine waves, and the integral of their product is also zero over the interval T (e.g., see Papoulis, 1965, pp. 219, 242).

In contrast, the crossplot of $\sin[10(2\pi)t]$ against itself yields a straight line (Fig. 12.3, right), which indicates a perfect linear correlation. The integral of $\sin^2[10(2\pi)t]$ over an integral number of cycles in time period T is equal to $.5T$. As a result of the lack of correlation between the other signals, the only nonzero integral in Equation 5 corresponds to the sine-squared term. Hence, the evaluation of the integral yields $.5B_{11}$. In a similar manner,

$$\left(\frac{1}{T}\right)\int_0^T [y(t) - u_y]\cos(\omega_1 t)dt = 0.5B_{21} \tag{6}$$

The remaining B_{1k}'s and B_{2k}'s can be calculated in a similar manner to the way B_{11} and B_{21} were calculated in Equations 5 and 6. It is possible to solve for the A_k's and θ_k's via Equations 4 and complete the sinusoidal approximation to the signal $y(t)$ corresponding to Equation 1. In summary, by determining how strongly $y(t)$ covaries with each of a set of mutually uncorrelated sine waves and cosine waves, it is possible to determine the appropriate weighting of each of them.

A few remaining practical details concern how to choose the range of frequencies ω_1 to ω_n to obtain a good approximation to a particular signal $y(t)$. Typically, researchers use enough frequencies so that $e(t)$ is negligible. How many frequencies is "enough" depends on how smooth the function $y(t)$ is. The smoother the changes in $y(t)$, the lower the highest frequency ω_n can be and still make $e(t)$ negligible. If $y(t)$ exhibits many rapid and abrupt changes, then researchers sometimes apply smoothing algorithms to $y(t)$ before approximating it with a set of sine waves. Then they can use fewer sine waves to represent $y(t)$. The lowest frequency, ω_1, will equal $2\pi/T$, and this will also be the spacing between successive frequencies in the Fourier spectrum. Therefore, to achieve a finer resolution among frequencies, it is necessary to use a longer measurement interval, T.

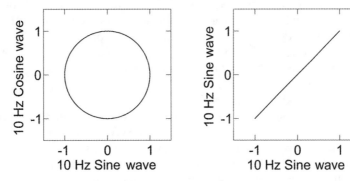

FIG. 12.3.   Crossplot of a sine wave and cosine wave of the same frequency over one cycle (left). Crossplot of a sine wave and itself (right).

Because most researchers use digital computers rather than analog computers for their computations, they sample a continuous signal $y(t)$ and use these samples in place of the continuous signal. If ω_n is the highest frequency necessary to make $e(t)$ negligible, then take samples at least at twice that frequency in order to detect the peaks and valleys of the trend associated with ω_n (one peak and one valley per cycle). More formally, this constraint is described in Shannon's Sampling Theorem (e.g., see Briggs & Henson, 1995). Practically, it is better to sample a bit faster than the minimal factor of twice the highest frequency of interest (Bendat & Piersol, 1966).

There are more efficient algorithms called Fast Fourier Transforms for estimating the B_{1k}'s and B_{2k}'s. However, the previous equations are useful for conceptual insight into Fourier analysis, and these equations have been used quite literally for computational purposes (e.g., Allen & Jex, 1972; McRuer & Krendel, 1957; Sheridan & Ferrell, 1974). Equations 1 to 3 are *Fourier series approximations* to the signal $y(t)$. The *Fourier transform* of $y(t)$ can also be defined in a manner similar to the Laplace transform (chap. 5). The complex variable s is replaced by $j\omega$, where ω is a continuous real variable representing frequency. The Fourier transform provides a value of amplitude and phase for each value of ω. The Fourier transform of a signal with limited duration is basically equivalent in amplitude and phase to its Fourier series approximation at those values of ω that correspond to an integer number of cycles over the signal duration (see Briggs & Henson, 1995). Fourier transforms also provide mathematical ease in calculating the response of linear dynamic systems in a manner similar to Laplace transforms (see chaps. 13, 14, and 19).

FOURIER SPECTRA OF SINUSOIDS: ONE, TWO, MANY

If a signal $y(t)$ consists of a single sine wave, $A_1\sin(\omega_1 t + \theta_1)$, then its Fourier spectrum will be zero at all but frequency ω_1, if that frequency coincides with one of the $k2\pi/T$ frequencies used to construct the spectrum. The amplitude of the spectrum will be A_1 (Fig. 12.4, top), and the phase will be θ_1 (not shown). If $y(t)$ consists of a single sine wave with a frequency ω_1 in between the $k2\pi/T$ frequencies, then its Fourier spectrum will consist of a band of nonzero frequencies in the neighborhood of ω_1 (Fig. 12.4, bottom; see Berge, Pomeau, & Vidal, 1986, for additional detail). The sum of the squares of the amplitudes in the Fourier spectrum will equal A_1^2; however, because $y(t)$ is spread over several of the $k2\pi/T$ frequencies, the amplitude of the highest component in the Fourier spectrum will be less than A_1. In other words, $y(t)$ has been "smeared" across several frequencies.

If the signal $y(t)$ consists of the sum of two sine waves of different frequencies, $A_1\sin(\omega_1 t + \theta_1) + A_2\sin(\omega_2 t + \theta_2)$, then the time history will look more complex (Fig. 12.5, top). Because the two frequencies are different, the phases of the two sine waves will increase at different rates. Sometimes the two sine waves will be in phase (i.e., their phases will differ by a multiple of 2π or an integer number of cycles). Around this occurrence, their sum will exhibit a larger amplitude than either of the individual sine waves. However, as time passes, they will drift out of phase. When they are one-

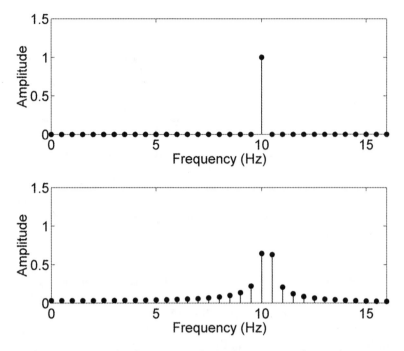

FIG. 12.4. Fourier amplitude spectrum of a single sine wave when its frequency coincides with one of the $k2\pi/T$ frequencies used to construct the spectrum (top) and when its frequency falls between two of these frequencies (bottom).

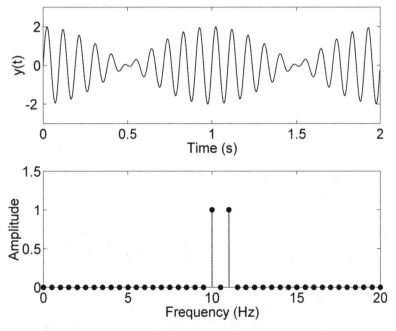

FIG. 12.5. The sum of two sine waves of different frequencies, 10 and 11 cycles per second (top), and the corresponding Fourier amplitude spectrum (bottom).

half cycle out of phase (i.e., their phases differ by an odd multiple of π), they will maximally cancel each other. Around this occurrence, their sum will have a smaller amplitude than either of the individual sine waves. This alternation between summing to produce a larger amplitude and summing to produce a smaller amplitude is often referred to as a *beat frequency*. Numerically, this frequency is equal to the difference between the two frequencies, $\omega_1 - \omega_2$. Namely, if the time between successive maximal cancellations is t^*, the increase in phase for the faster sine wave over this interval, $\omega_2 t^*$, is just 2π greater than the increase in phase in the slower sine wave, $\omega_1 t^*$, or $\omega_2 t^* - \omega_1 t^* = (\omega_2 - \omega_1)t^* = 2\pi$. In that manner, the two sine waves return to the same out-of-phase relation they had in the previous instant of maximal cancellation. Equivalently, t^* corresponds to one cycle of a beat frequency equal to $(\omega_2 - \omega_1)$. When people attempt manually to track a high frequency sinusoidal target, the frequency of their motion sometimes fails to match the frequency of the target. The resulting error then displays a beat frequency equal to the difference between the motion frequency of the person and target frequency (e.g., Noble, Fitts, & Warren, 1955; Fig. 12.6).

Although the time history is complicated by the presence of the beat frequency (Fig. 12.5, top), the Fourier spectrum for the sum of these two sine waves is still quite simple. Namely, it will be nonzero only at the two frequencies ω_1 and ω_2, if they coincide with the $k2\pi/T$ frequencies used to construct the spectrum (Fig. 12.5, bottom). The amplitudes will be A_1 and A_2, and the phases will be θ_1 and θ_2 (not shown). If ω_1 and ω_2 do not coincide with the $k2\pi/T$ frequencies, then the spectrum will consist of the sum of two bands of nonzero frequencies.

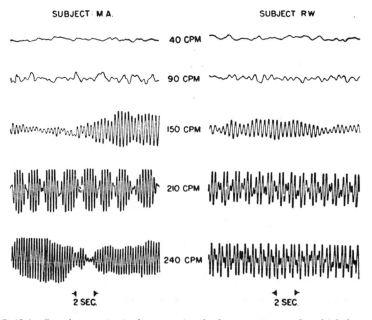

FIG. 12.6.   Beat frequencies in the error signal when a person tracks a high frequency sine wave . CPM = cycles per minute. From "The frequency response of skilled subjects in a pursuit tracking task" by M. Noble, P. M. Fitts, and C. E. Warren, 1955, *Journal of Experimental Psychology, 49*, pp. 249–256. Copyright 1955 by APA. Reprinted by permission.

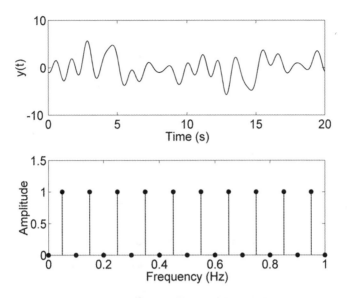

FIG. 12.7. An example of a rough approximation to white noise—the sum of 10 sine waves (top) and their corresponding Fourier amplitude spectrum (bottom).

If the signal $y(t)$ consists of the sum of a large number, n, of sine waves, then the time history will look very irregular (Fig. 12.7, top). When researchers want to generate a pattern that people cannot easily predict, they often use a sum of five or more sine waves. If the amplitudes of all the sine waves are equal, then this spectral pattern is a rough approximation to "white noise" (i.e., equal amplitude at all frequencies; Fig. 12.7, bottom). However, researchers investigating manual tracking often make the amplitudes of the higher frequency sine waves smaller, so that the tracking task does not become too difficult.

Figure 12.8 shows a portion of the Fourier amplitude spectrum of the error signal in a manual tracking task (Allen & Jex, 1972). Actually, the amplitudes have been squared, which makes this a "power spectrum." The power of a signal is often defined as its mean-squared value over time, and for a sinusoid, this is equal to one half its amplitude squared. More loosely, the factor of one half is often omitted for sinusoids, and in general the squared amplitudes of a Fourier spectrum are referred to as a power spectrum (e.g., Berge et al., 1986). Power is typically plotted on a logarithmic scale. (If the base 10 logarithm of power is multiplied by 10, then the unit is referred to as a decibel, abbreviated "db").[3] The input signal in this manual tracking task consisted of a sum of five sine waves, which made it relatively unpredictable. The error signal had larger power at the input frequencies, four of which correspond to the circled points in Fig. 12.8. The lower level of error power at noninput frequencies is typ-

---

[3]More technically, a decibel is defined as $10\log_{10}(A_1^2/A_0^2)$, that is, as 10 times the logarithm of the ratio of the power of a given signal with amplitude A_1 to the power of a reference signal with amplitude A_0. This unit of measurement is equivalent to $20\log_{10}(A_1/A_0)$. In nonauditory contexts, the reference signal is sometimes taken to have unity amplitude.

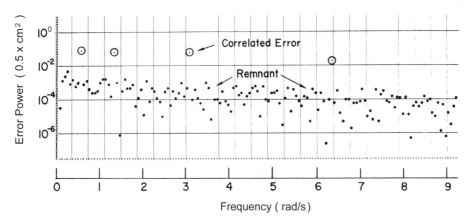

FIG. 12.8.   The Fourier power spectrum of the error signal in a manual tracking task. From "A simple Fourier analysis technique for measuring the dynamic response of manual control systems" by R. W. Allen and H. R. Jex, 1972, *IEEE Transactions on Systems, Man, and Cybernetics*, SMC-2, pp. 630–643. Copyright 1972 by IEEE. Adapted by permission.

ically referred to as "remnant," and its behavioral significance is discussed in a later chapter.

FOURIER SPECTRA: POSITION, VELOCITY, AND ACCELERATION

Suppose that the time history in Fig. 12.7 (top) represents a position signal, $y(t) = \sum_{k=1}^{n} [A_k \sin(\omega_k t + \theta_k)]$. By differentiating this signal, the velocity of $y(t)$ can be determined. Remembering that $d[\sin(u)]/dt = [du/dt]\cos(u)$ and that $\cos(u) = \sin(u + \pi/2)$, it is possible to calculate:

$$\text{velocity of } y(t) = dy(t)/dt = \sum_{k=1}^{n} A_k \omega_k \cos(\omega_k t + \theta_k) = \sum_{k=1}^{n} A_k \omega_k \sin(\omega_k t + \theta_k + \pi/2) \quad (7)$$

In other words, the amplitude of each of the sinusoidal components would be multiplied by ω_k, and the phase would be increased by $\pi/2$. The Fourier amplitude spectrum of the velocity of $y(t)$ would not have a uniform appearance. Instead, the higher frequencies would be amplified (Fig. 12.9, top).

If $y(t)$ were differentiated a second time to obtain its acceleration, then the same effect would occur again:

$$\text{acceleration of } y(t) = d[\text{velocity of } y(t)]/dt = \sum_{k=1}^{n} A_k \omega_k^2 \sin(\omega_k t + \theta_k + \pi) \quad (8)$$

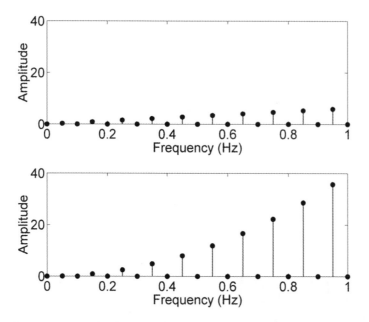

FIG. 12.9.   Fourier amplitude spectra for the velocity (top) and acceleration (bottom) of the signal at the top of Fig. 12.7.

Now each of the sinusoidal components of $y(t)$ would be multiplied by ω_k^2, and the phases would each be increased by π. The high frequency components in the Fourier amplitude spectrum are now amplified even more (Fig. 12.9, bottom).

In many measurement applications, the signal of interest may be at low frequencies, and in addition there may be high frequency measurement noise. In such cases, simply differentiating the noisy measurement of the signal to obtain its velocity and acceleration may unduly amplify the measurement noise and thereby produce an unreliable estimate of velocity and acceleration. For this reason, researchers typically use special techniques for filtering out the high frequency measurement noise before differentiating a signal. Better yet is to try to measure the velocity or acceleration signal more directly, rather than trying to differentiate a position signal (Winter, 1979).

In contrast, if the signal of interest is at high frequencies and an unwanted signal is at low frequencies, then measuring acceleration rather than position will emphasize the high frequency components of interest. This approach is used in the measurement of tremor (Elble & Koller, 1990). Tremor consists of a high frequency oscillation that occurs in addition to the intended movement. For most people, the magnitude of this high frequency oscillation is quite small. However, for some individuals, the magnitude is relatively large and is a severe impediment to skilled movement. In order to measure tremor, the lower frequency movement on which it occurs can be deemphasized by measuring acceleration (Fig. 12.10; Elble & Koller, 1990). It is possible to place a measurement device on a finger or other limb that is directly sensitive to the acceleration of the limb rather than to its position. Alternatively, limb position can be measured, high frequency measurement noise above the tremor frequencies can be filtered out, and the signal can be differentiated twice to approximate accelera-

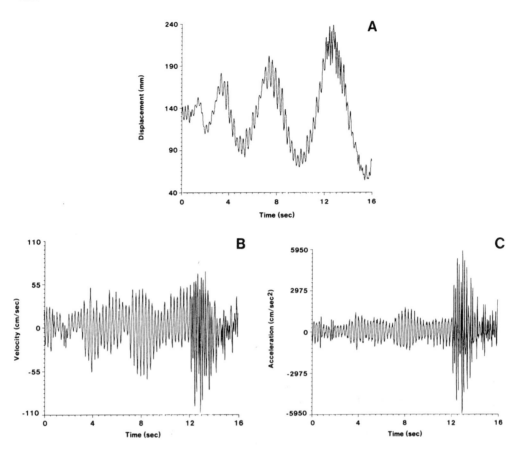

FIG. 12.10.   Horizontal position, velocity, and acceleration of a person with accentuated tremor attempting to draw a spiral. From *Tremor* (p. 17, Fig. 2.3) by R. Elble and W. Koller, 1990, Baltimore: Johns Hopkins University Press. Copyright 1990 by the Johns Hopkins University Press. Reprinted by permission.

tion. These measurement procedures emphasize the high frequency tremor. Figure 12.11 shows the power spectra (squared Fourier amplitudes) of the position, velocity, and acceleration of wrist movements by an individual with abnormal accentuated tremor (Elble & Koller, 1990).

IRREGULAR SIGNALS: SPEED, COMPLEXITY, AND NONSTATIONARITY

More generally, any continuous signal $y(t)$ with finite absolute value and limited duration can be approximated with a sum of sine waves even if it looks highly irregular. The relative weights of the various sine waves, the A_k's in Equation 1, provide an indication of how rapidly $y(t)$ changes over time. If the high frequency sine waves have

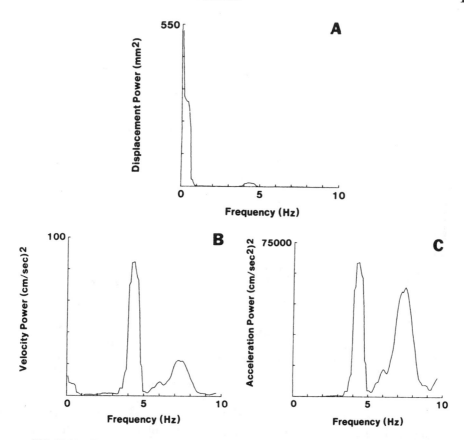

FIG. 12.11. Fourier power spectra for the signals in Fig. 12.10. From *Tremor* (p. 23, Fig. 2.5) by R. Elble and W. Koller, 1990, Baltimore: Johns Hopkins University Press. Copyright 1990 by the Johns Hopkins University Press. Reprinted by permission.

large weights [i.e., they play a major role in approximating $y(t)$], then $y(t)$ must be changing relatively rapidly. On the other hand, if only the low frequency sine waves have large weights, then $y(t)$ must be changing relatively slowly. Manual tracking data typically have most of their spectral power below 2 Hz or equivalently below about 12 radians/s (Fig. 12.8). Tremor is a higher frequency oscillation than is associated with most controlled motion, and its spectral power can range up to 40 Hz, depending on the part of the body involved (Elble & Koller, 1990). As another example, when measuring the heart rate of an individual at rest, it is not found to be perfectly steady. Variations tend to occur over a period of minutes, and the Fourier spectrum has most of its power below .05 Hz (e.g., Goldberger & Rigney, 1991, Fig. 12.12).

The shape of a Fourier spectrum can also provide some indication of the relative complexity of a signal. For example, Fig. 12.12 shows a heart rate time history and the corresponding Fourier amplitude spectrum for three individuals, one who was healthy and two who suffered heart failure. The time history for the healthy individual exhibits a relatively complex temporal structure, and the Fourier spectrum decreases inversely with frequency. The middle individual exhibits a more regular low

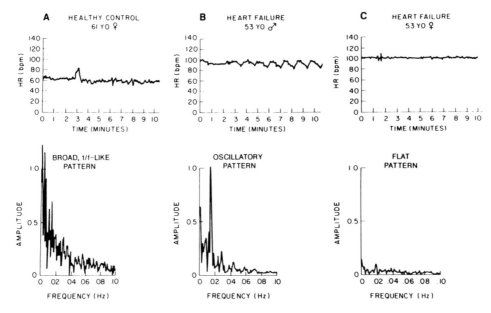

FIG. 12.12. Heart rate time histories and associated power spectra for a healthy individual (left) and two individuals with heart problems (center and right). From "Nonlinear dynamics at the bedside" by A. L. Goldberger and D. R. Rigney, 1991, in *Theory of heart* (p. 594, Fig. 22.10) by L. Glass, P. Hunter, and A. McCulloch (Eds.), New York: Springer-Verlag. Copyright 1991 by Springer-Verlag. Reprinted by permission.

frequency oscillation, and the Fourier spectrum is dominated by a corresponding peak at about .015 Hz. Finally, the third individual exhibits relatively little variation in heart rate. The amplitudes of the Fourier spectrum are therefore quite low, and the spectrum is flat, resembling white noise.

A final issue is stationarity (i.e., whether the temporal characteristics of some observed behavior are relatively invariant over time). A Fourier spectrum can be calculated for any continuous signal with finite absolute value and limited duration, regardless of whether it is relatively invariant (stationary) in its style of motion or whether it undergoes radical changes. However, if the observed process is not stationary, then care must be taken in interpreting the Fourier spectrum. For example, the tremor of the individual shown in Fig. 12.10 suddenly changed from a low frequency to a high frequency. The corresponding acceleration power spectrum (Fig. 12.11, lower right) shows two peaks, one for each of these frequencies (around 4.5 Hz and around 7.5 Hz). If researchers only saw this power spectrum and not the corresponding time history, then they might incorrectly conclude that the two frequencies were present simultaneously. Although the sine waves making up the Fourier spectrum extend over the entire duration of the time history, the phases are such that the low frequency sine waves dominate the resulting pattern early in the time history, and the high frequency sine waves dominate the resulting pattern late in the time history.

This contrast between the simultaneous presence of the component Fourier sine waves and the sequential nature of a corresponding nonstationary temporal pattern has led many researchers to calculate Fourier spectra over time spans for which a sig-

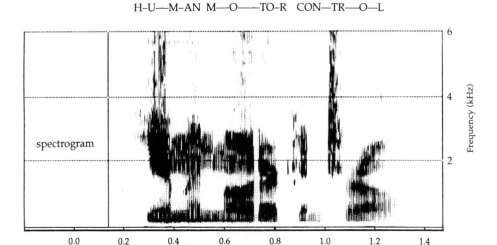

FIG. 12.13.   The spectrogram for someone saying the words, "human motor control." From *Human motor control* (p. 310) by D. A. Rosenbaum, San Diego: Academic. Copyright 1991 by Academic Press. Reprinted by permission.

nal is relatively stationary. For example, the power spectrum of speech varies markedly during the course of saying several words. Speech researchers portray the power spectrum of the speech signal over very brief intervals of time, and the two-dimensional graph of the power spectrum versus time is called a *spectrogram*. Figure 12.13 shows the spectrogram for someone saying the words, "human motor control" (Rosenbaum, 1991). The more highly weighted sine waves (i.e., larger A_k's) are represented more darkly. The frequencies are displayed up to 6,000 Hz and are much higher than those observed in manual control. Note that the "tr" in "control" consists primarily of frequencies above 2,000 Hz, whereas the subsequent "ol" consists of much lower frequencies. If a single Fourier spectrum had been calculated for this entire utterance, it would be impossible to appreciate how much the Fourier spectrum was varying over time.[4]

In summary, Fourier analysis provides an alternative representation of time histories in terms of an additive combination of sinusoidal components. The patterns revealed in the Fourier spectra often provide additional insight into the structure of the underlying process. Subsequent chapters examine how the responses of various dynamic systems can be characterized in terms of Fourier analysis.

REFERENCES

Allen, R. W., & Jex, H. R. (1972). A simple Fourier analysis technique for measuring the dynamic response of manual control systems. *IEEE Transactions on Systems, Man, and Cybernetics, SMC-2*, 630–643.

---

[4]A more advanced treatment of this issue is provided by Gabor functions (Delp & Crossman, 1972; Gabor, 1946).

Bendat, J. S., & Piersol, A. G. (1966). *Measurement and analysis of random data.* New York: Wiley.

Berge, P., Pomeau, Y., & Vidal, C. (1986). *Order within chaos: Towards a deterministic approach to turbulence* (L. Tuckerman, Trans.). New York: Wiley. (Original work published 1984.)

Briggs., W. L., & Henson, V. E. (1995). *The DFT: An owner's manual for the discrete Fourier transform.* Philadelphia: Society for Industrial and Applied Mathematics.

Delp, P., & Crossman, E.R.F.W. (1972, May). Transfer characteristics of human adaptive response to time-varying plant dynamics. In *Proceedings of the Eighth Annual Conference on Manual Control* (AFFDL-TR-72-92, pp. 245–271). University of Michigan, Ann Arbor, MI.

Elble, R. J., & Koller, W. C. (1990). *Tremor.* Baltimore: Johns Hopkins University Press.

Gabor, D. (1946). Theory of communication. *Journal of the Institute of Electrical Engineers, 93,* 429–457.

Goldberger, A. L., & Rigney, D. R. (1991). Nonlinear dynamics at the bedside. In L. Glass, P. Hunter, & A. McCulloch (Eds.), *Theory of heart* (pp. 583–605). New York: Springer-Verlag.

Hsu, H. P. (1970). *Fourier analysis.* New York: Simon & Schuster.

McRuer, D. T., & Krendel, E. S. (1957). *Dynamic response of human operators* (WADC-TR-56-524). Wright Patterson Air Force Base, OH.

Noble, M., Fitts, P. M., & Warren, C. E. (1955). The frequency response of skilled subjects in a pursuit tracking task. *Journal of Experimental Psychology, 49,* 249–256.

Papoulis, A. (1965). *Probability, random variables, and stochastic processes.* New York: McGraw-Hill.

Rosenbaum, D. A. (1991). *Human motor control.* San Diego: Academic Press.

Sheridan, T. B., & Ferrell, W. R. (1974). *Man–machine systems: Information, control, and decision models of human performance.* Cambridge, MA: MIT Press.

Winter, D. A. (1979). *Biomechanics of human movement.* New York: Wiley.

The Frequency Domain:
Bode Analysis

The Nyquist stability criterion . . . enables us to investigate both the absolute and relative stabilities of linear closed-loop systems from a knowledge of their open-loop frequency-response characteristics. In using this stability criterion, we do not have to determine the roots of the characteristic equation. This is one advantage of the frequency-response approach. Another advantage of this approach is that frequency-response tests are, in general, simple and can be made accurately by use of readily available sinusoidal signal generators and precise measurement equipments. Often the transfer functions of complicated components can be determined experimentally by frequency response tests. Such experimentally obtained transfer functions can be easily incorporated in the frequency-response approach. Also, the frequency-response methods can be applied to systems that do not have rational functions, such as those with transport lags. In addition, plants with uncertainties or plants that are poorly known can be handled by the frequency-response methods. A system may be designed by use of the frequency-response approach such that the effects of undesirable noises are negligible. Finally, frequency-response analyses and designs can be extended to certain nonlinear control systems.

—Ogata (1990, pp. 426–427)

This chapter introduces the Bode plot as a way of representing the performance of a dynamic system in the frequency domain. Patterns in the Bode space will allow inferences to be made about the order of control, the presence and magnitude of time delays, and the stability of a control system. The ability to represent stability constraints, in terms of phase or gain margins, is one reason why the Bode plot is an important analytic tool for evaluating dynamic control systems.

Earlier chapters characterized the Laplace transfer function of a linear system in terms of its response to a step input. The classical Fitts' Law task is an example of the use of a step input to generate response functions for humans. Characteristics of the time histories generated in response to the step input could be used to make infer-

ences about the underlying dynamics (e.g., is the system first, second, or higher order). This section discusses how sinusoidal inputs can be used as the test signals. The
next chapter discusses how tracking tasks, using sinusoidal signals, can be used to
make inferences about human performance.

THE FREQUENCY RESPONSE

A convenient feature of sine waves is that when a sine wave is input to a linear system, the steady state output of the system will be a sine wave of the same frequency.
However, the output may be altered in amplitude and phase relative to the input.
The degree of change in amplitude and phase will generally depend on the frequency
of the input sine wave (ω). The pattern of amplitude change and phase shift as a function of frequency is the "frequency response" for the system. Just as various systems
have distinctive step responses, systems also have distinctive frequency responses.
Thus, it is possible to use the pattern of change in amplitude and phase across frequency to make inferences about the dynamic properties of a linear system. The frequency response will be particularly useful for making inferences about the stability
of a linear system.

The steady state frequency response of a system can be analytically determined
from its Laplace transfer function by substituting ($j\omega$) for the Laplace operator (s) in
the transfer function (assuming zero initial conditions). The result of this substitution
is a complex number. The j in the $j\omega$ represents the pure imaginary number $\sqrt{-1}$.
Mathematicians tend to use i rather than j for $\sqrt{-1}$. Engineers use j. This inconsistency
emerged because engineers tend to use the i to represent electrical current. To avoid
confusion, they adopted the j convention for complex numbers. This book uses the
engineering convention. The ω is a variable that represents the frequency (typically
in radians) of a sinusoidal input.

Whereas a real number can be represented as a position along a line (one-
dimensional space), a complex number can be represented as a position in a plane (a
two-dimensional space). Thus, a complex number is an ordered pair (x, y), where x is
the real part, and y is the imaginary part of the number. One convention for representing the complex number is $x + jy$.

Complex numbers can be depicted as vectors in a two-dimensional plane. The horizontal axis (abscissa) of this plane represents the real part and the vertical axis (ordi-
nate) represents the imaginary part, as shown in Fig. 13.1. The vectors shown in Fig.
13.1 can also be expressed in terms of their polar form. That is, the vector is specified
by its length, r (Euclidean distance from the origin), and by an angle, θ (actually a set
of angles that are different by multiples of 2π radians). Note the similarity to the unit
circle referenced in the discussion of sine waves (Fig. 11.2).

$$r = \sqrt{x^2 + y^2} \qquad (1)$$

$$\theta = \tan^{-1}\left(\frac{y}{x}\right) \qquad (2)$$

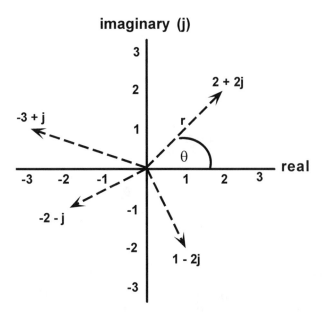

FIG. 13.1. Complex numbers are illustrated as vectors in a two-dimensional plane.

It is common to refer to the length of the vector as the *magnitude* of the complex number and to the angle as the *argument* of the complex number. These alternative values are the polar coordinates for the complex number. Using the trigonometric relations

$$x = r\cos\theta \text{ and } y = r\sin\theta \tag{3}$$

the complex number $x + jy$ can be rewritten in polar form as

$$r(\cos\theta + j\sin\theta) \tag{4}$$

An additional form for expressing a complex number can be obtained using the following identity involving the exponential function:

$$e^{j\theta} = \cos\theta + j\sin\theta \tag{5}$$

Thus, the complex number can be expressed in the exponential form:

$$re^{j\theta} \tag{6}$$

To compute the frequency response for a linear system from its Laplace transfer function, the first step is to substitute $(j\omega)$ for (s) in the transfer function. The result, also referred to as the Fourier transfer function, is a complex number with the real variable (ω) representing frequency. The magnitude of the resulting complex number is an index of the amplitude change for a given ω, and the argument of the resulting complex number is an index of the phase change for a given ω. For example, to compute the frequency response of an integrator:

$$\frac{1}{s} \Rightarrow \frac{1}{j\omega} = \left(\frac{-1}{\omega}\right)j \tag{7}$$

$$\left|\frac{1}{j\omega}\right| = \sqrt{\left(\frac{1}{j\omega} \times \frac{1}{-j\omega}\right)} = \frac{1}{\omega} \tag{8}$$

$$\arg\left(\frac{1}{j\omega}\right) = \tan^{-1}\left(\frac{-\dfrac{1}{\omega}}{0}\right) = \tan^{-1}(-\infty) = \frac{-\pi}{2} = -90° \tag{9}$$

Note that the magnitude was computed by multiplying the complex number by its conjugate [the conjugate of $x + jy$ is $x - jy$; the conjugate of $re^{+j\omega}$ is $re^{-j\omega}$]. This was obviously not necessary in this case, because there was a pure imaginary number (the magnitude is equal to the absolute value of the imaginary part). However, this is a general procedure for computing the magnitude (i.e., the magnitude is equal to the square root of the product of a complex number and its conjugate). Note that the magnitude of the frequency response of an integrator is inversely related to the frequency of the input (ω). As the frequency of the input sine wave (ω) increases, the amplitude of the resulting output sine wave will be reduced by a proportional amount. If the frequency of the input sine wave was .5 rad/s, then the amplitude of the output sine wave would be twice as great as the amplitude of the input sine wave. If the frequency of the input sine wave was 1 rad/s, then the input and output amplitudes would be equivalent. It the frequency of the input sine wave was 2 rad/s, then the output amplitude would be reduced by a factor of .5.

The argument for the integrator does not depend on ω. Thus, the phase shift from input to output will be the same for all frequencies of sine wave input to the integrator. The output sine wave will be one quarter of a cycle behind the input sine wave independent of its frequency. Note that the 90° phase shift is relative to the sinusoidal cycle (360°). It is not an absolute shift in time, but a shift of the sinusoidal waveform relative to itself. (See Fig. 11.3.)

The inverse relation between frequency and amplitude together with the one quarter cycle phase shift is the frequency signature for an integrator.

Consider another example. What is the frequency response for a first-order lag? Remember from earlier chapters that the transfer function for a first-order lag is:

$$\frac{k}{s+k} \tag{10}$$

where k is the gain in the forward loop. For this example, a first-order lag with unity gain ($k = 1$) is used. Thus, to compute the frequency response:

$$\frac{1}{s+1} \Rightarrow \frac{1}{j\omega+1} = \frac{1}{\omega^2+1} - \frac{j\omega}{\omega^2+1} \tag{11}$$

$$\left|\frac{1}{j\omega+1}\right| = \sqrt{\left(\frac{1}{j\omega+1}\right)\left(\frac{1}{-j\omega+1}\right)} = \sqrt{\frac{1}{\omega^2+1}} \tag{12}$$

$$\arg\left(\frac{1}{j\omega+1}\right) = \tan^{-1}\left(\frac{\dfrac{-\omega}{\omega^2+1}}{\dfrac{1}{\omega^2+1}}\right) = \tan^{-1}(-\omega) = -\tan^{-1}(\omega) \tag{13}$$

For the first-order lag, both the amplitude and the phase of the output are dependent on frequency. Table 13.1 shows the amplitude and phase response for a broad range of frequencies.

For very low frequencies, the first-order lag causes little change in amplitude from input to output. However, for high frequencies, the output amplitude is inversely proportional to the input frequency (as with the integrator). Because of its tendency to pass low frequencies without attenuation, but to attenuate high frequencies, the first-order lag is also referred to as a low-pass filter. The bandwidth of this filter is determined by the forward loop gain. The higher the gain (smaller the time constant), the greater the bandwidth of the filter. That is, a greater range of frequencies will pass through the system without significant attenuation. A good exercise to test your understanding would be to compute the frequency response for a first-order lag with less than unity (e.g., .1) and greater than unity (e.g., 10) forward loop gain. Note that the "break" between approximating a unity gain (no attenuation) and approximating an integrator (attenuation inversely proportional to frequency) shifts as a function of the change in forward loop gain.

Consider one last example: What is the frequency response for a time delay? The transfer function for a time delay is

$$e^{-as} \tag{14}$$

where a is the magnitude of the delay in seconds. Thus, the frequency response would be

TABLE 13.1
Amplitude Ratio and Phase Shift for a 1st-Order Lag

ω (rad/s)	Amplitude	Phase (rad)	Phase (degrees)
.001	1.000	−.001	−.057
.01	1.000	−.010	−.573
.1	.995	−.100	−5.710
1	.707	−.785	−45.000
10	.099	−1.471	−84.289
100	.010	−1.561	−89.427
1000	.001	−1.570	−89.942

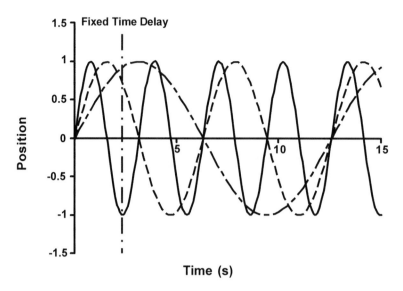

FIG. 13.2.   A fixed time delay will result in a phase lag that increases with the frequency of the sinusoidal signal. Note that a fixed time covers a larger portion of the cycle for a higher frequency sinusoid.

$$e^{-aj\omega} = 1e^{(-a\omega)j} \tag{15}$$

This is a complex number in exponential form. This makes it simple to determine the magnitude (1) and the argument ($-a\omega$). Note the magnitude does not depend on frequency. The amplitude of the output from a pure time delay will be equal to the amplitude of the input, independent of frequency. The phase response does depend on frequency. Phase lag will be directly proportional to frequency. The proportionality constant, a, is the magnitude of the delay in seconds. This makes good sense because the phase response is expressed in proportions of the sinusoidal cycle, as noted earlier. The frequency of a sine wave determines how many cycles are completed in a fixed time (e.g., cycles per second). A faster (higher frequency) sine wave will be a larger portion of the cycle behind the input after waiting a seconds than a slower (lower frequency) sine wave. This is illustrated in Fig. 13.2.

THE BODE PLOT

The Bode plot is a useful convention for representing the change in amplitude (output amplitude/input amplitude) and the phase shift (output phase – input phase) that is produced by a particular linear system. The Bode plot graphs the amplitude change, referred to as the gain curve, in a log-log space. The power of the amplitude ratio (i.e., magnitude squared) is plotted in decibels ($10 \log_{10}$) or equivalently $20 \log_{10}$(magnitude) [because the $\log(x^2) = 2 \log(x)$]. This is plotted against log frequency (in radians/s). The phase shift is plotted in degrees, also against log frequency (radi-

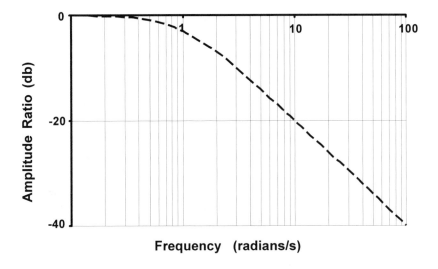

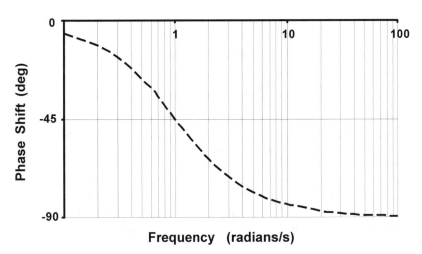

FIG. 13.3. This Bode plot shows the amplitude and phase response for a first-order lag.

ans/s). Figure 13.3 shows a Bode plot representation for the first-order lag presented in the last section of this chapter (see Table 13.1).

Note that there is a break in the gain curve at one radian/s. Below one radian/s, the gain curve is relatively flat at zero db (a one-to-one amplitude ratio, because log 1 = 0). Beyond 1 radian/s, the gain curve asymptotically approaches a line that slopes downward with a 20 db reduction in gain for each decade (factor of 10) increase in frequency. The 20 db per decade decrease in amplitude is the gain curve for an integrator:

$$20 \log\left(\frac{1}{\omega}\right) = -20 \log(\omega) \tag{16}$$

Thus, for every factor of 10 increase in frequency, the amplitude ratio (gain) will change by –20 db.

At one radian/s the gain response of the first-order lag (low pass filter) is approximately –3 db:

$$20\log\left(\sqrt{\frac{1}{\omega^2+1}}\right) = 20\log\left(\sqrt{\frac{1}{1+1}}\right) = 20\log(.707) = -3db \qquad (17)$$

The –3 db gain represents a .707 change in amplitude or a one half reduction in power ($.707^2 = .5$).

The phase response of the first-order lag asymptotically approaches 0° as frequency gets small and asymptotically approaches –90° as frequency increases. Again, note that the –90° phase shift is characteristic of a pure integrator. At 1 radian/s, the phase shift is –45°.

What would the Bode plot look like if the forward loop gain was different than unity (try plotting the gain and phase response that was computed for forward loop gains of .1 and 10)? How does the graph change? Notice that the overall pattern remains the same, but the location of the break changes. For a loop gain of .1, the break will happen at .1 radians/s. That is, there will be a gain of –3 db and a phase shift of –45° at .1 radians/s. This gain results in a low pass filter with a narrower bandwidth. For a forward loop gain of 10, the break will happen at 10 radians/s. Here the bandwidth is greater than with unity gain. Note that the shapes of the curves do not change. The only change is a lateral shift that is determined by an important parameter of this system, the forward loop gain (or time constant).

You should compare the frequency response with the step response of the first-order lags that were presented in chapter 4 (see Figs. 4.3 and 4.4). Note that as the forward loop gain increases, the bandwidth of the frequency response increases and the step response becomes sharper. That is, at higher bandwidth, the response to a step input is crisper; at lower bandwidth, the response to a step input is more rounded or gradual. This should help you to appreciate the frequency composition of the step signal. The higher the bandwidth, the better the resolution of the step response.

Figure 13.4 shows the amplitude response for systems that are simple gains (transfer function = K). The amplitude of the output is proportional to the magnitude of the gain (K). Gains of K = .1, 10, and 20 are shown in Fig. 13.4. Note that there is no phase lag associated with a simple gain.

Figure 13.5 shows the phase response for systems that are simple time delays (transfer function = e^{-as}). Time delays of .01, .1, and 1 s are plotted. Note that a time delay has no effect on gain, but there is a phase lag that increases linearly with frequency. Figure 13.5A shows the conventional Bode format, where the phase lags are curvilinear functions of log frequency. Figure 13.5B shows the phase lag plotted against frequency (not log frequency as in the conventional Bode plot). Note that when plotted against frequency, the phase lag curves in Figure 13.5B appear as straight lines whose downward slopes are equal to the time delay. This reflects the earlier analysis of the frequency response of a time delay:

$$e^{-aj\omega} = 1e^{(-a\omega)j} \qquad (18)$$

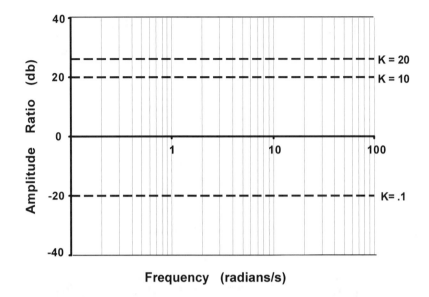

FIG. 13.4. The amplitude response for gains of 20, 10, and .1. Note that the gain does not affect the phase response (i.e., phase lag = 0).

Examine Fig. 13.6. What type of system would generate this pattern of Bode plot? The amplitude response looks like the pattern associated with a first-order lag. It is flat for a range of lower frequencies and asymptotically approaches a slope of –20 db per decade as frequency increases. But it is raised relative to the plot in Fig. 13.3 and the break frequency is shifted. The phase response is somewhat like the pattern for a time delay (increasing phase lag with increasing frequency), but it tends to break down more steeply at the lower frequencies.

The Laplace transfer function for the system whose frequency response is plotted in Fig. 13.6 is:

$$\frac{2e^{-.1s}}{.1 + s} = (20)(e^{-.1s})\left(\frac{.1}{.1 + s}\right) \tag{19}$$

The system whose Bode plot is shown in Fig. 13.6 is a combination of a pure gain, a time delay, and a lag. Here is the derivation for the frequency response equations for this system:

$$\frac{2e^{-.1j\omega}}{.1 + j\omega} = (20)(e^{-.1j\omega})\left(\frac{.1}{.1 + j\omega}\right) \tag{20}$$

$$\left|\frac{2e^{-.1j\omega}}{.1 + j\omega}\right| = \sqrt{\left(\frac{2e^{-.1j\omega}}{.1 + j\omega}\right)\left(\frac{2e^{.1j\omega}}{.1 - j\omega}\right)} = \sqrt{\frac{4}{.01 + \omega^2}} \tag{21}$$

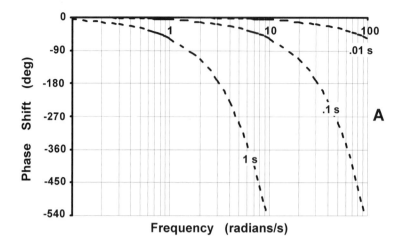

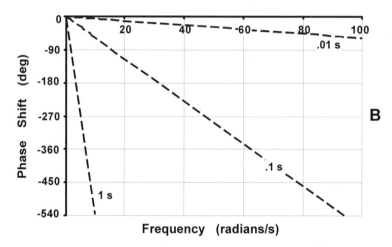

FIG. 13.5.   The phase response for time delays of .01, .1, and 1 s. The top plot (A) uses the standard Bode format in which a log scale is used for the frequency axis. The bottom plot (B) uses a linear frequency axis. Note that the time delay does not affect the amplitude response (i.e., gain = 0 db).

or

$$\left|\frac{2e^{-.1j\omega}}{.1+j\omega}\right| = |20| \times |e^{-.1j\omega}| \times \left|\frac{.1}{.1+j\omega}\right| = 20 \times 1 \times \sqrt{\frac{.01}{.01+\omega^2}} \qquad (22)$$

$$\arg\left(\frac{2e^{-.1j\omega}}{.1+j\omega}\right) = \arg(20) + \arg(e^{-.1j\omega}) + \arg\left(\frac{.1}{.1+j\omega}\right) \qquad (23)$$

$$\arg\left(\frac{2e^{-.1j\omega}}{.1+j\omega}\right) = 0 - .1\omega - \tan^{-1}(10\omega) \text{ radians} \qquad (24)$$

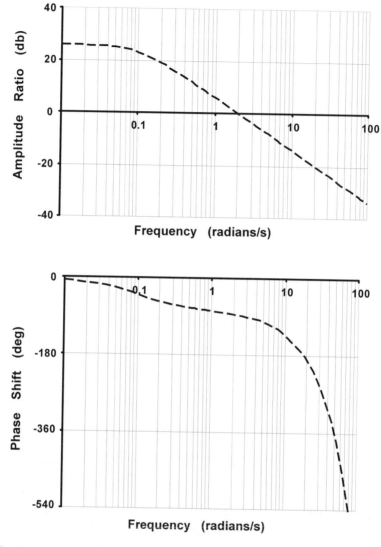

FIG. 13.6. The Bode plot shows the frequency response for a system composed of a gain, first-order lag, and time delay.

Note that the gain effects combine multiplicatively and the phase effects combine additively. This is why amplitude ratio or gain is plotted as a log value. A multiplicative combination is additive in log space [$\log (a \times b) = \log a + \log b$]. The result is that the joint effects of the three component linear systems combine additively within the Bode space. This is illustrated in Table 13.2, which shows the gain response (in db) and phase response (in degrees) for the components and for the total system. Note that the values at each frequency for the total system are just a simple sum of the values for the components. With practice, it is possible to learn to "see" the additive properties in the graphical Bode space. A complex system can be decomposed into

TABLE 13.2
Additive Components of Frequency Response

		Gain Response (db)		
ω (rad/s)	$20\ log\ |20|$	$20\ log\ |e^{-.1j\omega}|$	$20\ log\left|\dfrac{.1}{.1 + j\omega}\right|$	$20\ log\left|\dfrac{2e^{-.1j\omega}}{.1 + j\omega}\right|$
.01	26.02	0	−0.04	25.98
.05	26.02	0	−0.97	25.05
.1	26.02	0	−3.01	23.01
.5	26.02	0	−14.15	11.87
1	26.02	0	−20.04	5.98
2	26.02	0	−26.03	−0.01
3	26.02	0	−29.55	−3.53
4	26.02	0	−32.04	−6.02
5	26.02	0	−33.98	−7.96
6	26.02	0	−35.56	−9.54
7	26.02	0	−36.90	−10.88
8	26.02	0	−38.06	−12.04
9	26.02	0	−39.08	−13.06
10	26.02	0	−40.00	−13.98
15	26.02	0	−43.52	−17.50
20	26.02	0	−46.02	−20.00

		Phase Response (degrees)		
ω (rad/s)	$arg(20)$	$arg(e^{-.1j\omega})$	$arg\left(\dfrac{.1}{.1 + j\omega}\right)$	$arg\left(\dfrac{2e^{-.1jw}}{.1 + j\omega}\right)$
.01	0	−0.06	−5.71	−5.77
.05	0	−0.29	−26.56	−26.85
.1	0	−0.57	−45.00	−45.57
.5	0	−2.86	−78.69	−81.55
1	0	−5.73	−84.29	−90.02
2	0	−11.46	−87.14	−98.60
3	0	−17.19	−88.09	−105.28
4	0	−22.92	−88.57	−111.49
5	0	−28.65	−88.85	−117.50
6	0	−34.38	−89.04	−123.42
7	0	−40.11	−89.18	−129.29
8	0	−45.84	−89.28	−135.12
9	0	−51.57	−89.36	−140.93
10	0	−57.29	−89.43	−146.72
15	0	−85.94	−89.62	−175.56
20	0	−114.59	−89.71	−204.30

simpler components from simply looking at the pattern in a Bode plot, and the parameters of the components can be estimated from landmarks in the patterns.

Figure 13.7 shows the frequency response for a second-order system with a Laplace transfer function of the form:

$$\frac{O(s)}{I(s)} = \frac{\omega_n^2}{s^2 + 2\zeta s\omega_n + \omega_n^2}$$

or substituting $j\omega$ for s, to obtain the Fourier transfer function

$$\frac{O(j\omega)}{I(j\omega)} = \frac{\omega_n^2}{(j\omega)^2 + 2\zeta j\omega\omega_n + \omega_n^2} = \frac{\omega_n^2}{-\omega^2 + 2\zeta j\omega\omega_n + \omega_n^2}$$

As can be seen in the amplitude response shown in Fig. 13.7, there is a peak in the frequency response when the damping ratio (ζ) is small (.1 and .3 in the example). This

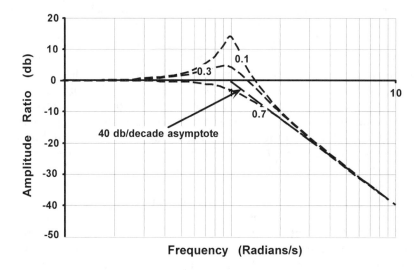

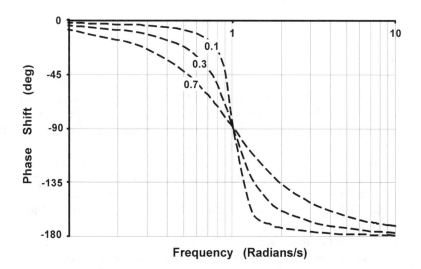

FIG. 13.7.   The Bode plot shows the frequency response for a second-order system with undamped natural frequency ω_n = 1 rad/s and with damping coefficients ζ = .1, .3, and .7. Note that the strongly underdamped second-order systems (ζ = .1 or .3) show a resonance peak (amplification) approximately at the undamped natural frequency.

peak reflects the fact that the second-order system will "resonate to," or amplify, inputs that are near the undamped natural frequency of the system (ω_n), which is equal to 1 rad/s in the example. When the damping ratio is higher (e.g., .7), the resonance peak is not evident.

Remember the step response of the second-order system shown in Fig. 6.3. When the damping ratios were small, the step response showed a damped sinusoidal oscillation at approximately the undamped natural frequency. The damped oscillation in response to a step input and the resonance peak in the Bode space are distinct signatures of underdamped second-order systems. At frequencies considerably below the undamped natural frequency, the second-order system behaves like a unity gain. At frequencies considerably above the undamped natural frequency, the second-order system asymptotes to a 40 db/decade attenuation of amplitude. The –40 db/decade slope is characteristic of double integration. Each integrator contributes –20 db/decade.

If the second-order system is heavily (or over-) damped ($\zeta > 1$), then there will be no overshoot in the step response (Fig. 6.3) and there will be no resonance peak in the frequency response. For the overdamped second-order system, the primary distinction between it and the first-order lag (Fig. 13.3) with regard to the amplitude ratio, is the rate at which the gain rolls off at the high frequencies. With a first-order lag, the roll off asymptotes at –20 db/decade; with a second-order system, the roll off asymptotes at a –40 db/decade rate.

The phase response of the second-order system asymptotes to zero degrees for very low frequencies and asymptotes to –180 degrees at higher frequencies. The phase lag is –90 degrees at the undamped natural frequency (ω_n). The roll off between zero and –180 degrees is steeper when the damping ratio is low. It should not be surprising to see the phase response approach –180 degree lag, as this represents the summation of two –90 degree phases lags (one for each integrator in the system).

STABILITY

A stable system is one that produces a bounded output to any bounded input. An unstable system will "explode" for some input (i.e., the output will be unbounded). In most cases, this unbounded output has catastrophic consequences for the system. For example, most fixed wing aircraft (except very high performance aircraft) have inherently stable aerodynamics. That is, most disturbances will be resisted, that is, damped out over time. For example, a paper plane will right itself. Sometimes, however, the incorrect responses of pilots can cause an inherently stable aircraft to become unstable (National Research Council, 1997). The pilot's responses tend to aggravate rather than correct the errors. This might be caused by control reversals (moving the control stick in the wrong direction) or by reaction time delays (responding too late to correct the error). The results of these errors are sometimes referred to as "pilot induced oscillations" or "pilot-involved oscillations," as described in chapter 2. The ultimate result might be an aircraft accident. On the other hand, rotary wing aircraft (e.g., helicopters) tend to be inherently unstable. The dynamics of the craft tend to exaggerate disturbances. A helicopter will not right itself. However,

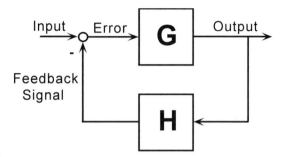

FIG. 13.8.   A canonical closed-loop system.

skilled pilots can compensate for the natural instability so that the human–machine system (pilot + helicopter) is generally stable. A principal motivation for applying frequency analysis to human performance is to be able to predict whether human–machine systems will be stable. A key point is that stability is a property of the human–machine system. That is, the pilot can cause a naturally stable aircraft to become unstable or the pilot can stabilize a naturally unstable aircraft. Stability is a property of the control system and many aircraft only become control systems when the loop is closed through the pilot.

To understand how Bode plots can help to analyze the stability of a system, a distinction must be introduced between the open-loop and the closed-loop transfer function. Figure 13.8 shows a canonical closed-loop system. The closed-loop Fourier transfer function of this system describes the output (O) relative to the input (I):

$$\frac{G(j\omega)}{1 + G(j\omega)H(j\omega)} = \frac{O(j\omega)}{I(j\omega)} \tag{25}$$

The open-loop (or "opened-loop") transfer function of the canonical system is:

$$G(j\omega)H(j\omega) = FS(j\omega)/E(j\omega) \tag{26}$$

It describes the feedback signal (FS) relative to the error (E). Suppose that

$$G(j\omega) = \frac{2e^{-.1j\omega}}{j\omega} \tag{27}$$

$$H(j\omega) = 1 \tag{28}$$

Then the frequency response for the open-loop transfer function is shown in Fig. 13.9. The frequency response for the closed-loop transfer function would be:

$$\frac{\dfrac{2e^{-.1j\omega}}{j\omega}}{1 + \dfrac{2e^{-.1j\omega}}{j\omega}} = \frac{2e^{-.1j\omega}}{j\omega + 2e^{-.1j\omega}} \tag{29}$$

$$\left| \frac{2e^{-.1j\omega}}{j\omega + 2e^{-.1j\omega}} \right| = \left[\left(\frac{2e^{-.1j\omega}}{j\omega + 2e^{-.1j\omega}} \right) \times \left(\frac{2e^{.1j\omega}}{-j\omega + 2e^{.1j\omega}} \right) \right]^{1/2}$$

$$= \left[\frac{4}{\omega^2 + 2j\omega(e^{.1j\omega} - e^{-.1j\omega}) + 4} \right]^{1/2}$$

This expression can be simplified using the following trigonometric identities:

$$e^{jx} = \cos x + j \sin x \quad \text{and} \quad e^{-jx} = \cos x - j \sin x \tag{30}$$

or

$$\sin x = \frac{e^{jx} - e^{-jx}}{2j} \quad \text{and} \quad \cos x = \frac{e^{jx} + e^{-jx}}{2} \tag{31}$$

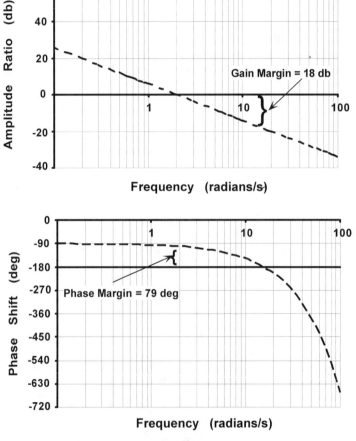

FIG. 13.9. The open-loop response (output relative to error) for a control system, which includes a gain, an integrator, and a time delay. The feedback signal equals the output for this system. The closed-loop response is illustrated in Fig. 13.10.

Thus, the magnitude for this closed-loop transfer function is:

$$\left| \frac{2e^{-.1j\omega}}{j\omega + 2e^{-.1j\omega}} \right| = \left[\frac{4}{\omega^2 - 4\omega \sin(.1\omega) + 4} \right]^{\frac{1}{2}} \tag{32}$$

The argument for the closed-loop transfer function can also be computed:

$$\arg\left(\frac{2e^{-.1j\omega}}{j\omega + 2e^{-.1j\omega}} \right) = \arg(2) + \arg(e^{-.1j\omega}) - \arg(j\omega + 2e^{-.1j\omega}) \tag{33}$$

The third term in this sum is the only difficult term. To solve for the argument of this term, it is useful to substitute for the e term using the previous trigonometric identities:

$$(j\omega + 2e^{-.1j\omega}) = j\omega + 2[\cos(.1\omega) - j\sin(.1\omega)]$$
$$= 2\cos(.1\omega) + [\omega - 2\sin(.1\omega)]j \tag{34}$$

Thus, the phase contribution of the third term can be added as the arctangent of the imaginary part divided by the real part:

$$\arg\left(\frac{2e^{-.1j\omega}}{j\omega + 2e^{-.1j\omega}} \right) = 0 - .1\omega - \tan^{-1}\left(\frac{[\omega - 2\sin(.1\omega)]}{2\cos(.1\omega)} \right) \text{ radians} \tag{35}$$

Figure 13.10 shows a Bode plot of the frequency response for the closed-loop system that has the open-loop response shown in Fig. 13.9. Compare these two figures. The point where the open-loop response goes through zero db is referred to as the Crossover frequency. Note that at the lower frequencies (well below the crossover frequency), where the open-loop gain is much greater than 1, the closed-loop gain is near unity and the phase lag is near zero. Thus, at these lower frequencies, the output of the closed-loop system will follow the input signal very closely. In other words, the closed-loop system will track frequencies that are below the crossover frequency well. Inputs at frequencies well above the crossover frequency will be attenuated in the closed-loop response. This result is consistent with the discussion in chapter 2 where the concept of closed-loop control was introduced. High gain in the forward loop results in a close match (i.e., 1 to 1 relation) between the input and output. The frequencies where the open-loop gain is considerably below unity are not tracked as well. These inputs are "filtered out." Thus, the crossover frequency of the open-loop transfer function determines the bandwidth of the response of the closed-loop system.

As was noted in the earlier discussion of closed-loop control, the situation is more complicated when there are time delays present (as in the system currently being discussed). When time delays are present, there are stability limits on the magnitude of the open-loop gain. Suppose that the gain of the forward loop is increased by a factor of 10. Figure 13.11 shows the open-loop response for this system. The gain increase

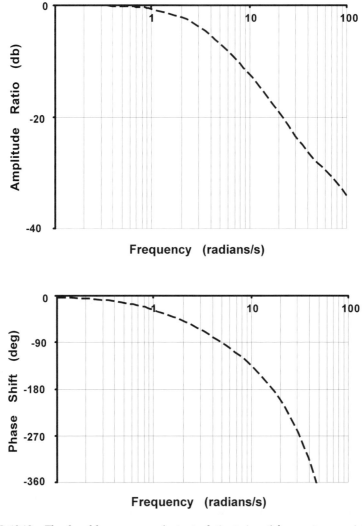

FIG. 13.10.   The closed-loop response (output relative to input) for a system consisting of a gain, an integrator, and a time delay. The open-loop response is shown in Fig. 13.9.

does not change the phase response, but the gain response is increased by 20 db. When computing the closed-loop response, the computation breaks down (there are negative values inside the square root). In other words, the system becomes unstable — the output becomes unbounded.

This example illustrates that when time delays are present, there are limits to how high the open-loop gain response can be and still produce a stable response for the closed-loop system. The boundary conditions can be readily seen in the Bode plot of the open-loop transfer function (Figs. 13.9 and 13.11). For a closed-loop system to be stable, the gain curve for its open-loop transfer function generally must go through 0 db, before the phase response of the open-loop transfer function goes through –180°, a relation referred to as the Nyquist stability criterion.

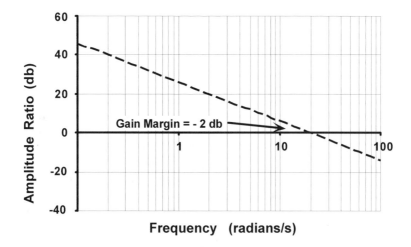

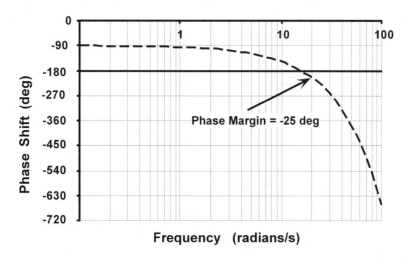

FIG. 13.11.   The open-loop response (output relative to error) when the gain for the system shown in Fig. 13.9 is increased by a factor of 10. Note that at crossover the phase has passed beyond –180°. This open-loop response will result in an unstable closed-loop response.

Stability requirements of closed-loop systems can be expressed as a function of the gain margin and the phase margin of the system. The *gain margin* is the magnitude of the reciprocal of the open-loop amplitude ratio, evaluated at the frequency at which the phase angle is –180°. This frequency is referred to as the *phase crossover frequency*. This value can be easily read off the Bode plots in Figs. 13.9 and 13.11. Simply identify where the phase reaches –180° and then trace up to read the gain value or amplitude ratio for that frequency. If the gain is less than 0 db, then the system will be stable. For the system in Fig. 13.9, the gain margin is approximately 18 db (i.e., 18 db less than 0), and for the system in Fig. 13.11, the gain margin is approximately –2 db (i.e., 2 db greater than 0). A stable system will have a positive gain margin. That is, the gain will be less than 0 db when the phase reaches –180°.

The *phase margin* is 180° plus the phase angle of the open-loop transfer function at the crossover frequency (unity gain). Again, this value can be easily read off the Bode plots. Identify the point of crossover for the gain plot and then trace down to the phase plot to see how far the phase is from –180°. If the phase response is less negative than –180°, then the system is stable. The phase margin is approximately 79° for the system shown in Fig. 13.9. The system in Fig. 13.11 has a phase margin of approximately –25°. A stable system will have a positive phase margin.

Whenever there are time delays in a system, the phase lag will reach –180° for some frequencies and, thus, there will be a U-shaped function relating open-loop gain and quality of performance. If gain is too low, then the system will be very "sluggish" (i.e., it will be slow to react to and correct errors). If gain is too high, then there is the danger of instability (i.e., the system will begin to exaggerate errors as in pilot induced oscillations). The relation between open-loop gain and time delay can be illustrated qualitatively in the *K-τ* Space (Jagacinski, 1977; see also Hill, 1970; Fig. 13.12). For short time delays, a wide range of gains will yield acceptable tracking and good tracking will result when gain is relatively high. As the time delay gets longer, the range of gains that will yield stable tracking diminishes. For very large time delays, no gain will yield acceptable tracking performance. In these situations, alterna-

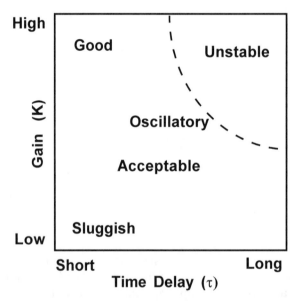

FIG. 13.12. The *K-τ* space illustrates the trade-off between gain and time delay with respect to system stability and tracking quality. When time delays are short, then high gain will produce good tracking (error will be eliminated quickly). However, if time delays are long, then high gains will lead to oscillations and/or instability. From "A qualitative look at feedback control theory as a style of describing behavior" by R. J. Jagacinski, 1977, *Human Factors, 19,* 331–347. Copyright by The Human Factors and Ergonomics Society, 1977. Adapted by permission.

tives to proportional control strategies (e.g., discrete control) will be necessary (e.g., Crossman & Cooke, 1974).

Thus, a Bode plot of the open-loop response (GH) for a closed-loop system $\left(\dfrac{G}{1+GH}\right)$ can provide important insights into the dynamic properties of the closed-loop system. The crossover frequency will provide a good approximation for the bandwidth of the closed-loop system. Signals at frequencies well below crossover will be tracked accurately. Signals at frequencies well above crossover will be attenuated (or filtered out). The gain and phase margins, which can be easily seen in the Bode space, indicate whether the closed-loop system will be stable. Positive gain and phase margins are indicative of a stable closed-loop response.

BODE ANALYSIS

In the earlier sections of this chapter, Bode plots were created for systems whose transfer functions were known. That is, the Bode plots were derived from the transfer functions. Bode analysis, however, can be used to discover the transfer function of a black box (i.e., a system component whose transfer function is unknown). The process of Bode analysis involves presenting sinusoidal inputs to the black box at different frequencies. If the black box is approximately linear (quasi-linear), then there will typically be an approximately sinusoidal output from the box at the same frequency as the input. However, there may be a change in amplitude and phase. In Bode analysis, inputs for a range of frequencies are input to the system and the amplitude ratios (gain) and phase shifts are recorded in the Bode space. The pattern in the Bode space corresponds to the Fourier transfer function if the black box is perfectly linear. The approximate transfer function for a quasi-linear black box is technically referred to as a *describing function* (e.g., Graham & McRuer, 1961) although the term "transfer function" is commonly used. In either case, the pattern can then be used to make inferences about the dynamic properties of the black box — what is its bandwidth? Will the associated closed-loop system be stable? With more exposure to the distinctive patterns associated with different dynamical elements, it is possible to learn to read the transfer function for the black box from the patterns in the Bode space. The next chapter explores how Bode analysis can be used to build quantitative models (i.e., transfer functions) for human operators.

REFERENCES

Crossman, E. R. F. W., & Cooke, J. E. (1974). Manual control of slow response systems. In E. Edwards & F. Lees (Eds.), *The human operator in process control* (pp. 51–66). New York: Plenum.

Graham, D., & McRuer, D. (1961). *Analysis of nonlinear control systems.* New York: Dover.

Hill, J. W. (1970). Describing function analysis of tracking performance using two tactile displays. *IEEE Transactions on Man–Machine Systems, MMS-11*, 92–101.

Jagacinski, R. J. (1977). A qualitative look at feedback control theory as a style of describing behavior. *Human Factors, 19*, 331–347.

National Research Council (1997). *Aviation safety and pilot control.* Washington, DC: National Academy Press.

Ogata, K. (1990). *Modern control engineering.* Englewood Cliffs, NJ: Prentice-Hall.

14

The Frequency Domain:
Describing the Human Operator

Formulations using the state space approach provide a good basis for developing general al-
gorithms that are well suited for computer-aided design purposes. On the other hand, the
methods of "classical" control theory are still widely used and their utility has been signifi-
cantly enhanced by modern computational techniques involving computer simulation and
computer graphics. Today's engineer must be familiar with the classical methods and the
new computational improvements.

—Palm (1986, p. VII)

Can the tools of classical linear control theory be used to model the human operator
as a continuous controller (i.e., a tracker or regulator)? There are two obvious reasons
why this question is interesting. First, quantitative models of the human operator
may provide insights into basic properties of human performance. Second, the ability
to derive transfer functions for human operators would greatly facilitate the ability to
predict the performance of human–machine systems. It was this second goal that motivated much of the early research on modeling human tracking. The goal was to
build describing functions, i.e., approximate transfer functions, of human performance using the same language (linear differential equations) that has been used to
model the performance of the machine systems (e.g., aircraft). The hope was that this
would allow prediction and evaluation of stability of the human–machine system
early in the design cycle.

There is good reason to be skeptical about linear models of human performance.
Clearly, humans exhibit nonlinear behaviors. For example, human perceptual systems have finite thresholds. Additionally, humans can learn and adapt to situations.
For example, they can detect and remember patterns when they are present. Thus, in
tracking a single sine wave, a human may not respond based on the moment-to-
moment input, as would a linear servomechanism. Rather, the human might respond

to parameters of the pattern (amplitude, frequency, phase). This is discussed in more depth in chapter 21. Thus, it is clear that human behavior can be nonlinear. However, it is also clear that, within certain contexts, humans do behave approximately linearly. Applying the analytical power of linear analysis to those behavioral contexts where they are appropriate may lead to important theoretical and practical insights about human performance. But care must be taken to differentiate the invariants that reflect the constraints of the linear mathematics from those invariants that reflect essential properties of the phenomenon of interest.

QUASI-LINEARITY

Although the human tracker is clearly nonlinear, linear analysis can still provide important insights into human performance and linear models may be able to give reasonable predictions for some situations. Thus, there have been significant efforts to develop a quasi-linear model of the human operator (Fig. 14.1). The quasi-linear model is an attempt to represent the human tracker as a constant coefficient linear differential equation (i.e., a linear transfer function) plus noise or remnant to account for those aspects of the output that cannot be accounted for by the linear transfer function (e.g., output at frequencies not in the input signal). It is generally assumed that the internal noise is random (uncorrelated with input signals). This noise is assumed to arise from perceptual and motor processes internal to the human operator.

The quasi-linear model has been developed with the realization that application of the model will be bounded. Although the model can provide momentary assessments of skill acquisition or adaptation (chaps. 15 & 26), the coefficients of the linear differential equations are most reliably estimated during stationary tracking by very highly trained performers. Also, the human behaves linearly generally when the input is unpredictable (i.e., no detectable pattern). In research, this is typically accomplished by creating disturbances from summing together a large number of sine waves (5 or more) and randomizing the phase relations within a trial and from trial to trial so that no pattern is ever repeated exactly. The result of this summation is a complex signal whose motions appear to be random. Elkind (1956), McRuer and Krendel (1959), and Stark, Iida, and Willis (1961) showed that when such random-appearing signals are used, human performance can be modeled fairly well using linear assumptions. Other boundary conditions of the quasi-linear model are considered throughout this chapter.

FIG. 14.1. A quasi-linear model of the human operator. Y_H is the linear transfer function; $u(t)$ is the linear response; $n(t)$ is internal noise (presumably reflecting noise in the perceptual and motor systems of the operator); and $u'(t)$ is the quasi-linear response. The noise is generally presumed to be uncorrelated with any input signal.

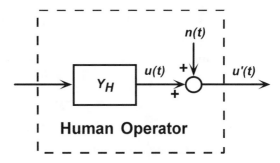

EXPERIMENT AND MEASUREMENT

Figure 14.2 illustrates a typical one-dimensional tracking experiment. The human operator [represented as Y_H and $n(t)$] is instructed to follow a quasi-random (sum-of-sines) input signal [$r(t)$]. The error [$e(t)$] is displayed in a compensatory tracking task, which is the topic of this chapter. Control responses [$u'(t)$] are typically made with a joystick or force stick and these control responses are input to a plant (e.g., vehicle) (Y_P). The output of the system [$y(t)$] is the response of the vehicle.

In order to build a describing function for the human operator, time histories are measured at three locations in the closed loop: input $r(t)$, error $e(t)$, and control $u'(t)$. These time histories are used to compute cross-correlation functions $\phi_{ru'}(\theta)$ and $\phi_{re}(\theta)$. The cross-correlation functions are Fourier transformed to $\Phi_{ru'}(j\omega)$ and $\Phi_{re}(j\omega)$. The ratio of these two transforms yields the open-loop frequency response for the human operator (see Sheridan & Ferrell, 1974, for a more detailed explanation of this measurement process). This mathematical procedure is conceptually analogous to calculating Fourier series approximations (chap. 12) to the error signal and the control signal, and then comparing the amplitudes and phases at each input frequency to infer the effective Fourier describing function for the human.

DESCRIBING FUNCTIONS

Figure 14.3 presents a Bode diagram that illustrates typical results of a one-dimensional compensatory tracking study. In this study, the "plant" was a simple gain ($Y_P = 5$). That is, the response of the simulated vehicle was proportional to the control input—it was a zero-order control system. Recognize this pattern? Can any inferences be made about the transfer function that would correspond to the Bode diagram? First, note the amplitude ratio exhibits a 20 db/decade slope at high frequencies, characteristic of integration, and a flatter slope at low frequencies. The phase response rolls off continuously with increased frequency suggesting a time delay. Thus, the human transfer function appears to consist of a gain, a lag, or an integrator at higher frequencies, and a time delay.

$$Y_H(j\omega) = \frac{Ke^{-\tau j\omega}}{(T_l j\omega + 1)}$$

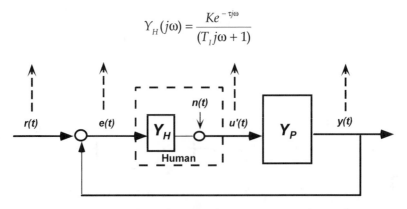

FIG. 14.2.   A typical one-dimensional compensatory tracking task.

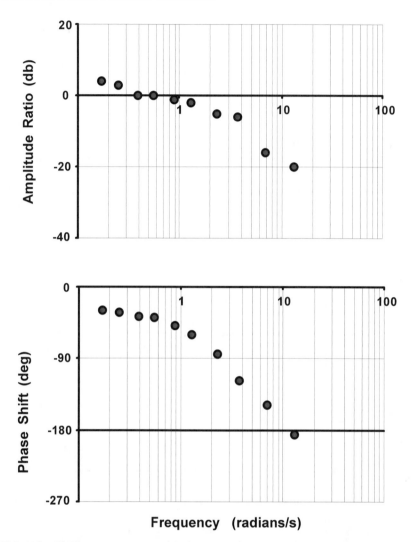

FIG. 14.3. The frequency response for a human operator controlling a zero-order system ($Y_P = 5$) (i.e., position control). The data were estimated from Figure 11.25 in *Engineering Psychology and Human Performance* (p. 456) by C. D. Wickens, 1992, New York: Harper Collins. The original source was McRuer & Jex (1967).

or

$$Y_H(j\omega) = \frac{Ke^{-\tau j\omega}}{j\omega} \tag{1}$$

This seems quite reasonable. The gain (K) is a scaling factor that influences the bandwidth of the control system. The time delay (τ) reflects human reaction time. Estimates from tracking studies like this find the time delay to be in the range of 100–250 ms, which overlaps with measures of reaction time in response to discrete stimuli. The lag $\dfrac{1}{(T_i j\omega + 1)}$ or integration $(1/j\omega)$ suggests that the human tracker has a low pass characteristic—that is, the human responds to low frequency components of error and ignores (i.e., filters out) the high frequency components of error.

Now that there is a transfer function for the human using a proportional control, it can be compared to performance with other types of plants. Suppose, that the human is asked to track using a first-order (velocity) control system ($Y_P = 4/j\omega$). The picture changes (Fig. 14.4). The amplitude pattern for the human is relatively flat. Here, the human seems to behave more like a simple gain and a time delay. The integral property is not evident.

The picture changes again (Fig. 14.5) when performance is measured with a second-order (acceleration) control system [$Y_P = 4/(j\omega)^2$]. With this plant, the frequency

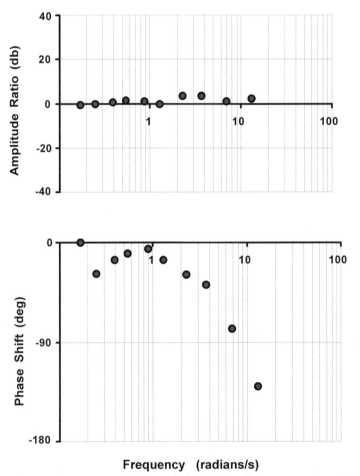

Frequency (radians/s)

FIG. 14.4.   The frequency response for a human operator using a velocity control system ($Y_P = 4/j\omega$). The data were estimated from Figure 11.25 in *Engineering Psychology and Human Performance* (p. 456) by C. D. Wickens, 1992, New York: Harper Collins. The original source was McRuer & Jex (1967).

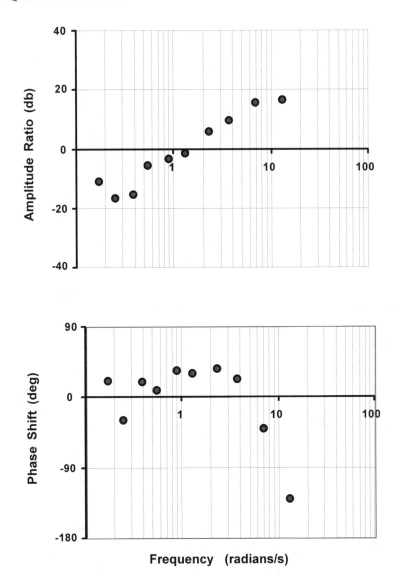

FIG. 14.5.   The frequency response for a human operator using an acceleration control system [$Y_P = 4/(j\omega)^2$]. The data were estimated from Figure 11.25 in *Engineering Psychology and Human Performance* (p. 456) by C. D. Wickens, 1992, New York: Harper Collins. The original source was McRuer & Jex (1967).

response of the human seems to show increasing gain with increasing frequency (up to a point). Also, note that the phase response is positive at some frequencies. That is, the controller is generating phase lead. This is characteristic of differentiation. A differentiator has a frequency response, $j\omega$, that is the inverse of an integrator; that is, gain increases 20 db/decade with frequency and there is a constant phase shift of +90°. Similarly, a lead, ($T_L j\omega + 1$), is the inverse of a lag. A lead resembles a unity gain at low frequencies and resembles a differentiator at high frequencies. The human describing function in Figure 14.5 appears to consist of a gain, a time delay, and lead, or differentiation at high frequencies.

Thus, the transfer function that is measured for the human seems to change as the dynamics of the control system change. Why is this? Do the changes make sense?

Figure 14.6 illustrates the adaptive nature of the human controller. The frequency response of the human element changes with changes in the plant dynamics. In each case, the skilled human tracker can be modeled fairly well using a constant coefficient differential equation (see Sheridan & Ferrell, 1974, for details). However, different equations must be used for each different plant. In one case, the human looks like an approximate integrator or lag, in another the human looks more like a gain, and in another the human looks more like an approximate differentiator or lead. In all cases, a time delay is evident. More insight into processing constraints within the human operator can be gained from analysis of nonlinear components of the human response as reflected in the remnant. This is discussed in chapter 19.

The third column in Fig. 14.6 gives a clue to the nature of the adaptation and to the source of invariance. In all three cases, the transfer function for the forward loop

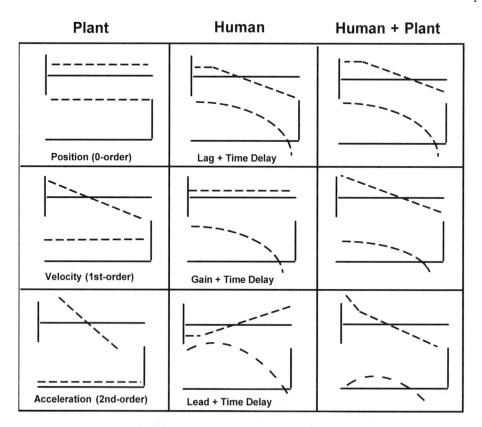

FIG. 14.6.   The schematic illustration shows the adaptive nature of the human controller. Note that the human transfer function (amplitude ratio and phase shift) changes depending on the system being controlled. The changes reflect the stability constraints on the total system (human + plant). The invariant form of the transfer function near the crossover frequency for the combination of human and plant satisfies the requirements for stable control.

$(Y_H Y_P)$ looks similar. That is, the combination of the human and the plant looks approximately like a gain, a time delay, and an integrator in the region of the crossover frequency (refer to chap. 13):

$$Y_H(j\omega)Y_P(j\omega) = \frac{\omega_c e^{-\tau j\omega}}{j\omega} \tag{2}$$

Note that at the crossover frequency, $\omega = \omega_c$, the net gain of the human + plant is 1.0, so the open-loop gain parameter in the numerator is equal to ω_c.

Remember the discussions of stability in the previous chapter. It appears that the human acts so that the transfer function for the forward loop satisfies the criteria for a "good" (i.e., stable) control system. The gain, ω_c, influences the bandwidth for the control system, and the integral property assures that the amplitude ratio will decrease and be less than 0 db before the phase shift goes through –180°. That is, there will be a positive phase margin. Thus, whereas the transfer function for the human varies with different control dynamics, the form of the transfer function for the full forward loop remains invariant in a way that reflects the constraints on stable control.

This reflects another aspect of quasi-linear models. For a given plant, the skilled human can be modeled using linear differential equations. These models can be quite stable (similar results are found in different laboratories using different subject populations) and quite powerful (accounting for large portions of the performance variability). However, different equations and different coefficients are required for different plants. These changes reflect the adaptive nature of the human controller.

The invariance at the level of the forward loop was recognized by McRuer and his colleagues (McRuer, Graham, Krendel, & Reisener, 1965; McRuer & Krendel, 1959) and the resulting model of the human operator is generally referred to as the McRuer crossover model. This model predicts that humans will act so that the total forward loop has the characteristics of a good servo. The gain (ω_c) of the system will be adjusted so that RMS error will be reduced. Ideally, open-loop amplitude ratio should be high over the bandwidth of signals that are to be followed (to insure that errors are responded to quickly), but should be below 1 (0 db) before the open-loop phase lag goes through –180° to avoid instability.

The following equation represents a general quasi-linear model of the human operator (Fig. 14.2):

$$U'(j\omega) = Y_H(j\omega)E(j\omega) + \text{Remnant} \tag{3}$$

where

$$Y_H(j\omega) = \left[\frac{e^{-\tau j\omega}}{T_N j\omega + 1}\right]\left[\frac{K(T_L j\omega + 1)}{T_I j\omega + 1}\right] \tag{4}$$

The terms in the first bracket represent inherent limitations of the human operator. The numerator reflects reaction time. That is, there is a finite time delay during which information is processed. This might be thought of as the time for an error signal from the receptor (e.g., eye in the case of simple visual tracking task) to be appropriately transformed and communicated to the effector (e.g., the hand on the control stick). Typically this is treated as a free parameter that is adjusted to fit the data. This parameter changes with the plant dynamics and with the bandwidth of the input signal. The denominator reflects neuromuscular lag. This reflects the inertial properties of the motor system. This lag reflects the time to get the muscles in motion. The time constant for this lag is typically relatively small (<.2 s). The frequency response of this lag will be relatively flat in the region of crossover. It will have a break frequency out beyond 5 radians/s. This might be expected to reflect invariant properties of the motor system, which should be independent of changes in the dynamics of the controlled system. However, there is some evidence that humans can reduce the effective neuromuscular lag, for example, by tightening the grip on the stick (e.g., see McRuer, Magdaleno, & Moore, 1968).

The terms in the second bracket represent the primary "strategic" parameters within the model. These terms represent a lead-lag control strategy for achieving good control. The parameters inside this bracket will take on different values depending on the plant dynamics. For controlling a zero-order plant, the lag term will dominate the frequency response in the crossover region. The human will behave like an approximate integrator, i.e., the rate of change of control movement will be proportional to error. This means that the human will be effectively responding to the low and middle frequencies and will be ignoring the high frequencies (i.e., rapid fluctuations). For controlling a first-order plant, the human will act like a pure gain (the lead and lag terms will effectively cancel out; see chap. 19). That is, the human's control response will be in direct proportion to the size of the error. For controlling a second-order plant, the lead term will dominate. That is, the humans will behave more like approximate differentiators. Or, in other words, their responses will reflect both the magnitude and the velocity of the errors. In sum, skilled trackers adopt strategies that make the overall system effective.

Note that the fact that different plants require different control strategies is consistent with discussions of discrete or step tracking. That is, different control inputs are necessary to produce a step output (acquire a target in a discrete movement task) for different types of control systems (zero, first, or second-order).

The different control strategies are not all equivalent in terms of the ease with which the human can adapt and maintain them (McDonnell, 1969). One way in which the difficulty (or workload) of tracking can be assessed is through subjective ratings of handling qualities (Cooper, 1957; Harper, 1961). There appears to be a most comfortable level for the gain parameter. That is, if the gain of the plant is too high or too low, then the handling qualities of the system will receive poor ratings. Also, requirements to generate "lead" tend to cause handling qualities to be rated poorer. The greater the lead (T_L)that is required, the poorer the rating of the handling qualities (Anderson, 1970; Ashkenas, 1965, cited in McRuer & Jex, 1967). This suggests that the requirement to take velocity into account (act like a differentiator) demands greater effort. Acceleration (second-order) and higher order control systems are generally significantly more difficult or demanding than lower order systems.

BEYOND THE CROSSOVER MODEL

The crossover model raises general questions about the appropriate unit of analysis when studying behavior. If the goal of behavioral scientists is to discover invariants of behavior, then what unit of analysis is appropriate? Is it realistic to expect to find invariant transfer functions of the human? That is, is it possible to build a model of the human operator that will generalize across behavioral contexts? The crossover model suggests that there is a greater likelihood of finding invariance at the unit of human–environment systems. That is, humans adapt to the constraints of their environment (e.g., plant dynamics). This does not mean there are no general, fixed constraints specific to the human. However, performance is generally not a simple product of such limitations. The control loop between perception and action is almost always closed through an environment. Thus, constraints such as stability reflect global properties of this control loop. Models of behavior are likely to include terms like the control strategy in the crossover model, whose parameters depend on the task context. This suggests that the search for behavioral invariants might be best focused at the level of human–environment systems. In other words, it suggests that behavior is "situated." Behavior is an adaptive response to situation constraints. Models of performance should give consideration to the contextual or ecological constraints that contribute to shaping performance (e.g., Flach, Hancock, Caird, & Vicente, 1995; Gibson, 1979; Hutchins, 1995; Suchman, 1987).

REFERENCES

Anderson, R. D. (1970). *A new approach to the specifications and evaluation of flying qualities* (Technical Report AFFDL-TR-69-120). Wright-Patterson Air Force Base, Ohio: Air Force Flight Dynamics Laboratory.

Cooper, G. E., (1957). Understanding and interpreting pilot opinion. *Aeronautical Engineering Review, 116,* 47–52.

Elkind, J. I. (1956). *Characteristics of simple manual control systems* (Tech. Rep. No. 111). Lexington, MA: MIT Lincoln Laboratory.

Flach, J. M., Hancock, P. A., Caird, J. K., & Vicente, K. J. (Eds.). (1995). *Global perspectives on the ecology of human–machine systems.* Hillsdale, NJ: Lawrence Erlbaum Associates.

Gibson, J. J. (1979). *The ecological approach to visual perception.* Boston: Houghton Mifflin.

Harper, R. P. Jr., (1961). *In flight simulation of the lateral directional handling qualities of entry vehicles.* (WADD-TR-147) U.S. Air Force.

Hutchins, E. (1995). *Cognition in the wild.* Cambridge, MA: MIT Press.

McDonnell, J. D. (1969). An application of measurement methods to improve the qualitative nature of pilot rating scales. *IEEE Transactions on Man–Machine Systems, MMS-10,* 81–92.

McRuer, D. T., Graham, D., Krendel, E. S., & Reisener, W., Jr. (1965). *Human pilot dynamics in compensatory systems: Theory, models, and experiments with controlled element and forcing function variations.* (AFFDL-TR-65-15). Wright-Patterson AFB, OH: Air Force Flight Dynamics Laboratory.

McRuer, D. T., & Jex, H. R. (1967). A review of quasi-linear pilot models. *IEEE Transactions on Human Factors, 8,* 231–249.

McRuer, D. T., & Krendel, E. S. (1959). The human operator as a servo system element. *Journal of the Franklin Institute, 267,* 381–403.

McRuer, D. T., Magdaleno, R. E., & Moore, G. P. (1968). A neuromuscular actuation system model. *IEEE Transactions on Man–Machine Systems, MMS-9,* 61–71.

Palm, W. J. (1986). *Control systems engineering.* New York: Wiley.

Sheridan, T. B., & Ferrell, W. R. (1974). *Man–machine systems: Information, control, and decision models of human performance.* Cambridge, MA: MIT Press.

Stark, L., Iida, M., & Willis, P. A. (1961). Dynamic characteristics of the motor coordination system in man. *Biophysical Journal, 1,* 279–300.

Suchman, L. (1987). *Plans and situated actions: The problems of human–machine interaction.* Cambridge, England: Cambridge University Press.

Wickens, C. D. (1992). *Engineering psychology and human performance.* New York: Harper Collins.

Additional Adaptive Aspects of the Crossover Model

> *The human pilot evolves, during a learning and skill development phase, a particular multiloop structure. The active feedback connections in this system will be similar to those which would be selected by a skilled controls designer who has available certain variable system characteristics to use for control of given fixed system characteristics; and who also has available a relative preference guide for the variables. System variables comprise sensing channels for each of the feedback possibilities available to the pilot, and possible equalization in each loop which is tailored from an adaptive, but limited, set of equalization forms.*
>
> —McRuer and Jex (1967, p. 244)

As noted in the previous chapter, the simplest version of the McRuer crossover model has two parameters, a gain, K, and a time delay, τ. Different values of these parameters result in different styles of tracking behavior ranging from sluggish to oscillatory. The major question of interest in this chapter is whether people can actively adjust these two parameters, and if so, how they accomplish that. In other words, can people behave as *adaptive* controllers.

One type of adaptivity has already been discussed. Namely, when the dynamics of the control system change from position control to velocity control to acceleration control, the behavior of the person changes so that the combination of person plus control system is to a first approximation described by the crossover model. In order to accomplish this invariance, individuals must be changing their behavior, and therefore demonstrate one kind of adaptivity. It can be additionally asked whether (a) the combination of a particular input signal, control system, and primary tracking display uniquely determines the behavior of the person, or (b) the person is capable of a range of behaviors for any combination of input signal, control system, and primary tracking display. If the answer is (b), then individuals are capable of an additional kind of adaptivity, and it becomes interesting to ask how people choose a particular style of behavior over the other styles that they are capable of in any situation.

ADAPTIVE RESPONSE TO DIFFERENT PERFORMANCE CRITERIA

The plausibility of this latter type of adaptivity can be argued by considering different styles of driving. Driving in heavy city traffic places different demands on a person than driving along a lightly traveled country road. In a city, a driver places a premium on steering accuracy, even if it entails a great deal of effort. On a country road, a driver is typically less concerned about deviations from the center of the driving lane, and less effort is devoted to steering. However, if drivers were motivated to do so, they could exert greater effort to steering even on a country road. The trade-off between accuracy and effort can be quantified in terms of different scoring criteria for good performance.

For example, Obermayer, Webster, and Muckler (1966) investigated whether people could vary their style of compensatory tracking performance when given different instructions. Among the various instructions they investigated, one group of performers was instructed to minimize mean-squared error. Another group was instructed to minimize a weighted sum of mean-squared error and mean-squared control stick displacement. Both groups received end-of-trial feedback. There were large performance differences between the two groups. The group with the weighted sum criterion had a much higher mean-squared error score and lower mean-squared control stick displacement than the group minimizing only mean-squared error. Describing function analyses revealed that the predominant differences between groups were in the performers' gains. The gains of the performers minimizing a sum of mean-squared error and mean-squared control stick displacement were approximately three times smaller than the group only minimizing mean-squared error. In a related study, Miller (1965) demonstrated that by providing a performer with a second-to-second display of mean-squared error plus mean-squared control stick displacement, finer gradations of performance could be obtained. Again, the primary describing function differences across conditions were in the performer's gain.

Additional evidence of adaptivity is provided in a study by Rupp (1974). In a compensatory tracking task, he asked performers to achieve certain target values of K and τ. In addition to seeing a visual display of instantaneous tracking error, performers also viewed a display that indicated deviations from the target values of K and τ. Rupp demonstrated that with the same input signal, control system, and primary tracking display, performers were capable of approaching different target values of $(\tau = .250\ s, K = 4.5\ s^{-1})$ and $(\tau = .175\ s, K = 7.5\ s^{-1})$. Like the Obermayer et. al (1966) results, these findings demonstrate considerable adaptivity by the human performer.

Rupp's experiment required that performers receive feedback of their gain and time delay as they were performing the task. In other words, estimates of these parameters had to be continuously provided during a trial, rather than after the trial was completed. The estimates of K and τ were provided by a procedure previously tested by Jackson (1969), and it is an example of a general procedure for parameter estimation that is part of "model reference adaptive control" (e.g., Astrom & Wittenmark, 1989; Narendra & Annaswamy, 1989). In that it provides an alternate methodology for measuring the crossover model parameters, it is presented in some detail. For a review of other examples of adaptive model matching for measuring tracking behavior, see Sheridan and Ferrell (1974, chap. 10.3).

MODEL REFERENCE ADAPTIVE CONTROL

By Computerized Gradient Descent

In order to understand the measurement procedure used by Jackson, first consider the situation in which there are two simulated crossover models (top half of Fig. 15.1). They are responding to the same input signal. One has fixed parameters K_1 and τ_1. The task is to adjust the parameters of the second crossover model to match the performance of the first. In other words, adjust K_2 and τ_2 to match K_1 and τ_1. This task might be called a *model reference adaptive control task* (e.g., Astrom & Wittenmark, 1989) because the goal is to adjust the parameters of one dynamic system to match the performance of another model system. The behavior of the model system is the desired style of performance. This technique is often used to adjust the performance of electromechanical systems.

One approach to this problem is to consider the difference between the outputs of the two systems to be the "matching error," $e_m(t) = y_1(t) - y_2(t)$. If the two systems behave in the same manner, then $e_m^2(t)$ will be zero. In Fig. 15.1, K_2 and τ_2 have to be adjusted to minimize the squared matching error. To perform this adjustment, at the very least it is necessary to know in which direction to adjust K_2 and τ_2. In other words, one needs to know how e_m^2 changes for a small change in K_2 and a small change in τ_2. Ratios of these small changes are more formally referred to as the partial derivatives of the squared matching error with respect to K_2 and τ_2 that is, $\partial e_m^2(t)/\partial K_2$ and $\partial e_m^2(t)/\partial \tau_2$.[1] These derivatives are the "gradient" of $e_m^2(t)$ with respect to K_2 and τ_2. This information can be used to adjust K_2 and τ_2 to match K_1 and τ_1, respectively.

---

[1]This differentiation can be done in the time domain, the Fourier domain, the Laplace domain, or even some combination of these domains. For example, recalling that $du^n/dx = nu^{n-1}(du/dx)$, the partial derivative of e_m^2 with respect to K_2 can be calculated by treating τ_2 as a constant (the symbol ∂ replaces d to indicate partial differentiation):

$$\partial e_m^2(t)/\partial K_2 = 2e_m(t)[\partial e_m(t)/\partial K_2]$$

Because $e_m(t) = y_1(t) - y_2(t)$, and because $y_1(t)$ does not depend K_2,

$$\partial e_m(t)/\partial K_2 = \partial[y_1(t) - y_2(t)]/\partial K_2 = -\partial y_2(t)/\partial K_2$$

Therefore,

$$\partial e_m^2(t)/\partial K_2 = -2e_m(t)[\partial y_2(t)/\partial K_2]$$

The gradient descent technique for model reference adaptive control adjusts each parameter in the direction of its negative gradient on the matching error. For example, if $\partial e_m^2(t)/\partial K_2$ is positive, that means that $e_m^2(t)$ increases as K_2 increases. Therefore, to decrease $e_m^2(t)$, K_2 should be decreased. In other words, K_2 should be adjusted in the direction of the *negative* gradient. One way of implementing this technique is simply to make the rate of change of the parameters K_2 and τ_2 proportional to the respective negative gradients:

$$dK_2/dt = c_k[-\partial e_m^2(t)/\partial K_2]$$
$$d\tau_2/dt = c_\tau[-\partial e_m^2(t)/\partial \tau_2] \tag{1}$$

To calculate $\partial[y_2(t)]/\partial K_2$, momentarily switch to the Laplace domain (e.g., see Astrom & Wittenmark, 1989, p. 112). The closed-loop response, $y_2(s)$, of a system with input $i(s)$, forward-loop transfer function $G(s)$, and feedback transfer function $H(s)$, is $y_2(s) = i(s)G(s)/[1 + G(s)H(s)]$, and the error is $e_2(s) = y_2(s)/G(s)$ $= i(s)/[1 + G(s)H(s)]$ (see chap. 13). For the second crossover model in Fig. 15.1, $G(s) = K_2 e^{-\tau_2 s}/s$, and $H(s) = 1$. Therefore,

$$\partial y_2(s)/\partial K_2 = \partial\{[i(s)K_2 e^{-\tau_2 s}/s]/[1 + K_2 e^{-\tau_2 s}/s]\}/\partial K_2$$

Remembering that $d(u/v)/dx = (du/dx)(1/v) + u(-dv/dx)(1/v^2)$,

$$\partial y_2(s)/\partial K_2 = [i(s)e^{-\tau_2 s}/s]/[1 + K_2 e^{-\tau_2 s}/s]$$
$$+ [i(s)K_2 e^{-\tau_2 s}/s][-e^{-\tau_2 s}/s]/[1 + K_2 e^{-\tau_2 s}/s]^2$$

Given that $e_2(s) = i(s)/[1 + K_2 e^{-\tau_2 s}/s]$,

$$\partial y_2(s)/\partial K_2 = e_2(s)[e^{-\tau_2 s}/s] - e_2(s)[e^{-\tau_2 s}/s][K_2 e^{-\tau_2 s}/s]/[1 + K_2 e^{-\tau_2 s}/s]$$
$$= e_2(s)[e^{-\tau_2 s}/s]/[1 + K_2 e^{-\tau_2 s}/s]$$

To perform this calculation, Jackson (1969) took the delayed error signal corresponding to $e_2(s)e^{-\tau_2 s}$ and passed it through a feedback loop with forward-loop transfer function $G(s) = 1/s$, and feedback $H(s) = K_2 e^{-\tau_2 s}$. The closed-loop transfer function of this loop was therefore

$$G(s)/[1 + G(s)H(s)] = [1/s]/[1 + K_2 e^{-\tau_2 s}/s]$$

which yields the desired transformation on $e_2(s)e^{-\tau_2 s}$. This calculation is illustrated in the bottom half of Fig. 15.1. The output of this auxiliary feedback loop corresponds to $\partial[y_2(t)]/\partial K_2$. Multiplying this quantity by $-2e_m(t)$ yields the desired partial derivative of squared matching error with respect to K_2, namely, $\partial e_m^2(t)/\partial K_2$. Using similar calculations, one can calculate the partial derivative of e_m^2 with respect to τ_2 while treating K_2 as a constant:

$$\partial e_m^2(t)/\partial \tau_2 = 2e_m(t)[\partial e_m(t)/\partial \tau_2]$$
$$= -2e_m(t)[\partial y_2(t)/\partial \tau_2]$$

Shifting to the Laplace domain again, one finds that $\partial y_2(s)/\partial \tau_2$ is related to the previously calculated $\partial y_2(s)/\partial K_2$:

$$\partial y_2(s)/\partial \tau_2 = -K_2 s[\partial y_2(s)/\partial K_2]$$

which is also implemented in the bottom of Fig. 15.1. The above calculations reveal how squared matching error, $e_m^2(t)$, changes as a function of K_2 and τ_2.

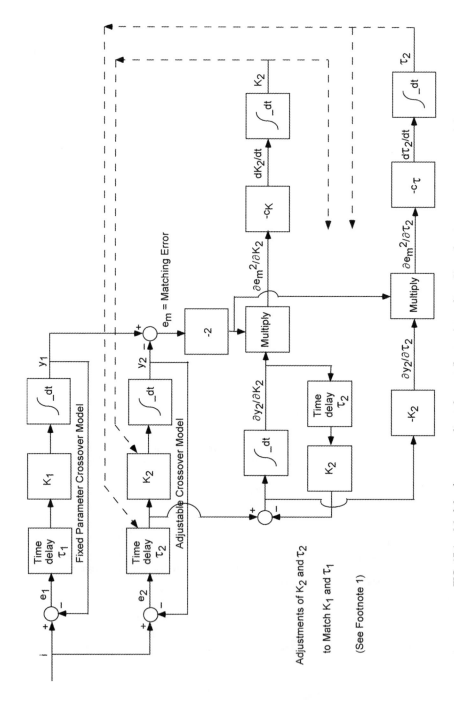

FIG. 15.1. Model reference gradient descent (bottom) to adjust K_2 and τ_2 in one crossover model to match the performance of a second crossover model with fixed parameters, K_1 and τ_1. From "A method for the direct measurement of crossover model parameters" by G. A. Jackson, 1969, *IEEE Transactions on Man–Machine Systems, MMS-10*, pp. 27–33. Copyright 1969 by IEEE. Adapted by permission.

172

where c_k and c_τ are small positive constants (Fig. 15.1, lower right). It is important that these constants be relatively small in order for K_2 and τ_2 to converge to K_1 and τ_1. Even so, this approach does not guarantee convergence for all systems. However, it has proved to be very useful in the present application.

Jackson (1969) tested the usefulness of this technique for measuring human performance by substituting a person for the simulated system with fixed parameters in Fig. 15.1. He found that the gradient descent technique (with the addition of some lag in the adjustment equations for K_2 and τ_2 to provide a little smoothing or averaging over time) led to convergence of K_2 and τ_2 to stable values. To determine whether these stable values were valid estimates of the behavior of the human performer, Jackson additionally calculated describing functions for the human performers using Fourier analysis. He found excellent agreement between the crossover frequency and the effective time delay derived from Fourier analysis and the values of K_2 and τ_2 generated by the gradient descent technique.

A measure of how well the simple version of the crossover model approximated the person's performance is the mean-squared value of the matching error (based on constant values of K_2 and τ_2) divided by the mean-squared output of the person's tracking system. If this value were zero, then the model would perfectly mimic the person's performance. Jackson found that the mean-squared matching error was 10% or less of the person's tracking system output with a velocity control system and 40% or less with an acceleration control system. The higher values of normalized matching error for each control system were obtained with a rapid input (4 rad/s bandwidth), whereas low values of normalized matching error below 10% were obtained for both control systems with a slow input (1 rad/s bandwidth).

By Human Performers

Given the validity of this measurement technique, Rupp (1974) utilized it to test the adaptivity of the human performer, as noted earlier. Basically, he required human participants to perform model reference adaptive control. In other words, they had to mimic a particular model dynamic system consisting of a crossover model with specified parameters. Their success in achieving these target values was measured by the computerized gradient descent technique already described. Thus, the person tried to match a reference model specified by the experimenter, and a computer simulation not seen by the participant tried to mimic the performance of the person in order to measure the participant's values of gain and time delay. Performers were generally quite successful at matching two very different styles of performance even though they were experiencing the same control system and input signal. Thus they exhibited adaptivity.[2]

The model reference adaptive control task can also be used to investigate perceptual and/or cognitive limitations in human performance. For example, Pew and

---

[2]Pew and Rupp (1971) also used the gradient descent measurement technique to quantify performance differences among 4th-, 7th-, and 10th-grade boys. They found marked decreases in the effective time delay with increasing age, which they interpreted as maturational differences. The 10th-grade boys also had much larger gains.

Jagacinski (1971; see also Jagacinski, 1973) studied how well human performers could use a set of pushbuttons to adjust the parameters K_2 and τ_2 in one simulated crossover model to match the fixed parameters K_1 and τ_1 in a second simulated crossover model (top half of Fig. 15.1). One pair of pushbuttons raised or lowered K_2 by .6 s^{-1} each time one was pressed; a second pair of pushbuttons raised or lowered τ_2 by .2 s each time one was pressed. Identical inputs to the two crossover models were displayed as two dots, one vertically above the other, moving horizontally in unison. The outputs of the crossover models were displayed as two horizontally moving cursors that tracked the movements of their respective inputs. The human performers could thus see how the responses of the two crossover models differed. That was the only information they had as the basis for adjusting the parameters. Whereas Rupp's task required participants to actually perform the tracking task as well as adaptively adjust their performance, this task only required participants to make adaptive adjustments to one of two computer simulations of crossover models. In this way, possible motor limitations were not a factor, but the task still involved adaptive adjustments to match a model system.

The initial values of K_2 and τ_2 were randomly varied across trials, and the pattern of adjustments was mapped out in the (K, τ) parameter space that has been previously used to describe the crossover model. The fixed model parameters corresponding to K_1 and τ_1 in Fig. 15.1 are represented as a circle in the center of the graph. At each point in the space, the number of adjustments in each of the four possible directions were tallied (+ and – K_2, + and – τ_2; Fig. 15.2a, top). These transition frequencies were represented as a set of vectors whose lengths summed to one at each point in the space (Fig. 15.2a, bottom). The individual vectors (light vectors) and their vector sum (bold) vectors are shown for three performers in Fig. 15.2a (bottom) and Fig. 15.2b.

Although there are large individual differences, all three performers were more successful in adjusting K_2 than in adjusting τ_2. The large dashed line (τ-Ambiguous line) and small dashed line (K-Ambiguous line) for each performer, respectively, indicate approximate boundaries between increasing and decreasing τ_2 adjustments and between increasing and decreasing K_2 adjustments. Across all three performers, the boundary between increasing and decreasing K adjustments is approximately perpendicular to the K axis and nearly goes through the model value (corresponding to the circle in the center of the graph), which indicates a high degree of success in making these adjustments. However, the corresponding boundaries between increasing and decreasing τ adjustments are diagonal in the (K, τ) space for two performers, and nearly perpendicular to the K axis rather than the τ axis for the third performer. These results indicate that the performers had difficulty adjusting τ.

Whereas the present task emphasized perceptual and/or cognitive aspects of adaptive control, it is interesting to note that active trackers also seem better able to adjust K than τ. For example, if the gain of a control system is varied over a wide range, skilled performers are able to alter their own gain in a compensatory manner so that the product of their own gain and the control system gain remains nearly constant (McRuer, Graham, Krendel, & Reisener, 1965). In contrast, additional time delays in a control system are not dealt with so readily (e.g., see Poulton, 1974, for a review). Behavioral mechanisms for decreasing the effective time delay in active

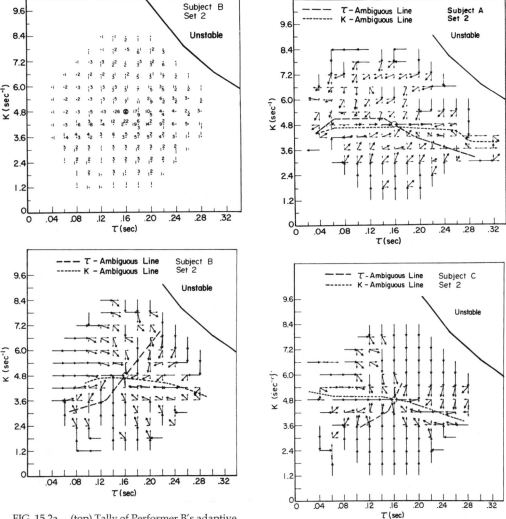

FIG. 15.2a. (top) Tally of Performer B's adaptive adjustments of K and τ in order to match a second crossover model with fixed parameters $K = 4.8/s$ and $\tau = .16$ s; (bottom) vector representation of the adjustment frequencies for Performer B.

FIG. 15.2b. Vector representations for two other performers, A and C.

tracking are presumed to be the reduction of neuromuscular lag (e.g., by tightening one's grip) and increasing high frequency lead generation (e.g., perceptually emphasizing velocity) (McRuer et al., 1965). The adaptive adjustment data in Fig. 15.2 raise the question of whether τ is a parameter by which tracker's intuitively characterize their performance and adjust accordingly, or whether some other parameter that is a function of both K and τ is more salient to the performer, e.g., closed-loop resonant frequency (see Figs. 15.4 and 15.5).

ADAPTIVE RESPONSE TO DIFFERENT INPUT BANDWIDTHS

Additional evidence for human performers' adaptive capability comes from the performance variations they exhibit with variations in the bandwidth of the input signal. If the input consists of a set of sinewaves with equal amplitudes and closely spaced frequencies, then the bandwidth refers to the highest frequency in the input. If the input signal consists of filtered noise, then the bandwidth roughly corresponds to the highest frequency at which there is significant power.

McRuer et al. (1965) investigated compensatory tracking performance with three different input bandwidths, 1.5, 2.5, and 4.0 rad/s with position, velocity, and acceleration control systems. Their experimental participants were aircraft pilots. The variations in the gain, K, and the time delay, τ, of the simple crossover model are shown in Fig. 15.3. The nature of the control system exerts a large effect on both K and τ. As the number of integrators in the control system increases (position control = 0; velocity control = 1; acceleration control = 2), K decreases and τ increases, resulting in poorer performance. K also exhibits a slight increase with input frequency, except for the acceleration control with the highest input frequency. τ decreases with input frequency, again with the exception of the acceleration control with the highest input frequency.

In order to understand the implications of the observed changes in K and τ, it is useful to consider the increase in input bandwidth as an experimental demand to respond more quickly. If the human performer's tracking of an input signal with a low bandwidth represents a desirable style of performance, then consider how a performer would have to alter K and τ to maintain approximately the same style of performance with a faster input having a higher bandwidth. If the higher input bandwidth was twice as large as the low one, then the highest sinusoidal component in the input signal with significant power would move twice as fast. In order to respond in approximately the same style, the human performer would have to respond more quickly. In terms of the crossover model parameters, this adjustment would correspond to decreasing τ by a factor of two, and increasing K by a factor of two.[3] However, that type of adjustment is not exhibited in Fig. 15.3, where the input bandwidth changes by nearly a factor of three (1.5 rad/s to 4.0 rad/s). τ generally decreases, but by less than an amount proportional to the ratio of input bandwidths; K exhibits a much smaller proportional change.

Some insight into the consequences of these less than proportional changes can be gained by considering the closed-loop frequency response. As noted in chapter 13, for the simple crossover model, the closed-loop describing function is $Ke^{-j\omega\tau}/(Ke^{-j\omega\tau} + j\omega)$. Figure 15.4 shows the open- and closed-loop describing functions for this simple model with the parameters corresponding to the three control systems with the

---

[3]The highest frequencies of the input spectrum containing significant power will have a dominant effect on overall performance. If the bandwidth is characterized by frequency ω_i, then the open-loop phase lag is $-\pi/2 - \tau\omega_i$, and the open-loop amplitude ratio is K/ω_i at this frequency. To maintain these same values of phase lag and amplitude ratio for a new bandwidth, $\omega_i' = 2\omega_i$, τ would have to decrease by a factor of 2, and K would have to increase by a factor of 2.

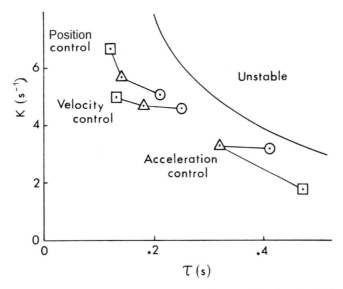

FIG. 15.3.   Crossover model parameters K, gain, and τ, time delay, for three different input bandwidths (circles = 1.5 rad/s; triangles = 2.5 rad/s; squares = 4.0 rad/s) and for position, velocity, and acceleration control systems. Data from McRuer et al. (1965). From "A qualitative look at feedback control theory as a style of describing behavior" by R. J. Jagacinski, 1977, *Human Factors*, *19*(4), pp. 331–347. Copyright 1977 by the Human Factors and Ergonomics Society. All rights reserved. Adapted by permission.

1.5 rad/s input bandwidth (circles in Fig. 15.3). The crossover frequency, which equals K rad/s, is more than twice the input bandwidth for all three control systems; however, it exceeds the input bandwidth by the greatest margin for the position control system, where it is approximately three times larger. This large ratio of crossover frequency to input bandwidth corresponds to the best tracking behavior among the three systems.

Note also that the closed-loop describing functions can roughly be characterized in terms of three regions: frequencies much below crossover, frequencies near crossover, and frequencies much above crossover:

1. At frequencies much below the crossover frequency, the closed-loop amplitude ratio is near 1.0 and the phase lag is small. In other words, the amplitude of the system output closely matches that of the input signal, and there is only a small amount of lag.

2. For frequencies near crossover, all three systems have resonant peaks (i.e., the amplitude ratio of the closed-loop response is greater than one). If there were significant components of the input signal in this region, then the system output at these frequencies would be larger than the input signal. However, if this resonance occurs at frequencies where there is not significant input power (as in Fig. 15.4), then these undesirable effects are not realized.

3. The amplitude ratio at frequencies much above crossover is much below 1.0. In other words, if there were significant components of the input signal in this region,

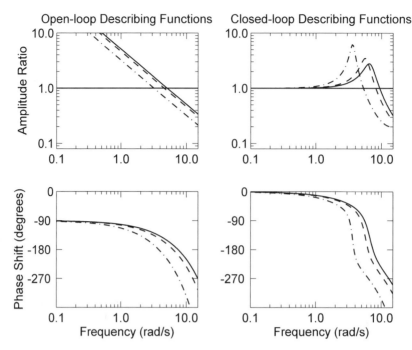

FIG. 15.4.   Open-loop (output relative to error) and closed-loop (output relative to input) describing functions for the crossover model for a low input bandwidth (1.5 rad/s, circles in Fig. 15.3) and a position control system (solid lines), velocity control system (dashed lines), and acceleration control system (dot-dashed lines).

then the system output would not be large enough. Also, the phase lag would be quite large. However, once again, if the input signal has no significant power in this region (as in Fig. 15.4), then these deleterious effects are not realized.

Although the crossover model was only intended to be accurate for the frequencies near crossover, variations in the open-loop describing function would typically not produce consequential variations in the closed-loop describing function in regions much below and much above crossover frequency, so the previous considerations with regard to the closed-loop describing function are fairly general.

As already noted, a person would have to increase the crossover frequency, K, in proportion to the input bandwidth in order to maintain approximately the same high quality tracking as exhibited with the low bandwidth input. However, K increases only slightly in Fig. 15.3, and other investigators have even found K to decrease somewhat as the input bandwidth increases (Jackson, 1969). Given the lack of proportional increase in the crossover frequency, the high frequency components of the input will approach the region of the closed-loop resonance as the input bandwidth increases. If the resonance remained as pronounced as for the low input bandwidth, then the tracking response in this region would be too large to match the input. Performers avoid this unwanted effect by reducing the closed-loop resonance. For example, Fig. 15.5 shows the open-loop and closed-loop describing functions for the sim-

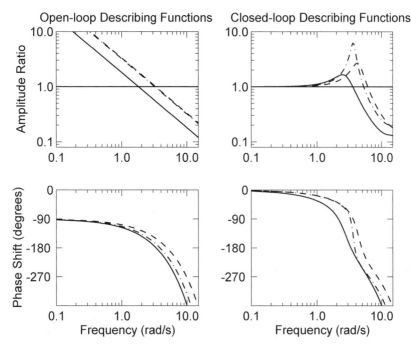

FIG. 15.5. Open-loop (output relative to error) and closed-loop (output relative to input) describing functions for the crossover model for an acceleration control with 1.5 rad/s (dot-dashed lines), 2.5 rad/s (dashed lines), and 4.0 rad/s (solid lines) input bandwidths (parameters are from Fig. 15.3).

ple crossover model for the parameters corresponding to the acceleration control with 1.5, 2.5, 4.0 rad/s input bandwidths in Fig. 15.3. With the 2.5 rad/s input (dashed lines), a decrease in τ has resulted in a smaller closed-loop resonance peak than with the 1.5 rad/s input (dot-dashed lines). Equivalently, a decrease in τ has resulted in less phase lag at the crossover frequency, which has increased the phase margin. (The phase lag due to the time delay, τ, at the crossover frequency, K, is equal to τK radians.) As the phase margin increases, the resonance peak diminishes (McRuer et al., 1965, p. 182). This effect is similar to increasing the damping in a second-order system (see chaps. 6 and 13) in that the system is less oscillatory.

Figure 15.5 also shows that as the input bandwidth is increased even more to 4.0 rad/s, the closed-loop resonance peak is decreased even more. In this condition, the phase lag at the crossover frequency, τK, has been reduced by decreasing K rather than by decreasing τ. The crossover frequency is actually less than the input bandwidth by more than a factor of two. This phenomenon, which has been replicated by other investigators (e.g., Jackson, 1969), has been termed *crossover frequency regression* (McRuer et al., 1965, p. 173). The quality of tracking is quite poor, because as can be seen in Fig. 15.5, the closed-loop amplitude ratio is well below 1.0 for the high frequency components of the input signal. However, this style of tracking requires less effort than producing the large amplitude movements that would be characteristic of a larger resonance peak. Crossover frequency regression can be considered as part of a general trend to decrease the closed-loop resonance peak by increasing the phase margin as the input bandwidth increases (see McRuer & Jex, 1967).

THE CRITICAL TRACKING TASK: ADAPTIVE RESPONSE TO INCREASING INSTABILITY

In chapter 4, the first-order lag was introduced as an elementary negative feedback loop (Fig. 4.2). The output signal was subtracted from the command input to generate the error signal; the error signal was passed through a gain and an integrator to generate the output signal. If there was a sudden step change in command input, then the output exponentially approached the new value of the command input with a time constant that was inversely proportional to the gain. In other words, the system responded to the change in the input so as to reduce the error signal. Error reduction is a characteristic response of a negative feedback loop.

If instead of subtracting the output from the command input as in Fig. 4.2, the output was added to the command input, the negative feedback loop would be changed into a positive feedback loop. Such loops amplify errors rather than diminish them. For example, if the command input and the system output summed to zero, then such a system would be at equilibrium. However, if the command input suddenly increased to a new constant value, the system output would also increase, which would lead to an increase in the sum of the command input and system output, which would lead to an increase in the system output, and so on. The system output would exhibit an exponential runaway in the direction of the step change in the command input. The difference between the command input and the system output would thus become greater over time. The higher the gain of the positive feedback loop, the faster the runaway would occur.

Jex, McDonnell, and Phatak (1966) investigated how well a person could stabilize a positive feedback loop that became more difficult to control. As shown in Fig. 15.6, the dynamic system that the person had to control consisted of a gain, λ, and an integrator in a positive feedback loop. Such a system will run away exponentially with a time constant that is inversely proportional to λ, unless the human performer exerts appropriate control. The actions of the performer can be thought of as a negative feedback loop that is competing with the positive feedback loop. This task can be better understood by considering another task with different dynamics, but that also involves stabilization of an unstable system, namely, balancing a broomstick in a vertical position (Cannon, 1967; Geva & Sitte, 1993; Pew, 1973). The force of gravity acts to amplify small deviations of the broomstick from vertical and makes it fall over, and people move their hand to reduce small deviations from vertical. A long broomstick is easier to balance than a short one. If the broomstick were to gradually become shorter, then the difficulty of the balancing task would increase until the person could no longer control it, and the broomstick would fall. In an analogous manner, the speed with which the positive feedback loop shown in Fig. 15.6 runs away can be gradually increased by increasing λ until it reaches a "critical" value, and the person finally loses control, and the cursor runs off the display screen. This critical value, λ_c,

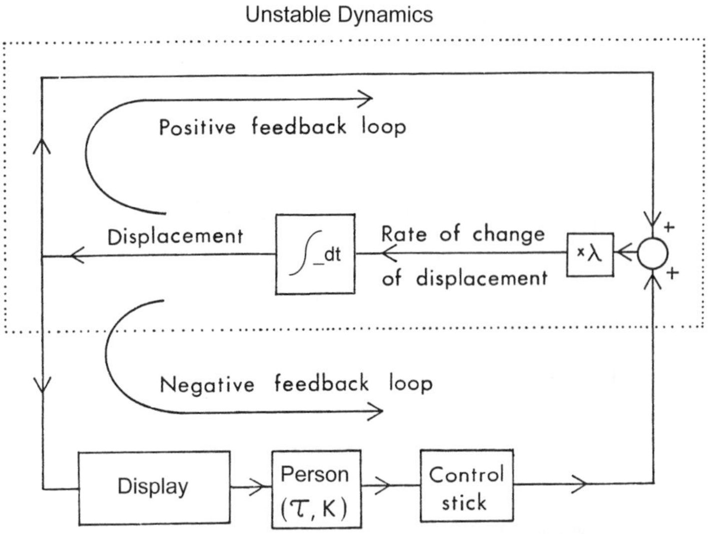

Unstable Dynamics

FIG. 15.6. The critical tracking task. From "A 'critical' tracking task for manual control research" by H. R. Jex, J. D. McDonnell, and A. V. Phatak, 1966, *IEEE Transactions on Human Factors in Electronics, HFE-7*, pp. 138–144. Copyright 1966 by IEEE. Adapted by permission.

is linearly related to the inverse of the person's effective time delay. The shorter the time delay, the higher the value of λ_c.

To understand why this simple relation exists, it is helpful to consider the control exerted by the person. As shown in Fig. 15.6, the person can be considered to be in a negative feedback loop that is competing with the positive feedback loop. The person's negative feedback loop contains one integrator and therefore resembles a rate control system. Previous discussions of the McRuer crossover model suggest that the person can be approximated as a gain, K, and a time delay, τ, which is empirically found to be so. As shown in Fig. 15.7, there are both upper and lower constraints on K. K cannot be too low or the positive feedback loop will win out. In particular, the gain around the positive feedback loop is λ, and the multiplication of gains around the negative feedback loop is $K\lambda$. $K\lambda$ must be at least as large as λ, so K must at least be equal to 1.0. This constraint is shown toward the bottom of Fig. 15.7. The second constraint is that $K\lambda$ cannot be too large or the combination of high gain and time delay τ will lead to unstable oscillations as discussed in the preceding chapter on the crossover model. Therefore, as λ increases, the maximum stable value of K decreases. This constraint is shown by the solid, dashed, and dotted curves in the upper right of Fig. 15.7. These curves represent decreasing maximal values of K for stable behavior as λ is increased. The person adapts to this ever more restrictive upper bound by lowering their time delay, τ, to the limited degree possible, and by lowering their gain, K (dots 1, 2, 3 in Fig. 15.7). Eventually, K is driven close to 1.0 (dot 3 in Fig. 15.7). At that point, the person is caught between the upper and lower constraints, and then there is no longer any latitude for further adjustment. Therefore, the person loses control (Jex et al., 1966). The value of λ_c that forces the person's gain to 1.0 is determined by

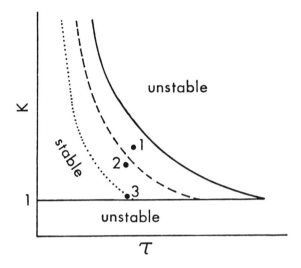

FIG. 15.7. The curved upper stability boundary moves leftward as λ increases. Individuals adjust their gain, K, (dots 1, 2, 3) to try to stay below this upper stability boundary and above the lower stability boundary of 1.0.

the person's time delay, τ. The smaller the value of τ, the larger the value of λ_c.[4] Thus λ_c provides a measure of τ.

To measure the relation between τ and λ_c, λ can be increased until it is just short of λ_c, and then for this fixed value of λ measure the person's describing function and examine the linear trend in the phase as a function of frequency to estimate τ. The inverse of the values of τ derived from describing functions were found by Jex and Allen (1970) to be linearly related to λ_c. Namely, $1/\tau = 2.88 \text{ s}^{-1} + 0.93\lambda_c$, and the correlation was .90. The success of the critical tracking task in providing a measure of τ derives from the person's ability to adjust their gain toward 1.0 as λ increases. Thus, the critical tracking task is another example of adaptive adjustment of crossover model performance. This task has been used as a general test of fitness for duty in industrial settings and also as part of a system that deters drunk driving by requiring a person to achieve a certain level of critical instability before starting a car (see Allen, Stein, & Miller, 1990, for a review). To the extent that alcohol, drugs, and fatigue can increase a person's effective time delay, the value of the critical instability that person can achieve will be lowered. The critical tracking task has also been used to evaluate and compare different controls and displays (e.g., Jagacinski, Miller, & Gilson, 1979; Jex, Jewell, & Allen, 1972).

In summary, both this chapter and the preceding one presented a number of ways that performers are able to adapt their feedback control behavior to different performance criteria, control systems, and input signals. To a first-order approximation,

---

[4]Based on the simple crossover model, the theoretical point at which the person loses control corresponds to $Ke^{-j\omega\tau}\lambda_c/(j\omega - \lambda_c) = -1$. Namely, the open-loop gain is 1, and the phase shift is $-180°$. Because K is equal to 1, and $e^{-j\omega\tau} = \cos(\omega\tau) - j\sin(\omega\tau)$ (see chap. 13), this equation can be rewritten as: $[\cos(\omega\tau) - j\sin(\omega\tau)]\lambda_c = -(j\omega - \lambda_c)$. Equating the imaginary parts of both sides of the equation, $\lambda_c\sin(\omega\tau) = \omega$. Assuming $\omega\tau$ is small enough that $\sin(\omega\tau)$ is approximately equal to $\omega\tau$, one obtains: $\lambda_c\omega\tau = \omega$. Equivalently, $\lambda_c = 1/\tau$.

the human performer's ability to make such adjustments involves an impressive degree of adaptation of gain and to a lesser degree effective time delay within broad limitations on feasible values of these parameters.

REFERENCES

Allen, R. W., Stein, A. C., & Miller, J. C. (1990, October). *Performance testing as a determinant of fitness-for-duty*. (Society for Automotive Engineering Technical Paper 901870). Paper presented at the Aerospace Technology Conference and Exposition, Long Beach, CA.

Astrom, K. J., & Wittenmark, B. (1989). *Adaptive control*. Reading, MA: Addison-Wesley.

Cannon, R. H., Jr. (1967). *Dynamics of physical systems*. New York: McGraw-Hill.

Geva, S., & Sitte, J. (1993, October). A cartpole experiment benchmark for trainable controllers. *IEEE Control Systems, 13*, 40–51.

Jackson, G. A. (1969). A method for the direct measurement of crossover model parameters. *IEEE Transactions on Man–Machine Systems, MMS-10*, 27–33.

Jagacinski, R. J. (1973). *Describing multidimensional stimuli by the method of adjustment*. Unpublished doctoral dissertation, University of Michigan, Ann Arbor, MI.

Jagacinski, R. J. (1977). A qualitative look at feedback control theory as a style of describing behavior. *Human Factors, 19*, 331–347.

Jagacinski, R. J., Miller, D. P., & Gilson, R. D. (1979). A comparison of kinesthetic-tactual and visual displays via a critical tracking task. *Human Factors, 21*, 79–86.

Jex, H. R., & Allen, R. W. (1970). Research on a new human dynamic response test battery. In *Proceedings of the Sixth Annual Conference on Manual Control* (pp. 743–777). Wright-Patterson AFB, OH.

Jex, H. R., Jewell, W. F., & Allen, R. W. (1972). Development of the dual-axis and cross-coupled critical tasks. In *Proceedings of the Eighth Annual Conference on Manual Control* (AFFDL-TR-72-92, pp. 529–552). University of Michigan, Ann Arbor, MI.

Jex, H. R., McDonnell, J. D., & Phatak, A. V. (1966). A "critical" tracking task for manual control research. *IEEE Transactions on Human Factors in Electronics, HFE-7*, 138–144.

McRuer, D., Graham, D., Krendel, E., & Reisener, W., Jr. (1965). *Human pilot dynamics in compensatory systems* (Air Force Flight Dynamics Laboratory Tech. Rep. No. 65-15). Wright-Patterson Air Force Base, OH.

McRuer, D., & Jex, H. R. (1967). A review of quasi-linear pilot models. *IEEE Transactions on Human Factors in Electronics, HFE-8*, 231–249.

Miller, D. C. (1965). The effects of performance-scoring criteria on compensatory tracking behavior. *IEEE Transactions on Human Factors in Electronics, HFE-6*, 62–65.

Narendra, K. S., & Annaswamy, A. M. (1989). *Stable adaptive systems*. Englewood Cliffs, NJ: Prentice-Hall.

Obermayer, R. W., Webster, R. B., & Muckler, F. A. (1966). Studies in optimal behavior in manual control systems: The effect of four performance criteria in compensatory rate-control tracking. In *Proceedings of the Second Annual Conference on Manual Control* (NASA Special Rep. No. 128, pp. 311–324). MIT, Cambridge, MA.

Pew, R. W. (1973). Performance assessment via the critical task and dowel balancing. In *Proceedings of the Ninth Annual Conference on Manual Control*. MIT, Cambridge, MA.

Pew, R. W., & Jagacinski, R. J. (1971). Mapping an operator's perception of a parameter space. In *Proceedings of the Seventh Annual Conference on Manual Control* (NASA Special Rep. No. 281, pp. 201–206). University of Southern California, Los Angeles, CA.

Pew, R. W., & Rupp, G. L. (1971). Two quantitative measures of skill development. *Journal of Experimental Psychology, 90*, 1–7.

Poulton, E. C. (1974). *Tracking skill and manual control*. New York: Academic Press.

Rupp, G. L. (1974). *Operator control of crossover model parameters*. Unpublished doctoral dissertation, University of Michigan, Ann Arbor, MI.

Sheridan, T. B., & Ferrell, W. R. (1974). *Man–machine systems*. Cambridge, MA: MIT Press.

Driving Around in Circles

> *A nontrivial part of any control problem is modeling the process. The objective is to obtain the simplest mathematical description that adequately predicts the response of the physical system to all anticipated inputs.*
>
> —Kirk (1970, p. 4)

"Driving around in circles" typically means that drivers have lost their way. However, in this chapter it refers to the geometry of the automobile steering system. Namely, if you hold the steering wheel of a car at a fixed angle, the tire angle[1] will be equal to some constant value θ, and the car will go around in a circle. Consider some of the details of the geometry of this situation. As shown in Fig. 16.1, the right front tire will go around a slightly larger circle than the right rear tire (e.g., Gillespie, 1992). The turn radius of the right front tire, R_f, is perpendicular to the right front tire, which is tangent to the circular path.[2] It is possible to construct a right triangle in which R_f is the hypotenuse, L is the distance between the front and rear tires (also referred to as the "wheel base"), and R_r is the turn radius of the rear right tire. Note that the angle between R_f and L is equal to 90° − θ, because the sum of several angles whose outer radii form a straight line is 180°. Additionally, the angle between R_f and R_r is equal to θ, because the sum of the internal angles of a triangle is 180°. Knowing these angles, it can be concluded that R_f is equal to $L/\sin\theta$, and R_r is equal to $R_f\cos\theta$. In other words, the turn radius of the front right tire is inversely related to the sine of the front tire angle. If the steering wheel is held at just a small angle away from its neutral position,

---

[1] In the automotive literature, the tire angle is referred to as the "steer angle."

[2] This discussion assumes that the car is traveling at a slow enough speed or that the steering wheel deflection is sufficiently small that side slip (i.e., motion of the tires perpendicular to their plane of rotation) can be ignored. For more elaborate modeling of the vehicle response, see Gillespie (1992) or Weir, Shortwell, and Johnson (1966).

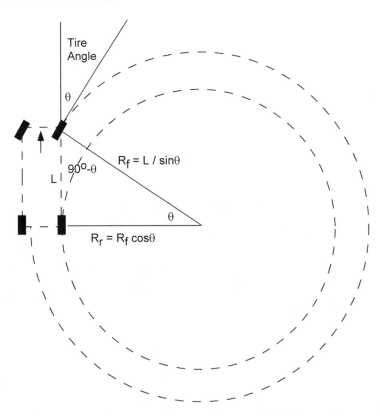

FIG. 16.1. Holding the steering wheel at a constant angle causes the front and rear tires to follow circular paths with different radii and a common center. From "Man–machine steering dynamics" by J. G. Wohl, 1961, *Human Factors, 3*, pp. 222–228. Copyright 1961 by the Human Factors and Ergonomics Society. All rights reserved. Adapted by permission.

then the radius of the circle will be very large. If the steering wheel is held at a larger angle away from its neutral position, then the radius of the circle will be smaller. Making a tight turn means traveling on a circle with a small radius, which requires a large steering wheel angle. Whatever the turn radius of the front right tire, the rear right tire turn radius is smaller by a factor of the cosine of the front tire angle. Although these geometric observations seem straightforward enough, they have strong implications for the task of steering a car.

THE CAR STEERING SYSTEM: POSITION, VELOCITY, OR ACCELERATION CONTROL?

Previous chapters have discussed position, velocity, and acceleration control systems. These are simplified abstractions of many control systems that may be encountered, and their simplification helps to emphasize important relations in control theory. Which control system describes the steering system of a car? Actually, all of them do depending on what is considered to be the output of the steering control system. The following derivation parallels that of Wohl (1961).

If tire angle is considered to be the output of the steering control system, then the steering system can be approximated as a position control with gain K_s. As shown in Fig. 16.2, the tire angle, θ, is directly proportional to the steering wheel angle, δ_{sw}. Power steering systems may introduce a bit of lag, but to a reasonable approximation the driver changes tire angle relative to the body of the car by a position control system.

A second possible way of thinking about the steering system is as a means of changing the direction in which the car is pointing. In other words, the output of the steering system is the heading angle of the car, ψ_c. The right front tire travels in a circular path with radius R_f (Fig. 16.1). If the magnitude of the velocity of the right front tire is V meters per second, and R_f is the turn radius in meters, then it will take R_f/V seconds to cover a distance R_f along the circumference of the circle. In other words, the front tire will travel one radian (57.3°) in R_f/V seconds. Over that interval, the heading angle of the car will change by 57.3° or one radian. Equivalently, the rate of change of the heading angle, $d\psi_c/dt$, is equal to 1 radian/(R_f/V) second = V/R_f radians/second. Substituting the relation between turn radius and tire angle, the following equation is obtained:

$$d\psi_c/dt = V/R_f = V/(L/\sin\theta) = V\sin\theta/L \tag{1}$$

In other words, the sine of the tire angle determines the rate of change of heading angle for a given velocity and wheel base. Equivalently, the tire angle and wheel base determine the curvature of the car path, $1/R_f$, which in turn determines the rate of change of heading angle. Substituting the proportional relation between tire angle and steering wheel angle:

$$d\psi_c/dt = V\sin(K_s\delta_{sw})/L \tag{2}$$

For small angles, $\sin\theta$ is approximately equal to θ, so that the previous relation simplifies to:

$$d\psi_c/dt = \delta_{sw}(K_s/L)V$$

or equivalently,

$$V\int[\delta_{sw}(K_s/L)]dt = \psi_c \tag{3}$$

In other words, the steering system can be considered a rate control system if the heading angle is considered to be the output. The gain of the system is proportional to the velocity of the front tire. The faster the car is traveling, the more quickly the heading angle will change for a given steering wheel angle. These relations are depicted on the right side of Fig. 16.2. The difference between the angle of the driving lane, ψ_i, and the heading angle of the car, ψ_c, is considered an error in heading angle, ψ_e, which is shown in the lower left of Fig. 16.2.

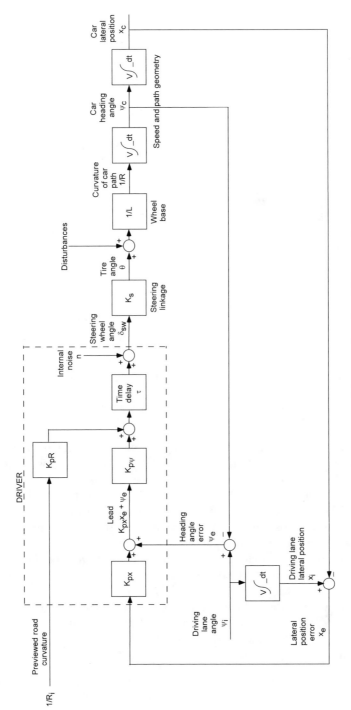

FIG. 16.2. Steering control for a car (based on McRuer et al., 1977).

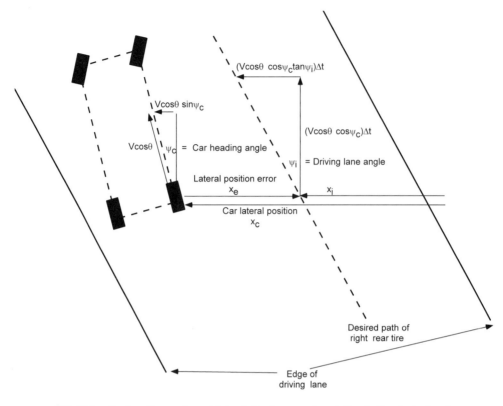

FIG. 16.3.   Car heading angle and lateral displacement in relation to the desired path.

A third way of thinking about the steering system is as a means of changing the lateral position of the car in the driving lane. Let x_c be the lateral position of the right rear tire. If the desired lateral position of the right rear tire is x_i, then the difference between x_i and x_c is the lateral position error, x_e. The subscript "e" indicates that this displacement is an error or deviation from the desired position (Fig. 16.3).[3]

Consider how these positions are related to the velocity of the right rear tire. Given that the rate of change of heading angle is the same for both the front and rear tires as the car drives around in a circle (Fig. 16.1), the right rear tire must be traveling at a slower velocity than the right front tire, because the rear tire is going around a slightly smaller circle. More precisely, if the rear tire travels with velocity $V\cos\theta$ around a circle with radius $R_f\cos\theta$, then its rate of change of heading angle is $V\cos\theta/R_f\cos\theta = V/R_f$, which is the same as the right front tire. If the velocity of the right rear tire is $V\cos\theta$, then the component of that velocity in the lateral direction is $V\cos\theta\sin\psi_c$ (Fig. 16.3). Therefore,

$$dx_c/dt = V\cos\theta\sin\psi_c \tag{4}$$

_____

[3]If lateral error were measured perpendicular to the edge of the road, then it would be equal to $x_e\cos\psi_i$ in Fig. 16.3. However, $x_e\cos\psi_i$ approximately equals x_e for small values of ψ_i.

For small angles, use the approximations $\cos\theta = 1$ and $\sin\psi_c = \psi_c$, so that

$$dx_c/dt = V\psi_c$$

or equivalently,

$$V\int\psi_c dt = x_c \qquad (5)$$

This relation is depicted on the far right side of Fig. 16.2. In other words, the car's heading angle determines the rate of change of lateral position. Also, the faster the car is moving, the more quickly the lateral position changes.

Note that there are two integrations between the steering wheel angle and the lateral position. Substituting into Equation 5 the expression for heading angle (Equation 3):

$$V\int\psi_c dt = V\int\{V\int[\delta_{sw}(K_s/L)]dt\}dt = V^2\iint[\delta_{sw}(K_s/L)]dt\,dt = x_c \qquad (6)$$

The steering system is thus an acceleration control system, if lateral position is considered the output. Note also that the effective gain between steering wheel angle and lateral position error is proportional to the square of the velocity. In other words, the system becomes much more sensitive as the car travels faster (Wohl, 1961). Another way of thinking about this result is that the centripetal acceleration of an object following a circular path is proportional to V^2/R, where V is its velocity and R is the radius of curvature. For a small angle approximation, R_f and R_r are approximately equal, because the cosine of the tire angle, $\cos\theta$, is approximately 1.0. Therefore, a common term $1/R$, may be used to refer to them both. The last two transfer functions on the right of Fig. 16.2 multiply $1/R$ by V^2 and integrate twice to transform acceleration into position. Thus, the lateral acceleration of the car is V^2/R, as would be expected of circular motion (Weir, Shortwell, & Johnson, 1966).

The implication of this derivation is that if people drive a car, then they have already experienced an acceleration control system, probably without thinking about it in those terms. If they have doubts about the aforementioned derivation, then there is a simple test that can be performed. As noted previously, to produce a step output from an acceleration control system, a double pulse must be executed with the control—a pulse to the right followed by a pulse to the left, or vice versa. A step response in the lateral position of a car corresponds to a lane change. Therefore, if individuals feel how they move the steering wheel to perform a lane change, then they should feel a double pulse. (It is strongly recommended that this self-observation not be performed with other cars nearby so as not to risk having an accident while lost in analytic thought.) Figure 16.4 shows some experimental data for two successive lane changes. The form of the lateral acceleration time history closely mimics the steering wheel angle time history, as would be expected for an acceleration control system (see also Tijerina & Hetrick, 1997).

Many people find this surprising. Namely, when asked how they manipulate the steering wheel to perform a lane change, they often report using a single pulse. However, that is not the appropriate control for an acceleration control system, and it is not what these people feel when they are asked to monitor their own performance. In

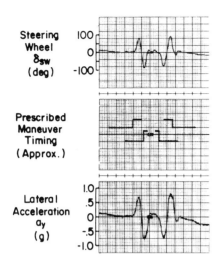

FIG. 16.4. Experimental data for two successive lane changes. The dark vertical lines are spaced at 1-second intervals. From "New results in driver steering control models" by D. T. McRuer, R. W. Allen, D. H. Weir, and R. H. Klein, 1977, *Human Factors*, 19(4), pp. 381–397. Copyright 1977 by the Human Factors and Ergonomics Society. All rights reserved. Reprinted by permission.

other words, many people cannot correctly report from memory how they perform this task.

Next consider the desired path of the right rear tire, which is assumed to be a constant offset away from the edge of the driving lane. Just as the rate of change of the car's lateral position is related to its heading angle, the rate of change of the *desired* lateral position of the right rear tire is related to the driving lane angle. The longitudinal component of the right rear tire velocity is $V\cos\theta\cos\psi_c$ (Fig. 16.3). In a short amount of time, Δt, the right rear tire will travel a longitudinal distance $(V\cos\theta\cos\psi_c)\Delta t$. The *desired* lateral position will similarly change by $(V\cos\theta\cos\psi_c\tan\psi_i)\Delta t$. In other words, in the limit of very small Δt,

$$dx_i/dt = V\cos\theta\cos\psi_c\tan\psi_i \qquad (7)$$

Assuming that all three angles in the previous expression are small, then a simpler approximation is

$$dx_i/dt = V\psi_i$$

or equivalently,

$$V\int\psi_i dt = x_i \qquad (8)$$

This relation is shown in the bottom left of Fig. 16.2.

DESCRIBING THE DRIVER

After approximately characterizing the response of the car to variations in steering wheel angle, the next step is to consider the mathematical characterization of the driver. To maintain the car's rear wheel at the desired distance from the edge of the driving lane, the driver needs to keep the lateral position error, x_e, near zero. Given that the steering control system is an acceleration control, previous discussions of the McRuer crossover model in chapter 14 indicate that the driver certainly will introduce some lead or anticipation in order to stably close the feedback loop. If only the lateral position error of the car was visible, then the driver would have to perceive its rate and form an additive combination of error and error rate. However, the heading angle error is also visible, and it is proportional to the rate of change of lateral position error. Therefore, lead can be generated by a summation of lateral position error and heading angle error. This strategy should be easier than if the driver could only observe lateral position error. At the left side of Fig. 16.2, this summation is represented as $K_{px}x_e + \psi_e$.

One interpretation of this lead term is that the driver extrapolates the motion of the car forward over a duration T, and tries to null out this predicted lateral deviation error (e.g., McRuer, Weir, Jex, Magdaleno, & Allen, 1975). The approximate predicted lateral deviation is

$$x_e(t + T) = x_e(t) + T dx_e(t)/dt \tag{9}$$

Also,

$$x_e = x_i - x_c$$

$$dx_e/dt = dx_i/dt - dx_c/dt = V\psi_i - V\psi_c = V\psi_e$$

Substituting for $dx_e(t)/dt$,

$$x_e(t + T) = x_e(t) + TV\psi_e \tag{10}$$

The lead term shown in Fig. 16.2 can be rewritten as:

$$K_{px}(x_e + \psi_e/K_{px}). \tag{11}$$

Comparing the portion of this expression in parenthesis with the previous expression for $x_e(t + T)$, it is evident that $1/K_{px}$ corresponds to TV, which is the distance the car travels in T seconds (i.e., the "look ahead distance"). If the driver follows such a strategy, then the lead term might be estimated from separate perceptions of x_e and ψ_e, or the weighted summation of these two variables might be present as a feature of the optical flow field in relation to the visual image of the road (e.g., see Lee & Lishman, 1977). Optical flow is considered in chap. 22.

The previous description of the driver addresses compensatory error nulling behavior such as would be needed to nullify the effects of disturbances due to wind gusts and road surface irregularities, as well as inadvertent variability or noise in the operator's response (see Fig. 16.2). However, given that preview of the road is available, it would be expected that the driver would use feedforward (i.e., pursuit) control to respond directly to the effective input signal (see also chaps. 7, 20, and 24), as well as feedback control to respond to deviations from desired system performance. McRuer et al. (1975) suggested that the driver uses the perceived curvature of the road (i.e., 1/turn radius), available via preview of the upcoming road, in order to generate feedforward control (Fig. 16.2, upper left). Monitoring of drivers' eye movements suggests that they closely monitor the side of the road at the points at which the road appears to reverse its lateral bend (Land & Lee, 1994) from left to right or vice versa. These points can be used to judge the curvature of the road.

If the feedforward response is K_{pR}/R_i, where R_i is the radius of curvature of the road, then K_{pR} should approximate L/K_s. When this steering wheel command, $L/(K_sR_i)$, is passed through the steering linkage to generate wheel angle, it will effectively be multiplied by K_s and $1/L$ to generate path curvature (see Fig. 16.2). The result will be $1/R_i$, and the path curvature of the car will equal the path curvature of the road. In other words, the gain K_{pR} should partially invert the car dynamics in order to provide effective feedforward control. Any mismatch can then be nulled out by the compensatory loop. However, the better the feedforward control, the less the compensatory loop has to do.

Another way of putting this latter point is that the *effective* describing function between lateral position error, x_e, and the car's lateral position, x_c, should be improved with feedforward control. McRuer, Allen, Weir, and Klein (1977) calculated the open-loop describing function for drivers compensating for disturbances due to wind gusts on a straight road and also drivers following a curved path, both in a fixed base driving simulator. The wind gust condition presumably involved only feedback control; the curved path condition presumably involved both feedforward and feedback control. The data are shown in Fig. 16.5. Note that the amplitude ratio of car position to lateral position error is considerably greater for the curved path condition, reflecting the use of feedforward (i.e., pursuit) control. Also, note that in the region where the amplitude ratio for the wind gust condition passes through 0 dB, the amplitude ratio drops by approximately 20 dB as the frequency increases by a factor of 10, as would be expected by the crossover model. At higher frequencies, the amplitude ratio levels out. The curves in Fig. 16.5 represent fits of a model like that shown in Fig. 16.2, with an additional high frequency lead added to the driver's response to match the flattening of the amplitude ratio at high frequencies. Not all drivers exhibit this additional high frequency lead (McRuer et al., 1975). It might arise from drivers' sensitivity to the rate of change of heading angle error. In summary, this classical control theory model characterizes the driver as generating low frequency lead from an extrapolation of the car's motion and additionally providing feedforward control via the previewed curvature of the road.

Another early effort at modeling of the use of roadway preview was performed by Wierwille, Gagne, and Knight (1967). They measured the effective time-averaged weighting of previewed roadway positions at various distances in front of the vehicle in determining steering wheel feedforward control. They did not find evidence for a single well-defined preview distance, but instead found a broad band of effective information usage. In a theoretical analysis of the use of previewed position information from the perspective of optimal control theory, Miller (1976) noted that the relative weighting of positional information at various distances in front of a car would depend on a driver's trade-off between maintaining accurate position in the driving

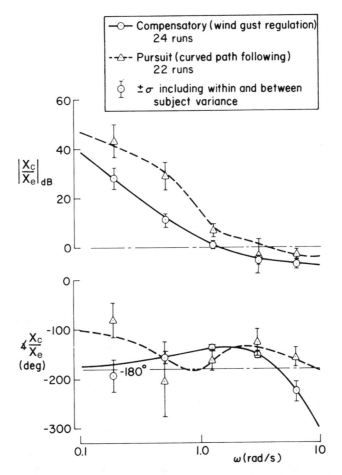

FIG. 16.5. Open-loop describing functions for a driver with wind gusts and with preview of the road curvature. From "New results in driver steering control models" by D. T. McRuer, R. W. Allen, D. H. Weir, and R. H. Klein, 1977, *Human Factors, 19*(4), pp. 381–397. Copyright 1977 by the Human Factors and Ergonomics Society. All rights reserved. Adapted by permission.

lane versus control effort. Greater emphasis on maximizing positional accuracy would correspond to more heavily weighting positional information closer to the car. Greater emphasis on minimizing control effort would correspond to more heavily weighting positional information farther from the car. How such positional information is in turn related to the features of the roadway and the motion patterns perceived by the driver is a topic of active research. For a brief review of additional steering models, see Hildreth, Beusmans, Boer, and Royden (2000).

Any theory of steering control must at least implicitly be a theory of perception as well as a theory of motion generation. The present discussion has used a very simplified description of the automobile response in order to present an introduction to this style of modeling. Examples of more elaborate models of the vehicle response typical of this literature can be found in McRuer et al. (1975), Gillespie (1992), and Weir et al.

(1966). As a first approximation, think of the problem of steering a car as one of coming to terms with the geometry of circles that is imposed by the structure of the car. When attempting to apply control theory to different tasks, understanding the geometry that emerges from the physics of the controlled object is often an important step in exploring different control strategies.

REFERENCES

Gillespie, T. D. (1992). *Fundamentals of vehicle dynamics.* Warrendale, PA: Society of Automotive Engineers.

Hildreth, E. C., Beusmans, J. M. H., Boer, E. R., & Royden, C. S. (2000). From vision to action: Experiments and models of steering control during driving. *Journal of Experimental Psychology: Human Perception and Performance, 26,* 1106–1132.

Kirk, D. E. (1970). *Optimal control theory: An introduction.* Englewood Cliffs, NJ: Prentice-Hall.

Land, M. F., & Lee, D. N. (1994). Where we look when we steer. *Nature, 369,* 742–744.

Lee, D. N., & Lishman, R. (1977). Visual control of locomotion. *Scandinavian Journal of Psychology, 18,* 224–230.

McRuer, D. T., Allen, R. W., Weir, D. H., & Klein, R. H. (1977). New results in driver steering control models. *Human Factors, 19,* 381–397.

McRuer, D. T., Weir, D. H., Jex, H. R., Magdaleno, R. E., & Allen, R. W. (1975). Measurement of driver–vehicle multiloop response properties with a single disturbance input. *IEEE Transactions on Systems, Man, and Cybernetics, SMC-5,* 490–497.

Miller, R. A. (1976). On the finite preview problem in manual control. *International Journal of Systems Science, 7,* 667–672.

Tijerina, L., & Hetrick, S. (1997). Analytical evaluation of warning onset rules for lane change crash avoidance systems. In *Proceedings of the Human Factors and Ergonomics Society 41st Annual Meeting* (pp. 949–953). Albuquerque, NM: Human Factors in Ergonomics Society.

Weir, D. H., Shortwell, C. P., & Johnson, W. A. (1966). *Dynamics of the automobile related to driver control* (Systems Technology Tech. Rep. No. 157-1). (Contract CPR-11-2770, Bureau of Public Roads, Department of Commerce, Washington, DC.) Hawthorne, CA.

Wierwille, W. W., Gagne, G. A., & Knight, J. R. (1967). An experimental study of human operator models and closed-loop analysis methods for high-speed automobile driving. *IEEE Transactions on Human Factors in Electronics, HFE-8,* 187–201.

Wohl, J. G. (1961). Man–machine steering dynamics. *Human Factors, 3,* 222–228.

Continuous Tracking: Optimal Control

Classical control system design is generally a trial-and-error process in which various methods of analysis are used iteratively to determine the design parameters of an "acceptable" system. Acceptable performance is generally defined in terms of time and frequency domain criteria such as rise time, settling time, peak overshoot, gain and phase margin, and bandwidth. Radically different performance criteria must be satisfied, however, by the complex, multiple-input, multiple-output systems required to meet the demands of modern technology. For example, the design of a spacecraft attitude control system that minimizes fuel expenditure is not amenable to solution by classical methods. A new and direct approach to the synthesis of these complex systems, called optimal control theory, has been made feasible by the development of the digital computer.

—Kirk (1970, p. 3)

When examining performance in tracking experiments such as those described in previous chapters, it would be reasonable to ask about how good humans are as trackers. Of course, this is a relative question, how good are they relative to what norms? Modern control theory provides the tools (e.g., linear algebra and variational calculus) for developing normative models of "optimal control" that provide an important basis against which to measure human tracking performance. In addition, modern control analyses associated with the design and implementation of optimal controllers may provide insights into the information processes involved in human tracking (e.g., Pew & Baron, 1978; Sheridan & Ferrell, 1974).

THE OPTIMAL CONTROL PROBLEM

There are three sets of constraints that must be explicitly addressed when approaching an optimal control problem: the dynamic constraints of the process being controlled (e.g., the vehicle dynamics), the physical constraints (e.g., the initial conditions and the range limits of control — maximum output), and the value constraints or performance criteria (e.g., fuel consumption or the measure of error in a tracking study).

The dynamic constraints of the process, together with the physical constraints, bound the set of possible paths through a state space. For example, the problem might be to fly from New York to Amsterdam. The *physical constraints* include the initial (New York) and final (Amsterdam) conditions, the control limits of the aircraft (e.g., maximum thrust), or any other external restrictions that bound performance (e.g., FAA flight restrictions in terms of minimum and maximum altitudes). The *dynamic constraints* could be expressed as the equations of motion of the plane (e.g., the plane is an inertial system that cannot instantaneously change altitude or instantaneously stop without slowing down). Together, the physical and dynamic constraints bound the set of possible paths between New York and Amsterdam. Although there may be an infinite number of paths between New York and Amsterdam, it is a bounded set (i.e., there are many paths that are not possible).

The *performance criteria* provide a value system for ranking the possible paths — for identifying an optimal, or best, path. It addresses the question implied earlier: "good relative to what?" Generally, *optimal*, or *best*, is defined in terms of minimizing or maximizing some measure or combination of measures. Typical criteria include time (Which path would reach Amsterdam in the least time?), distance (What is the shortest path between New York and Amsterdam?), or resource consumption (What path requires the least amount of fuel?).

Modern control theory provides the tools to express the control problem mathematically and in some cases to analytically solve for the "optimal" control solution. For example, it provides analytical tools that can be used to determine the control inputs (or control law) that would take the plane from New York to Amsterdam in the least time, using the least fuel, or traveling the minimum distance.

LINEAR ALGEBRA

Modern control theory heavily depends on analytic techniques involving the manipulation of matrices. Thus, before describing the optimal control model, it might be useful to review a bit of matrix algebra. The most important operations are multiplication and addition of matrices.

To add two matrices, they must be of the same dimension. The sum of two matrices is the sum of the corresponding elements within each matrix:

$$\begin{vmatrix} a & b \\ c & d \end{vmatrix} + \begin{vmatrix} e & f \\ g & h \end{vmatrix} = \begin{vmatrix} a+e & b+f \\ c+g & d+h \end{vmatrix} \tag{1}$$

To multiply two matrices, the number of columns in the first matrix must equal the number of rows in the second matrix (note that unlike scalar multiplication, order matters). The product of the multiplication of two matrices will be a matrix with the same number of rows as the first matrix and the same number of columns as the second matrix. Each element in the new matrix will be the result of multiplying an element from the row in the first matrix with the corresponding element in the column of the second matrix and then summing the products:

$$\begin{vmatrix} a & b \\ c & d \end{vmatrix}\begin{vmatrix} x_1 \\ x_2 \end{vmatrix} = \begin{vmatrix} ax_1 + bx_2 \\ cx_1 + dx_2 \end{vmatrix} \tag{2}$$

$$\begin{vmatrix} a & b \\ c & d \end{vmatrix}\begin{vmatrix} e & f \\ g & h \end{vmatrix} = \begin{vmatrix} ae + bg & af + bh \\ ce + dg & cf + dh \end{vmatrix} \tag{3}$$

In Equation 2, a 2 × 2 matrix is multiplied by a 2 × 1 matrix to yield a 2 × 1 matrix. This example, in which a vector (single column matrix) is multiplied by a square matrix to produce a second vector, is a typical operation for modeling dynamic systems. The elements in the first vector will be used to represent an "input," or a "current state," and the elements in the second vector will be used to represent an "output," or "future state." The square matrix can be thought of as a map between these two vectors (between input and output or between present and future states). The elements of the square matrix can be considered "weights" that map (or transform) the inputs to the outputs. Thus, in Equation 2, the product (output vector) is a new state vector in which each element is a weighted sum of the elements from the first (input) vector. The elements in the square matrix are the weights.

When representing dynamic systems in matrix equations, the dimension of the state vector will be directly related to the order of the control system. A system that is second order will have two rows in the state vector. In other words, there will be two state variables (typically position and velocity). A system that is third order will have three rows in the state vector (typically representing position, velocity, and acceleration). As is discussed in the next section, the second-order differential equation used to represent a second-order system can be written as a system of first-order differential equations. The matrix equation provides an alternative way for representing this system of first-order relations.

Another matrix operation used in this chapter is the *transpose*. The transpose of a matrix is a new matrix in which the rows and columns are interchanged:

$$\begin{vmatrix} a \\ b \end{vmatrix}^T = \begin{vmatrix} a & b \end{vmatrix} \tag{4}$$

$$\begin{vmatrix} a & b \\ c & d \end{vmatrix}^T = \begin{vmatrix} a & c \\ b & d \end{vmatrix} \tag{5}$$

A final property of matrices to consider is the inverse, A^{-1}, where

$$A\ A^{-1} = I$$
$$A^{-1}\ A = I$$

and I is an identity matrix (1's on the diagonal and 0's in all other positions). Note that a matrix must be a square matrix ($n \times n$) in order to have an inverse. However, not all square matrices have inverses. The examples in this chapter will not require the inversion of a matrix. To learn about the conditions a matrix must meet to have an inverse and to learn about the operations involved in computing an inverse consult a general text on linear algebra or linear systems (e.g., Luenberger, 1979; Shields, 1980; Wiberg, 1971).

An Analytic Approach to Optimality

A key to the mathematics of modern control theory is the representation of the dynamics of a process in terms of *state variables*. In chapter 7, state space diagrams were used when discussing discrete strategies for arm positioning. The dimensions of this state space were the position and velocity of the arm, which was being modeled as a second-order (e.g., inertial) system.

In general, the state of a system is specified by a minimum set of variables that if known for any particular time together with the input from that time forward would allow specification of the behavior of the system. Inertial (second-order) systems are typically described in terms of two state variables (position and velocity) as reflected in the state space diagram used earlier. Thus, if the position and velocity of a simple inertial system (e.g., an arm) at a specific time are known, and the forces affecting that arm from that point in time forward (i.e., the inputs) are known, then the path of the arm (i.e., all of its future states) can be fully specified. To say it another way, the future behavior of the system is fully determined by its current state together with all future input. In a nontrivial sense, the state summarizes the contribution of the "past" to future behavior. From the perspective of calculus, the state reflects the "initial conditions" that must be specified in order to find a particular solution to a differential equation. From the perspective of classical control, the state reflects the order of control. A zero-order system has no state variables; the output is fully determined by the input. A first-order system has one state variable; to determine the output, it is necessary to know the initial position in addition to the input. Finally, as already noted, a second-order system has two state variables; to determine the output, the initial position and velocity must be specified in addition to future input. This logic can be extended to higher order systems. For a vehicle, the state variables of motion will reflect both the translational and rotational axes of motion (degrees of freedom) and the control order for each axis. So, the state variables for specifying the motion of an aircraft would include at least 12 state variables—the position and velocity for each of the six axes of motion (three dimensions of translation plus pitch, roll, and yaw).

In a very real sense, the choice of state variables is a hypothesis about the meaningful dimensions of a problem. They specify what differences make a difference. For some systems, such as simple springs, physicists and engineers have a good sense about the critical dimensions. However, as systems get more complex (e.g., high performance aircraft, nuclear power plants), there is less confidence that it is possible to identify all the relevant dimensions. Thus, the choice of state variables is a guess (usually a very educated guess) about what dimensions are important. Again, remember that a system is a description of a natural phenomenon. The choice of state variables is a critical component of that description. It is important to always remember that the state variables are properties of the description, not absolute dimensions of a natural phenomenon. Thus, the choice of state variables is influenced as much by the goals of the description as it is by the phenomenon being described.

Modern control theory uses the analytic power of linear algebra to draw inferences about properties of dynamic systems. Models are constructed in the form of matrix equations and then these models can be manipulated as mathematical objects. The properties of the mathematical objects can then be used to make inferences about the dynamic systems they represent. Two important properties of dynamic systems are *controllability* and *observability*. A system is *controllable* if every state in the state space can be reached from any other state in the space with finite duration control inputs. This is sometimes called *complete reachability*. If there is a state that cannot be reached from another state (i.e., you cannot get there from here), then the system is not completely controllable. A system is *observable* if the initial state of the system can be determined by observing its outputs over a finite time interval. In other words, it is possible to use the outputs to trace back to the initial state of the system. If the initial state of the system cannot be determined from observation of the outputs (i.e., the behavior) with known control inputs, then the system is not completely observable. The controllability and observability of a dynamic system can be inferred from properties of the matrices used to represent the system (see Luenberger, 1979, for a more complete discussion of this issue). A controllability matrix will show whether the controls are completely "connected" to the states, and an observability matrix will show whether the outputs are unambiguously "linked" to the states of the system. Although the mathematics of modern control theory provide analytic tests of observability and controllability, Luenberger (1979) observed that "it is rare that in specific practical applications controllability or observability represent cloudy issues that must be laboriously resolved. Usually, the context, or one's intuitive knowledge of the system, makes it clear whether the controllability and observability conditions are satisfied" (p. 289).

The Controlled Process. The dynamics of a system can be mathematically expressed as a system of first-order differential equations (i.e., state equations). These equations specify the rate of change for each state variable as a function of the other state variables. Figure 17.1 shows a second-order system. This system has two state variables: the position (x_1) and the velocity (x_2). The rate of change of position (x_1) is equal to the velocity (x_2):

$$\dot{x}_1(t) = x_2(t) \tag{6}$$

The rate of change of velocity (x_2) is determined by the difference between the control input (u) and the negative feedback from position ($-x_1$) and velocity ($-x_2$):

$$\dot{x}_2(t) = -x_1(t) - x_2(t) + u(t) \tag{7}$$

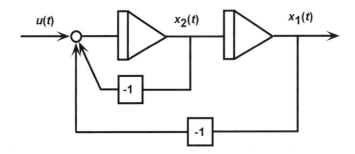

FIG. 17.1.   A simple second-order system. The x's represent the state variables of this system (x_1 = position and x_2 = velocity).

Thus, the dynamics for the system can be expressed as a system of two state equations:

$$\dot{x}_1(t) = x_2(t) \tag{8}$$

$$\dot{x}_2(t) = -x_1(t) - x_2(t) + u(t) \tag{9}$$

This system of equations can be expressed in matrix form:

$$\begin{vmatrix} \dot{x}_1(t) \\ \dot{x}_2(t) \end{vmatrix} = \begin{vmatrix} 0 & 1 \\ -1 & -1 \end{vmatrix} \begin{vmatrix} x_1(t) \\ x_2(t) \end{vmatrix} + \begin{vmatrix} 0 \\ 1 \end{vmatrix} u(t) \tag{10}$$

where $\begin{vmatrix} x_1(t) \\ x_2(t) \end{vmatrix}$ is the state vector;

$\begin{vmatrix} 0 & 1 \\ -1 & -1 \end{vmatrix}$ is the system matrix;

and $\begin{vmatrix} 0 \\ 1 \end{vmatrix}$ is the control or input matrix.

Conventionally, a capital A is used to represent the *system* matrix and a capital B is used to represent the *input* matrix for the general case. Thus, the general equation for the state dynamics of a system is:

$$\bar{x}(t) = A\bar{x}(t) + B\bar{u}(t) \tag{11}$$

(Bars indicate that these variables represent vectors. An alternative convention is to use bold type fonts to differentiate vectors from scalars.)

A second set of equations is included to express the system output as a function of system state and control inputs. For the system in Fig. 17.1, the equations would be:

$$y(t) = \begin{vmatrix} 1 & 0 \end{vmatrix} \begin{vmatrix} x_1(t) \\ x_2(t) \end{vmatrix} + |0| u(t) \tag{12}$$

or simply

$$y(t) = x_1(t)$$

For the general case, the matrix multiplying the state vector is typically represented as a capital C, and the matrix multiplying the control (u) is represented as a capital D. Thus, the dynamics of a process are modeled using two systems of equations:

$$\dot{\bar{x}}(t) = A\bar{x}(t) + B\bar{u}(t) \tag{13}$$

$$\bar{y}(t) = C\bar{x}(t) + D\bar{u}(t) \tag{14}$$

Figure 17.2 shows a general flow diagram that illustrates the role of the various matrices in relating the states and controls to the state derivatives and to the system output.

Note that in the example the input $u(t)$ and output $y(t)$ were scalars, however, the optimal control model can be used with multiple inputs and multiple outputs so that the input and output functions in the general model (Equations 13 & 14) are vectors. Also, controlability of the system represented by Equations 13 and 14 will depend on joint properties of the system matrix (A) and the input matrix (B). These two matrices determine the mapping from one state to another as a function of the control input. Observability of the system represented by Equations 13 and 14 will depend on joint properties of the system matrix (A) and the output matrix (C). These two matrices determine the mapping from the states to the output or conversely the mapping from the output back to the initial states. Finally, for most situations (particularly for tracking tasks), there will typically be an additional disturbance input $w(t)$. Thus, the complete general linear systems dynamic model can be represented as follows:

$$\dot{\bar{x}}(t) = A\bar{x}(t) + B\bar{u}(t) + E\bar{w}(t) \tag{15}$$

$$\bar{y}(t) = C\bar{x}(t) + D\bar{u}(t) \tag{16}$$

Physical Constraints. The physical constraints are typically expressed as boundary conditions on either the state variables, the control inputs, the outputs, or

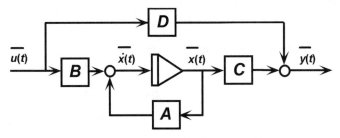

FIG. 17.2. A general flow diagram illustrating the relations between the state vector (x) and its derivative and the control (u) and the output of the system. The letters in the boxes represent matrices and the time variables are vectors.

time. In a discrete movement experiment, the boundary conditions on position would be the starting position and the potential ending positions (target region). For example, if the participant was required to move from a zero position to a 3 pixel target that is 11 pixels away, as is typical for the positioning task discussed in earlier chapters, then the constraints would be:

$$x_1(t_0) = 0 \tag{17}$$

$$10 \le x_1(t_f) \le 12 \tag{18}$$

The range of output from the control stick would be another significant physical constraint. If the control system were second order, this would set limits on the maximum acceleration (e.g., maximum control input of ±10 pixels per second squared):

$$-10 \le u(t) \le 10 \tag{19}$$

Performance Measures. The performance criteria are typically expressed as a function to be minimized (a capital J is conventionally used to represent the performance measure). For example, in Fitts' experiments, the subjects were asked to capture the target in minimum time. Thus, the function to be minimized in this discrete positioning task would be:

$$J = t_f - t_0 \tag{20}$$

In continuous tracking tasks, performance is typically scored in terms of integrated or average mean squared error. Thus, the cost funtional might be written in terms of the tracking error:

$$J = \frac{1}{t_f - t_0} \int_{t_0}^{t_f} e_1^2(t) dt \tag{21}$$

or assuming that $x_1 = 0$ is the target in a compensatory tracking task:

$$J = \frac{1}{t_f - t_0} \int_{t_0}^{t_f} x_1^2(t) dt \tag{22}$$

A general form for the performance measure is:

$$J = h(\bar{x}(t_f), t_f) + \int_{t_0}^{t_f} g(\bar{x}(t), \bar{u}(t), t) dt \tag{23}$$

The first term in this function reflects the value (or cost) of the final state and final time. The second, integral term represents a cumulative function of the states, the control actions, and time.

A very important performance function is the quadratic cost funtional:

$$J = \frac{1}{t_f - t_0} \int_{t_0}^{t_f} (\bar{x}^T Q \bar{x} + \bar{u}^T R \bar{u}) dt \tag{24}$$

Remember that the "T" indicates the transpose. The Q matrix represents a weighting on the squared values of the state variables (x), and the R matrix represents a weighting on the squared values of control actions (u). Using a cost functional of this form, it is known that J will be minimized in a steady state tracking task, i.e., when t_f is "very long," when the following "contol law" is utilized:

$$u(t) = -R^{-1} B^T K \bar{x}(t) \tag{25}$$

The matrix K is the solution of the Riccati equation (e.g., Kirk, 1970, pp. 90–93):

$$KA + A^T K + Q - KBR^{-1} B^T K = 0 \tag{26}$$

For example, consider the control of a second-order system (Fig. 17.3). The state equation for this system and a quadratic cost functional in which equal costs are associated with position (mean squared error) and control are:

$$\bar{x}(t) = \begin{vmatrix} 0 & 1 \\ 0 & 0 \end{vmatrix} \bar{x}(t) + \begin{vmatrix} 0 \\ 1 \end{vmatrix} u(t) \tag{27}$$

$$J = \frac{1}{t_f - t_0} \int_{t_0}^{t_f} \left(\bar{x}^T \begin{vmatrix} 1 & 0 \\ 0 & 0 \end{vmatrix} \bar{x} + \bar{u}^T |1| u \right) dt \tag{28}$$

or

$$J = \frac{1}{t_f - t_0} \int_{t_0}^{t_f} \left(x_1^2 + u^2 \right) dt \tag{29}$$

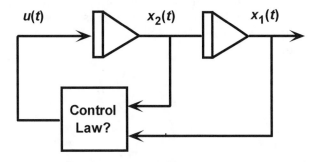

FIG. 17.3.   What would be the optimal control law for a simple second-order process?

The Riccati Equation would be:

$$K\begin{vmatrix} 0 & 1 \\ 0 & 0 \end{vmatrix} + \begin{vmatrix} 0 & 0 \\ 1 & 0 \end{vmatrix}K + \begin{vmatrix} 1 & 0 \\ 0 & 0 \end{vmatrix} - K\begin{vmatrix} 0 \\ 1 \end{vmatrix}1|0 \quad 1|K = 0 \tag{30}$$

$$K = \begin{vmatrix} \sqrt{2} & 1 \\ 1 & \sqrt{2} \end{vmatrix} \tag{31}$$

To solve this equation, you must first realize that K will be a matrix:

$$K = \begin{vmatrix} k_1 & k_2 \\ k_3 & k_4 \end{vmatrix}$$

If this matrix is substituted into the Riccati equation and matrix operations are carried out, then four equations (one for each position in the matrix) in four unknowns will result. It turns out that the solution always will be a symmetric matrix (i.e., $k_2 = k_3$).

Once the values for the K matrix have been found, then the optimal control can be determined using Equation 25:

$$u(t) = -1[0 \quad 1]\begin{bmatrix} \sqrt{2} & 1 \\ 1 & \sqrt{2} \end{bmatrix}\bar{x}(t) \tag{32}$$

or

$$u(t) = [-1 \quad -\sqrt{2}]\bar{x}(t) = -x_1(t) - \sqrt{2}x_2(t) \tag{33}$$

This optimal control law is a set of gains that maps the states of the system into the control action. In this case, the values are the weights given to the instantaneous position and velocity of the output in determining the current control input. In other words, the control $u(t)$ will be proportional to a weighted sum of the extent and rate of change of the position (error relative to zero). This is a control strategy that will minimize a weighted sum of mean-squared error and mean-squared control for a linear regulator problem of bringing the output toward its desired value of zero.

Suppose the cost associated with position (error) is increased to four and the weight on control is left at one. How would this affect the control law? The resulting control law would be:

$$u(t) = -1[0 \quad 1]\begin{bmatrix} 4 & 2 \\ 2 & 2 \end{bmatrix}\bar{x}(t) \tag{34}$$

or

$$u(t) = [-2 \quad -2]\bar{x}(t) = -2x_1(t) - 2x_2(t) \tag{35}$$

Note that as the cost on squared position error is increased, the control gains for both states increase. But, the absolute magnitude of the control weight on position is increased more than the weight on velocity so that now equal weight is given to position and velocity. To check whether this is a general trend, try increasing the weight on position to nine and see how this affects the optimal control law. Thus, the values for Q and R in the cost functional will determine how aggressive the controller will behave.

To summarize, the analytic approach to optimal control depends on a state-based representation of a process. Linear algebra and variational calculus can be used to determine the path through state space (or the control law) that minimizes (or maximizes) the performance criterion.

THE OPTIMAL CONTROL MODEL

With training, human trackers perform some tasks in a manner that is consistent with the assumptions of optimal control. This observation has prompted researchers to model the human as an optimal controller (e.g., Kleinman, Baron, & Levison, 1971). Figures 17.4 and 17.5 show two views of the optimal control model (OCM) for human tracking. The components are segregated to explicitly separate the human information-processing system from the situation or task constraints. There are two fundamental components of this system: the *perceptual component* and the *decision or control component*.

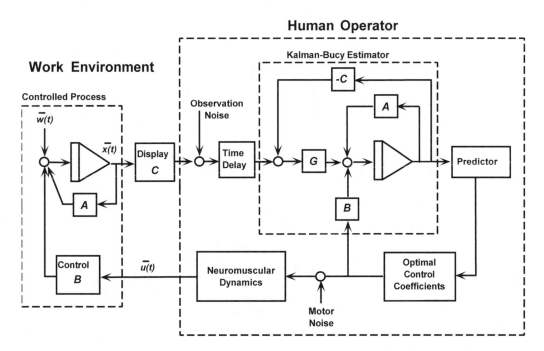

FIG. 17.4.   The Baron-Kleinman-Levison optimal control model. From "*Man-Machine Systems: Information, Control, and Decision Models of Human Performance*," (p. 254) by T. B. Sheridan and W. R. Ferrell, 1974, Cambridge, MA: MIT Press. Adapted by permission.

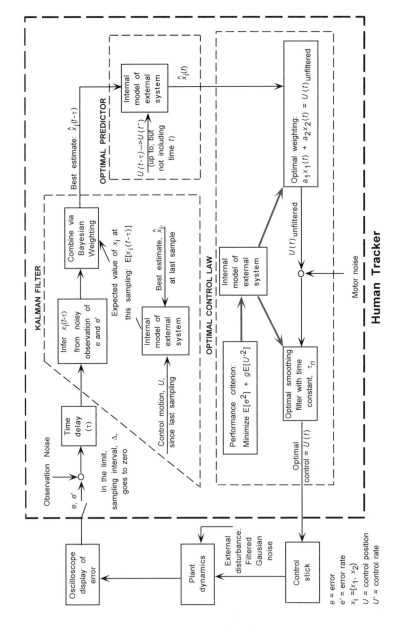

FIG. 17.5. The Baron-Kleinman-Levison optimal control model (after Jagacinski, 1973).

The perceptual component is designed to solve the observation problem, that is, to estimate the state of the process based on the behavior of the system. The simple examples that are used to illustrate optimal control in this chapter tend to trivialize the observer problem. However, in natural control problems, there can be many state variables, and the mapping of these state variables to output behavior (e.g., displays) may not be explicit. Thus, when faced with controlling a new system (or with designing an automated control system), identifying and estimating the state variables can be the major difficulty. In fact, one of the developers of the optimal control model is reported to have said that he regrets calling the model the optimal "control" model. He said that it should be called the optimal "estimation" model, because much of the variance from a human performance perspective is associated with parameters of the estimation process.

The perceptual component of the optimal control model includes observation noise, a time delay, a Kalman filter, and a predictor. The model assumes that humans must estimate the current state of the system based on noisy delayed observations of the output. The parameters for the perceptual component are the signal-to-noise ratios for each of the state variables and the processing time delay. The time delay is placed within the perceptual component, but its value reflects the total information-processing delay.

The Kalman filter is designed to reduce the uncertainty resulting from observation and motor noise. The Kalman filter uses the noisy observations of the output to estimate the values of the system state (x_i in Fig. 17.5). In Fig. 17.5, two state variables are assumed (this reflects a second-order control system). Think of the Kalman filter as sampling output every Δ seconds in order to update its estimate of state variables. The Kalman filter typically used to model human performance acts continuously, rather than using discrete observations of the display. However, for the purpose of gaining an intuitive understanding of the filter, it is easier to think in terms of discrete observations. The continuous filtering process can then be considered as the limiting case in which the sampling interval, Δ, shrinks toward zero.

The Kalman filter takes its best estimate from the previous sampling instant and also takes the value of the control motion over the interval between samples, and using an internal model of the external process being controlled, it generates an expected value of the state variables for the next sampling instant. Figure 17.4 illustrates the internal model as a veridical model that includes A, B, and C matrices corresponding to the controlled process. Typically, the expected value based on the internal model will not equal the estimate from the noisy observation at that next instant. The lack of agreement will reflect the observation noise on one hand, and the limits of the estimation process on the other hand. The Kalman filter then acts like a Bayesian decision-maker with two pieces of evidence. It weights these two inputs (a noisy observation and an expected value based on an internal model) to reflect their relative uncertainty. For example, if the observation noise was high, then relatively more weight would be given to the expected value and less weight would be given to the observation. However, if the observation noise was low, then greater weight would be given to the current observation relative to the expected value (Fig. 17.5).

An alternative way of thinking about the estimation process is illustrated in Fig. 17.4. In the Kalman–Bucy Estimator, the expectation of the model is compared with the observation at the summing junction. The weights in the G matrix represent the forward-loop gain in a closed-loop system around this "error" signal. The values of G are chosen so that the mean-squared error of the state estimate is minimized. If the forward-loop gain is low, then the system will be slow to revise its assessment based on new observations. If the gain is higher, then the estimator will quickly revise its estimate based on new observations. Thus, intuitions about the behavior of a first-order lag to a step input as a function of the gain can be applied to understanding the estimation process (see Fig. 4.4). An estimator with a high gain will be responsive to new information. An estimator with a low gain will be sluggish. In making an estimation based on an internal model of the process, the Kalman filter is essentially integrating over the past history of the process and combining this "history" with the current noisy observation to produce a more accurate estimate of the current state of the process. The gains in the G matrix reflect the relative confidence that this process gives to each new observation relative to an expectation derived from the history of the process. The estimation process is discussed in more detail in chapter 18 (also see Analytic Sciences Corporation, 1974, for a more complete treatment of the observer problem). The main point to understand here is that it may be possible for the values of the G matrix to be set so that the error in estimating the state from the noisy observations is minimized. To the extent that this is possible, the Kalman filter behaves as an "ideal observer (perceiver)."

The predictor element of the optimal control model compensates for the effects of the processing time delay (return to Fig. 17.5). The predictor takes the best estimate provided by the Kalman filter and then extrapolates this estimate over the interval of the time delay. This extrapolation is based on a knowledge of the system dynamics (i.e., an internal model—implemented in the OCM using the state transition matrix for the controlled process) and a knowledge of the control history over the estimation interval (up to, but not including, the instantaneously current action).

Note that both the Kalman filter and the predictor include internal models of the controlled process. Together the estimator and predictor model "the (cognitive) process whereby the operator constructs a set of expectancies concerning the state of the system on the basis of his understanding of the system and his incomplete knowledge of the moment-to-moment state as accessible to him from limited and noisy observations" (Baron & Corker, 1989, p. 205). An important assumption of the optimal control model (as typically applied) is that the operator has a veridical internal model of the process being controlled. This assumption seems to generally be justified in situations where experts are tracking familiar processes.

The decision or control component is implemented as an analytically determined optimal control law. For simple tracking tasks, Kleinman et al. (1971) used the following cost functional:

$$J(u) = \int [e^2(t) + g\dot{u}^2(t)]dt \tag{36}$$

The error term reflects position relative to the track. Note that control velocity is used instead of control. The constant g is a parameter that is typically determined based on empirical matches with the data. It reflects the cost associated with fast movements. With larger values of g, the system would tend to resist fast corrections. For many laboratory tracking studies, the g parameter is thought to reflect the neuromuscular dynamics of the arm. This corresponds to the neuromuscular lag (T_N) and reflects the natural inertial dynamics of the arm that filters out high frequency responses. The g parameter might also reflect other factors (e.g., the desire for smooth tracking or for a comfortable ride). For example, a driver who is very aggressive in correcting errors with little concern for the comfort of passengers would be placing a low weight on the cost of control velocity (i.e., low g value). A driver who was more concerned about his passenger's comfort might be less aggressive in correcting errors (high cost on control velocity or high g) in order to give his passengers a smooth ride. Thus, the optimal control reflects both the performance measure (mean squared error, control velocity) and the dynamic constraints of the arm. The control component also includes a source of motor noise that reflects the signal-to-noise ratio for the control signal. In other words, this reflects the variability inherent in the motor control system.

The derivation of the optimal control weights is accomplished analytically using the Riccati equation. As illustrated earlier in the chapter, the Riccati equation includes the system matrix. Thus, the derivation of the optimal weights assumes that the operator has an internal model of the process. Ironically, whereas the computation of the optimal weights may be an elegant feature of the mathematics of modern control theory, it can be a mundane aspect of formally implementing the optimal control model, particularly when applied to single axis tracking tasks. In part, this may be an illusion created by the partitioning of the problem into separate information components (e.g., a perceiver and a controller). Note that the system (A) matrix is a critical component of both the perceiver and controller. In identifying the "states" of the process (dimensionality of the A matrix) and the constraints among the states (the structure of the A matrix), consideration must be given both to the dynamic constraints (order of control and degrees of freedom) and the information (feedback) constraints of the control task. In texts on optimal control the A matrix is often a "given," but for biological systems discovering the dimensionality of the control problem (solving the degrees of freedom problem on the action side and the information problem on the perception side) may be the heart of the problem. That is, the search to discover the appropriate dimensionality of the control problem may account for the biggest step up the learning curve. Once the dimensionality of the control task is discovered, tuning the Kalman filter and the control matrix gains to optimal levels may only be reflected in the asymptotic regions of the learning curve. Thus, the parameters of the optimal control problem may reflect the search space of the modeler, but these parameters may not reflect the most interesting aspects of the "search" space for a biological system learning to control their actions (e.g., learning to walk). Discovering the system structure as reflected in the A and B matrices may be the difficult hurdle for a biological system learning a new skill.

Thus, the parameters of the optimal control model are the time delay, control velocity weight (g), signal-to-noise ratio for observation, and signal-to-noise ratio for control. The optimal control model is generally implemented as an analytic function of its parameters, which are iteratively adjusted to get the best match to empirical data. Although this is a large number of free parameters, converging operations can be used and the parameters can be set to reflect previous research on the component models. For example, the observation signal-to-noise ratios should be consistent with psychophysical studies of the perceptual system.

Sheridan and Ferrell (1974) reported some typical values for the parameters in the optimal control model. The time delay is typically about 150 ms. The neuromuscular lag has a time constant of about 100 ms. The noise on observation is typically about 3% and the noise on control is about 1% of their respective signal variances.

The adaptations of the human transfer function to changes in the vehicle or plant dynamics that motivated the crossover model (see chapter 14) are consistent with the optimal control model. These adaptations reflect the differing control laws required to minimize the cost functional. Thus, Fig. 14.6 (in the chapter describing the crossover model) provides a means for visualizing the analytical solutions to the different optimal control problems posed by the changing plant dynamics. Adjustments of the lead and lag constants within the strategic portion of the crossover model are related to changes of the control law (gains in the control matrix derived to minimize the cost functional) within the optimal control model. Increasing lead in a classical control solution will show up in an optimal control solution as relatively more weight on the velocity variable.

Both the crossover model and the optimal control model provide good matches to the frequency response of the well-trained human operator in laboratory tracking tasks. However, an advantage of the optimal control model is that the simulation is more flexible for testing hypotheses and generating predictions targeted to specific components of the information-processing system. For example, the behavioral effects of changes in observation noise can be modeled directly. Also, the Kalman filter generally assumes that the operator has a valid internal model of the system dynamics. However, suppose the internal model is not valid. The optimal control model can be implemented so that the internal model of the estimator is inaccurate (i.e., the estimator has an inaccurate state description of the process). Thus, experiments with a simulation implemented as an optimal control model have the potential for providing insights to the performance of novices and to the development of tracking skill (Baron & Berliner, 1977). Finally, the optimal control model is designed to handle multiple inputs and outputs. Thus, the optimal control model has greater potential as a tool for modeling more complex control environments.

CONCLUSION

Modern control theory adds a new set of analytic tools to the arsenal. This chapter provides only a small glimpse of those tools. A good source for a more comprehensive introduction to modern control theory is Luenberger (1979). This new set of tools should not be viewed as a replacement of classical analytic methods. Rather, the new tools should be considered as an addition that complements the older tools of classical control theory to broaden the perspective on human performance.

The optimal control model provides a unique perspective on the modeling process because it supports both a wholistic view that supports appreciation of the global situational constraints that bound performance and it also supports a decomposition or reduction of the system into simpler information-processing components. The importance of the global situational constraints is reflected in the fact that essentially every cognitive component of the process (e.g., the Kalman filter, the predictor, the control law) incorporates the system matrices (principally the A matrix) as an internal constraint. Thus, each component is "situated." In fact, these situated constraints can be thought of as the threads that link the components into a coherent process. Decomposition is supported by localizing the modeling parameters within distinct components (e.g., Kalman filter, predictor, control matrix, neuromuscular lag, cost functional). These components match well with other decompositional or stage models of information processing. Thus, the optimal control model can support both a top-down, situated view of cognition and a bottom-up, information-processing perspective for modeling human performance. Perhaps the biggest danger of the optimal control model is the temptation to equate the modeler's search space (e.g., Kalman filter gains, control gains) with the human's learning process. It is likely that much of the human's learning process involves discovering the dimensionality of the control problem. This process is generally not reflected in the mathematics of optimal control, but in the analytic processes in which the modeler reasons from first principles (e.g., aerodynamics) to define the control system (e.g, specification of the state variables).

REFERENCES

Analytic Sciences Incorporation. (1974). *Applied optimal estimation* (A. Gelb, Ed.). Cambridge, MA: MIT Press.

Baron, S., & Berliner, J. E. (1977). The effects of deviate internal representations in the optimal model of the human operator. In *Proceedings of the Thirteenth Annual Conference on Manual Control* (pp. 17–26). Cambridge, MA: MIT.

Baron, S., & Corker, K. (1989). Engineering-based approaches to human performance modeling. In G. R. McMillan, D. Beevis, E. Salas, M. H. Strub, R. Sutton, & L. Van Breda (Eds.), *Applications of human performance models to system design* (pp. 203–217). New York: Plenum.

Jagacinski, R. J. (1973). *Describing multidimensional stimuli by the method of adjustment.* Unpublished doctoral dissertation, the University of Michigan, Ann Arbor, MI.

Kirk, D. E. (1970). *Optimal control theory: An introduction.* Englewood Cliffs, NJ: Prentice-Hall.

Kleinman, D. L., Baron, S., & Levison, W. H. (1971). A control theoretic approach to manned-vehicle systems analysis. *IEEE Transactions in Automatic Control, AC-16,* 824–832.

Luenberger, D. G. (1979). *Introduction to dynamic systems: Theory, models, and applications.* New York: Wiley.

Pew, R. W., & Baron, S. (1978). The components of an information processing theory of skilled performance based on an optimal control perspective. In G. E. Stelmach (Ed.), *Information processing in motor control and learning* (pp. 71–78). New York: Academic Press.

Sheridan, T. B., & Ferrell, W. R. (1974). *Man–machine systems: Information, control, and decision models of human performance.* Cambridge, MA: MIT Press.

Shields, P. C. (1980). *Elementary linear algebra.* New York: Worth.

Wiberg, D. M. (1971). *State space and linear systems.* New York: McGraw-Hill.

Estimating and Predicting
the State of a Dynamic System
With Lag-Like Calculations

> *An optimal estimator is a computational algorithm that processes measurements to deduce a minimum error estimate of the state of a system by utilizing: knowledge of system and measurement dynamics, assumed statistics of system noises and measurement errors, and initial condition information. Among the presumed advantages of this type of data processor are that it minimizes the estimation error in a well defined statistical sense and that it utilizes all measurement data plus prior knowledge about the system. The corresponding potential disadvantages are its sensitivity to erroneous a priori models and statistics, and the inherent computational burden.*
>
> — Analytic Sciences Corporation (1974, p. 2)

As seen in the chapter on optimal control, controlling or monitoring a dynamic system typically requires estimation of some parameters and variables describing the system's behavior. Such quantities might range from the mean of a noise source to the position, velocity, and acceleration of a vehicle. The process of estimation involves integrating information over time, and effective ways of performing this process have much in common with the behaviors involved in actually exerting control. This chapter explores some simple examples to perform such calculations.

ESTIMATING THE MEAN FROM A SERIES
OF MEASUREMENTS

A very general problem is to observe some process that continues over time and to attempt to predict the next observation to be encountered based on past observations. Common examples might be predicting tomorrow's weather, predicting next year's level of the stock market, or predicting the trajectory of an approaching car. These are all very complicated prediction problems. In order to understand some of the issues involved in the prediction process, first consider a much simpler example.

Suppose it is possible to observe successive samples, y_i, from a Gaussian noise source. These observations are normally distributed with mean equal to m, variance var_y, and zero correlation between successive samples. The task is to predict each sample as accurately as possible. In order to minimize the mean squared difference between each of the predictions, y_i', and the actual observation, y_i, let the prediction be equal to the best estimate of the mean of the random process. Given that successive observations are uncorrelated, a high observation may be followed by either a high or low next observation. Therefore, it is best to simply predict the mean of the process.

The prediction task has thus become a task of determining the best estimate of the mean. This task is a basic problem in elementary statistics. Given a random sample of n observations from a Gaussian distribution about which there is no prior information, an unbiased estimate of the mean of the process can be obtained by calculating the sample mean (i.e., adding up the n observations and dividing by n). The mean of the n observations provides an estimate of m, and the error variance associated with this estimate is var_y/n. Namely, the variability associated with the estimate of the mean is smaller than the variability associated with each individual observation. In fact, the estimate of the mean can be made arbitrarily accurate by taking a very large sample size (i.e., choosing a large n). Therefore, the longer the Gaussian process is observed, the better the estimate of the mean should become and hence the better the prediction of the next observation.

Next consider how to go about improving the estimate of the mean of the process after each successive observation. You could remember all n individual observations you had experienced, add these up, and then divide by n. However, as n gets large, this procedure becomes cumbersome. An equivalent way of performing the same calculation without having to remember all the individual observations is to use the previous estimate of the mean prior to the latest observation. Namely, the previous estimate of the mean based on $n - 1$ observations was the prediction of y_n, namely y_n'. y_n' can be multiplied by $n - 1$ to get the sum of all the previous $n - 1$ observations. Simply add the latest observation, y_n, to this sum and divide by n to get the updated estimate of the mean and the prediction of y_{n+1}.

$$y_{n+1}' = [(n - 1)y_n' + y_n]/n \tag{1}$$

In other words, y_{n+1}' is a weighted average of the previous prediction and the latest observation.

Note that if you only had your latest observation, y_n, that would also provide an estimate of the mean. However, the uncertainty associated with this estimate would be var_y, because it is only based on one observation. In contrast, the estimate of the mean y_n' is based on $n - 1$ previous observations and has a much smaller uncertainty, $var_y/(n - 1)$. Equation 1 averages these two sources of information about the true mean by weighting them in inverse proportion to their uncertainties. The estimate with the smaller uncertainty is weighted more heavily. The ratio of the two weights,

$(n - 1)/1$, is the inverse of the ratio of their uncertainties. It is also the ratio of the two sample sizes in this example where all the samples are drawn from the same distribution. y_n is based on a sample size of 1, and y_n' is based on a sample size of $n - 1$. Equation 1 minimizes the expected squared deviation between y_{n+1}' and the true value of the mean. This expected squared deviation is var_y/n, which is smaller than $var_y/(n - 1)$. The estimate of the mean has thus been made a little more accurate by incorporating information from another observation.

THINKING OF VARIANCES IN TERMS OF EFFECTIVE SAMPLE SIZES

In the previous example, y_n' was the result of $n - 1$ samples from the Gaussian distribution. However, suppose the same information was given without knowing the details of the sampling process. In other words, suppose that before measurement y_{n+1} is available, an expert reports that the best previous estimate of the mean has the same numerical value as y_n', and the variance of this estimate is $var_y/(n - 1)$. From a Bayesian perspective, calculate the estimate y_{n+1}' in exactly the same manner as before. Namely, prior knowledge of that degree of precision should be treated as though it were an "equivalent sample size" of $n - 1$ (Hays & Winkler, 1971). In other words, if the measurement y_n has variance var_y and the prior estimate y_n' has variance $var_{y_n'}$, then the equivalent sample size (*ss*) of the prior estimate is calculated as $var_y/var_{y_n'} = n - 1$. Then the combination of these two pieces of information can proceed as before. Namely,

$$\text{equivalent prior sample size} = ss_{\text{prior}} = var_y/var_{y_n'} = var_y/[var_y/(n - 1)] = n - 1$$

$$y_{n+1}' = [ss_{\text{prior}} y_n' + y_n]/[ss_{\text{prior}} + 1] \tag{2}$$

Equation 2 is more general than Equation 1. For this particular example, $ss_{\text{prior}} = n - 1$, and substituting this value into Equation 2 results in Equation 1.

Another way of rewriting Equations 1 is:

$$y_{n+1}' = (n/n)y_n' - (1/n)y_n' + (1/n)y_n = y_n' + (1/n)(y_n - y_n') \tag{3}$$

Another way of rewriting Equation 2 is:

$$y_{n+1}' = [ss_{\text{prior}} y_n' + y_n' - y_n' + y_n]/[ss_{\text{prior}} + 1] = y_n' + [1/(ss_{\text{prior}} + 1)](y_n - y_n') \tag{4}$$

Note that the term $y_n - y_n'$ corresponds to the difference between the last observation and the previous prediction (i.e., the error of the previous prediction). The new prediction is equal to the previous prediction updated by some fraction of the previous error. If the previous prediction was too low, then the next prediction will be higher. However, the adjustment will be tempered by how long you have been observing the process. The fraction is equal to one divided by the new effective sample size, $ss_{prior} + 1$. If ss_{prior} is large, then y_n' is already a reasonably good estimate of the mean, so the adjustment due to the previous prediction error will be small.

THE RECURSIVE ESTIMATION PROCEDURE RESEMBLES A LAG

Equations 3 and 4 look like discrete versions of a lag (chap. 4). Namely, if we draw a flow diagram of this recurring (or recursive) calculation (Fig. 18.1) and assume that the time between samples is fixed, it resembles a discrete implementation of a lag. It looks like a low pass filter with an input y_n, an output y_{n+1}', and n (or $ss_{prior} + 1$ more generally) is analogous to an increasing time constant (see Fig. 4.2). The box labeled "Sum" (Fig. 18.1) accumulates the successive weighted deviations between observation and prediction, $(y_n - y_n')/n$, to generate each new prediction. In other words, $(y_n - y_n')/n$ is added to the previous Sum, namely, y_n', to generate y_{n+1}'. The response of this system will become more sluggish as n increases, and the system becomes more sure of the value of the mean. This prediction scheme can be considered a very ele-

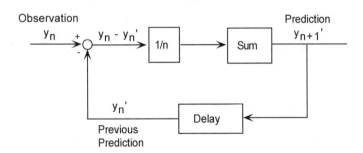

n	y_n	$[\, y_n' + (y_n - y_n')/n \,]$	$=$	Sum	$=$	y_{n+1}'
0	----	---- ----		0	$=$	y_1'
1	y_1	$0 + (y_1 - 0)/1$	$=$	y_1	$=$	y_2'
2	y_2	$y_1 + (y_2 - y_1)/2$	$=$	$y_1 + y_2/2 - y_1/2$	$= (y_1 + y_2)/2 =$	y_3'
3	y_3	$(y_1 + y_2)/2 + [y_3 - (y_1 + y_2)/2]/3 = y_1/2 + y_2/2 + y_3/3 - y_1/6 - y_2/6 = (y_1 + y_2 + y_3)/3 = y_4'$				

etc.

FIG. 18.1. Recursive estimation of the mean of a Gaussian process.

mentary form of a *Kalman filter*. It is a recursive estimation procedure that weights multiple sources of information in inverse proportion to their uncertainties.[1]

If the mean and variance of the Gaussian process remain constant, then the aforementioned scheme provides a very efficient estimate of that mean. However, if the mean were to suddenly change after a long period of observation, then it would take a long time for the system to change its estimate of the mean. This slowness results because the scheme is effectively averaging any new observations with all of its previous observations (e.g., Slotine & Li, 1991, chap. 8). The situation is analogous to the time constant of a continuous filter getting longer, so the system responds more sluggishly.

An alternate approach that would allow the system to detect changes more rapidly would be to avoid letting the multiplier, $1/n$, become too small (Fig. 18.1). Rather than changing the gain in the feedback loop according to a preset schedule of $1/($the number of observations), the gain could be set to a constant. One justification for using such a procedure is that it should gradually reduce the squared prediction error, $(y_n - y_n')^2$. Namely, if the derivative of the squared prediction error is calculated with respect to y_n', the result is:

$$d(y_n - y_n')^2/dy_n' = -2(y_n - y_n') \tag{5}$$

$-2(y_n - y_n')$ represents the "gradient," or rate of increase, of the squared prediction error as y_n' is increased. Adjusting y_n' in a direction that is opposite to this gradient should decrease the squared prediction error. One possible implementation of that strategy is to adjust the prediction by an amount proportional to $(y_n - y_n')$. This adjustment procedure can be accomplished with the adjustment scheme:

$$y_{n+1}' = y_n' + (1/b)(y_n - y_n') \tag{6}$$

Unlike the previous adjustment scheme, $(1/b)$ is a constant, and b is analogous to the time constant of a lag. If b is small (but greater than 1), then $(1/b)$ is a relatively large fraction, and the system will alter its estimate of the mean of the process relatively quickly if there is a sudden change in the mean of the Gaussian process being observed (e.g., Brown, 1963). This adjustment scheme can be considered to be a simple example of gradient descent (see chaps. 15 and 24).

This scheme is also referred to as exponential smoothing, although "geometric smoothing" would be a more appropriate term, because the weighting of past observations drops off as their contributions to y_n' are effectively multiplied by the fraction $(1 - 1/b)$ after each new observation (e.g., Brown, 1963). Previous observations have less impact than more recent observations. Because it effectively ignores observations in the distant past, the uncertainty associated with the estimate of the mean from this

---

[1]See, for example, Analytic Sciences Corporation (1974, ch. 4), Bozic (1994), Bryson and Ho (1969), du Plessis (1967), and Maybeck (1979, ch. 1) for similar discussions. From the perspective of adaptive control theory, Equation 3 can also be considered an example of "gain scheduling" in that the gain of the feedback loop is altered on a preset schedule, namely, $1, 1/2, 1/3, \ldots$ etc. See Astrom and Wittenmark (1989, ch. 9).

scheme will not be as small as with the previous scheme. However, this new scheme will respond more quickly to changes in the mean of the underlying Gaussian process (see Brown, 1963, for details).

This second scheme can also be considered a very simple example of an observer (Luenberger, 1979). Very roughly speaking, an observer is a process that updates the estimate of the state of the system on the basis of available observations, but without using time varying gains based on the uncertainties in the measurements and previous state estimates as in the Kalman filter.

ESTIMATING THE STATE OF A FIRST-ORDER LAG WITH AN ELEMENTARY KALMAN FILTER

In the previous example, the quantity being estimated was simply a mean. Suppose, instead, that the problem was to estimate the output of a continuous first-order lag, x, at times 0, T, $2T$, and so on, given some noisy observations, y, at those intervals (Fig. 18.2). The noise, v, is assumed to have zero mean and a Gaussian distribution. The additional complication in this example is that, between observations, the output of the lag will vary on its own in a manner determined by the input to the lag, u, and the lag time constant, $1/a$ (see chap. 4). Suppose for simplicity that the input to the lag, u, is zero, but that the initial output of the integrator, $x(0)$, might have any finite value. If the initial observation of the output, $y(0)$, is measured with some uncertainty var_y at time 0, then $y(0)$ will be the initial estimate of x. After an amount of time T, a new esti-

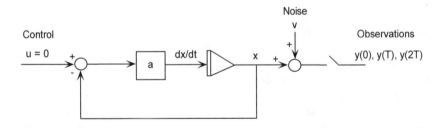

Time	Observation	Prediction	Variance of Prediction
0	$y(0)$	$x'(T) = e^{-aT}y(0)$	$e^{-2aT}var_y$
T	$y(T)$	$x'(2T) = e^{-aT}\{e^{-aT}y(0) + [1/(e^{2aT} + 1)] [y(T) - e^{-aT}y(0)]\}$	$e^{-2aT}var_y / (e^{2aT} + 1)$
2T	$y(2T)$	etc.	

FIG. 18.2. Recursive estimation of the output of a first-order lag based on noisy observations at discrete intervals (based on Maybeck, 1979).

mate or prediction, $x'(T)$, that takes into account the dynamics of the lag will be $e^{-aT}y(0)$ with a variance $e^{-2aT}var_y$. Namely, the output is expected to decay exponentially. (One could obtain this extrapolated estimate, $x'(T)$, by making a copy of the original system, setting its initial output to $y(0)$, and letting the system run for an interval T.) If a random variable is multiplied by some constant (e.g., e^{-aT}), then its variance, which is in squared units of measurement, is multiplied by the square of that constant (e.g., e^{-2aT}).

Next suppose that an observation $y(T)$ with the same uncertainty as before, var_y, is made at time T. What will be the new best estimate of x? The extrapolated prediction based on previous information, $x'(T)$, and the observation, $y(T)$, are both estimates of $x(T)$, and they can be combined in the usual manner. Namely, analogous to the computation used in the earlier example to estimate the mean of a sequence of measurements, the effective prior sample size is

$$ss_{prior} = var_y/(e^{-2aT}var_y) = e^{2aT} \tag{7}$$

The new sample size is $ss_{prior} + 1$. Therefore, the new estimate of x is:

$$x'(T) + [1/(ss_{prior} + 1)][y(T) - x'(T)] = e^{-aT}y(0) + [1/(e^{2aT} + 1)][y(T) - e^{-aT}y(0)] \tag{8}$$

Note that this looks very similar to Equation 4 for recursively estimating a mean. The major change is the inclusion of the exponential effect of the lag over the interval T between successive observations. Also, the new variance of the best estimate will be:

$$var_y/(ss_{prior} + 1) = var_y/(e^{2aT} + 1) \tag{9}$$

To make a prediction of what x will be after another interval T has passed, namely $x'(2T)$, it is necessary to extrapolate the effects of the lag over the time interval from T to $2T$ on both the latest estimate of x and its variance (i.e., multiply by e^{-aT} and e^{-2aT}, respectively; Fig. 18.2). Then another observation is made at time $2T$, namely $y(2T)$, and the extrapolated prediction, $x'(2T)$, is combined with the weighted deviation between observation and prediction in the usual fashion. Then the whole process repeats itself again. These alternating dynamic extrapolations and averagings of predictions and weighted deviations are the basic idea behind a Kalman filter based on discrete observations. As shown in chap. 17, the Kalman filter has been a building block for some models of human performance.

ESTIMATING THE STATE OF A FIRST-ORDER LAG WITH AN ELEMENTARY OBSERVER

In the previous example, the weighting of the difference between the observation and the estimated output of the lag will gradually decrease over time, in analogy with the first recursive calculation of the mean (Fig. 18.1). As the variance of the esti-

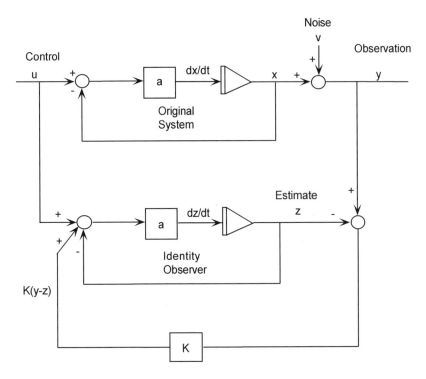

FIG. 18.3. The output, z, of an identity observer is an estimate of the output, x, of the original system (based on Luenberger, 1979).

mate becomes smaller and smaller, the effective prior sample size becomes larger and larger, so the impact of any new observation becomes small.[2]

Just as there was an alternate approach to recursively calculating the mean of a Gaussian process by using a constant weighting of the difference between the observation and the estimated value, there is a similar constant weighting approach to estimating the output of a first-order lag. Suppose that it is possible to continuously observe the output of a first-order lag that has some known input u. Assume that u is known exactly, but the continuous observation, y, consists of the true output of the lag, x, plus some zero mean Gaussian observation noise, v. As shown in Fig. 18.3, it is possible to build a copy of the original system and call its output z. It is desired that z be an estimate of x, the output of the original system. However, if the setting of the initial value of z differs from the unknown initial value of x [i.e., $z(0) = z_0$ is not equal to $x(0) = x_0$], then z will not be a good estimate of x at subsequent times until the effects of this initial mismatch exponentially decay to negligible levels. In order to overcome this difficulty more quickly, one can continuously calculate the difference between y and z, multiply this difference by a constant K, and make this signal an additional input to the z system. This latter system is called an *identity observer*, be-

---

[2]If the input to the lag, u, consisted of Gaussian noise, the effective weighting of new information would asymptotically approach a constant. See, for example, Bryson and Ho (1969, p. 367) and Maybeck (1979, pp. 220–226) for details.

cause it has the same structure as the original system (Luenberger, 1979). The equations governing the original system and the observer are:

$$dx/dt = a(u - x) \tag{10}$$

$$dz/dt = a[u - z + K(y - z)] = a(u - z) + aK(y - z) \tag{11}$$

Subtracting these two equations, one obtains:

$$d(x - z)/dt = -a(x - z) - aK(y - z) \tag{12}$$

Taking the expected value of both sides of Equation 12 (i.e., the mean value across different possible noisy observations):

$$E[d(x - z)/dt] = E[-a(x - z) - aK(y - z)]$$

$$d[x - E(z)]/dt = -a[x - E(z)] - aK[x - E(z)] = -a(1 + K)[x - E(z)] \tag{13}$$

The expected value of y is simply x, because $y = x + v$, and the expected value (or mean) of v is zero at all times. Equation 13 implies:

$$x - E(z) = (x_0 - z_0)e^{-a(1 + K)t} \tag{14}$$

Therefore, the difference between x and the expected value of z behaves like the output of a lag, and the lag time constant is $1/[a(1 + K)]$. K should be chosen to be large enough such that the difference between x and the expected value of z can be made to go to zero quickly. For example, if $a = 3$ and $K = 3$, the time constant of the original system is $1/a = 1/3$ s, but the time constant of $[x - E(z)]$ is much smaller (i.e., $1/[a(1 + K)] = 1/[3(1 + 3)] = 1/12$ s). If the difference between x and $E(z)$ becomes negligible, then z will become an unbiased estimate of x. On the other hand, K should be small enough that there is sufficient averaging of observations to reduce the effects of v on the variability of the estimate. Note that u is not explicitly represented in Equation 14. Therefore, for the special case of $u = 0$, the problem is analogous to the previous example with the Kalman filter.

This example, which is adapted from Luenberger (1979), illustrates an observer that uses a constant weighting of the difference between the observation, y, and the estimated state, z. The estimator involves a model of the original system as well as appropriate coupling to effectively match the observer to the original system. Note that as the difference between x and $E(z)$ becomes negligible, the velocity of x can be estimated as $a(u - z)$, the weighted difference between the input and the estimate of the output position. This velocity estimation procedure avoids amplifying the noise, v, as could occur with more direct attempts at approximate differentiation of the noisy observed signal, y. For more complex systems, the observer can provide estimates of other internal state variables based on observation of the system output (e.g., see Luenberger, 1971).

If this same problem were approached using a Kalman filter, then Fig. 18.3 would still be an appropriate representation of the calculation. However, instead of being a constant, K would be a variable that would get continuously smaller as the variance of the estimate became smaller. Kalman filters and identity observers are closely related in that they both involve a copy of the original system in order to extrapolate the effects of the system dynamics. If the parameters (e.g., a) of the original system are not known, then these parameters have to be estimated in order to implement the copy of the original system. Estimation of the system parameters is a problem in adaptive control (see chaps. 15 and 20). For a discussion of adaptive observers that estimate both system parameters and state variables, see Narendra and Annaswamy (1989).

REFERENCES

Analytic Sciences Corporation (1974). *Applied optimal estimation* (A. Gelb, Ed.). Cambridge, MA: MIT Press.

Astrom, K. J., & Wittenmark, B. (1989). *Adaptive control*. Reading, MA: Addison-Wesley.

Bozic, S. M. (1994). *Digital and Kalman filtering*. New York: Halsted Press.

Brown, R. G. (1963). *Smoothing, forecasting, and prediction of discrete time series*. Englewood Cliffs, NJ: Prentice-Hall.

Bryson, A. E., & Ho, Y. (1969). *Applied optimal control*. Waltham, MA: Blaisdell.

du Plessis, R. M. (1967). *Poor man's explanation of Kalman filtering or how I stopped worrying and learned to love matrix inversion*. Anaheim, CA: Autonetics Division of North American Aviation.

Hays, W. L., & Winkler, R. L. (1971). *Statistics: Probability, inference, and decision*. New York: Holt, Rinehart & Winston.

Luenberger, D. G. (1971). An introduction to observers. *IEEE Transactions on Automatic Control, AC-16,* 596–602.

Luenberger, D. G. (1979). *Introduction to dynamic systems*. New York: Wiley.

Maybeck, P. (1979). *Stochastic models, estimation, and control* (Vol. 1). New York: Academic Press.

Narendra, K. S., & Annaswamy, A. M. (1989). *Stable adaptive systems*. Englewood Cliffs, NJ: Prentice Hall.

Slotine, J. J. E., & Li, W. (1991). *Applied nonlinear control*. Englewood Cliffs, NJ: Prentice Hall.

Varieties of Variability

> With the goal of understanding the nature of human movement, our thesis is that the analysis of variability in movement provides a powerful tool in pointing to the underlying functional organization of the motor system.
>
> —Wing (1992, p. 709)

If a person is asked to perform the same manual control task several times in exactly the same manner, then the measured performance will not be identical across repeated performances. In other words, there is variability in performance. Mathematical descriptions of this variability are an important aspect of models of human manual control. The present chapter concentrates on variability in continuous tracking tasks; however, it also compares the pattern of variability in continuous tracking with other tasks, such as discrete movements and rhythmic tapping.

REMNANT

Linear describing functions are a mathematical approximation to the regularities in a person's response in a manual control task (chap. 14). However, there is also inconsistency or variability in the person's response as well. The magnitude of this variability limits the accuracy of the linear describing function as an approximation to the person's behavior; however, this variability also provides some structural insight into the underlying behavioral processes.

One way to generate a random-appearing input for experimental purposes is to add together numerous sine waves of different frequencies and randomize the initial phases across frequencies and from trial to trial. Although such a signal is perfectly predictable in principle, it will appear random to the person. A Fourier analysis of the

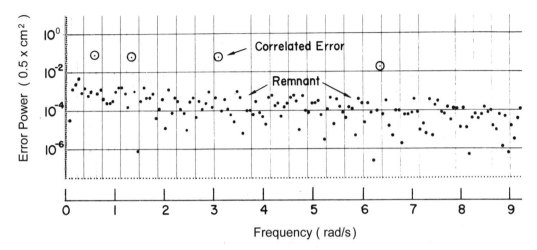

FIG. 19.1. A typical error spectrum when the input signal consists of a sum of sine waves. Peaks occur in the power spectrum (amplitude squared) at the input frequencies. Nonzero power at other frequencies is the remnant. From "A simple Fourier analysis technique for measuring the dynamic response of manual control systems" by R. W. Allen and H. R. Jex, 1972, *IEEE Transactions on Systems, Man, and Cybernetics*, SMC-2, pp. 630–643. Copyright 1972 by IEEE. Adapted by permission.

person's performance will reveal peaks in the movement and error spectra at the input frequencies (Fig. 19.1). However, at noninput frequencies, there will also be some small to moderate response. If a linear describing function provided a perfect description of the person's behavior, then the response at the noninput frequencies would be zero if the input frequencies are carefully chosen so they go through an integral number of cycles over the measurement period (see chap. 12). The linear transformations performed by the person's describing function and by the linear system being controlled would not change the frequencies of the input sinusoids; only their amplitudes and phases would be altered. The observed response at noninput frequencies indicates variability and/or nonlinear aspects of the person's behavior. It is often referred to as *remnant*, meaning it is the part of a person's response not captured by the linear describing function (e.g., McRuer & Krendel, 1957).

There are two immediate questions regarding the remnant: How large a fraction of the person's response is it? What is the shape of its Fourier spectrum? In describing how large the remnant is, it is typical to use an approach analogous to linear regression problems (see any introductory statistics text). Namely, the total error variance (i.e., the mean squared deviation of the error signal from its mean value, typically near zero) is separated into two components, one that is linearly correlated with the input signal and one that is not.

$$\sigma_e^2 \quad = \quad \sigma_{e_i}^2 \quad + \quad \sigma_{e_r}^2$$

Error Var$_{\text{Total}}$ = Error Var$_{\text{Linear Describing Function}}$ + Error Var$_{\text{Remnant}}$ (1)

The total error variance can simply be calculated from the observed experimental time signal generated by the person's performance. The variance of a signal is the average squared deviation from its mean value. If a linear describing function is calculated from the person's performance (see chap. 14), then a simulated tracking trial can be conducted with the linear describing function substituted for the person. The variance of the resulting error signal is the middle term in Equation 1. The difference between the variance from this linear simulation and the variance of the person's (total) error signal is equal to the variance of the remnant (Allen & Jex, 1972).

The ratio of the error variance captured by the linear describing function to the total error variance is typically represented by the symbol ρ^2, which is analogous to the square of the Pearson correlation coefficient. The ratio of the remnant variance to the total error variance is therefore $1 - \rho^2$. Researchers often report the analogously calculated ρ^2 for the person's control movements rather than for the error signal (e.g., see Elkind, 1956; McRuer & Krendel, 1957). The values of ρ^2 for error and for control movement will not typically be identical, although they may be close in value (e.g., see Jex & Magdaleno, 1969). Typical values for ρ^2 may vary from near unity for easy tracking tasks to below .5 for difficult tracking tasks.

Whereas the previous relation refers to the variance of signals over time, an analogous relation can be described in the frequency domain. Namely, if the Fourier transform of the error signal is taken, its power spectrum consists of the square of its amplitude at each frequency. The sum of these squared values is proportional to the total error power, which can be partitioned into two components:

$$\text{Error Power}_{\text{Total}} = \text{Error Power}_{\text{Linear Describing Function}} + \text{Error Power}_{\text{Remnant}} \qquad (2)$$

ρ^2 is also equal to the error power captured by the linear describing function divided by the total error power.

The connection between this equation and the previous equations involving variances in the time domain is a relation known as *Parseval's formula*. This formula states that the variance of a signal in the time domain is proportional to its total power in the frequency domain. To more easily understand this relation, consider a signal $y(t)$ that is composed of the sum of two sine waves:

$$y(t) = A_1\sin(\omega_1 t + \theta_1) + A_2\sin(\omega_2 t + \theta_2) \qquad (3)$$

Because the mean value of $y(t)$ is zero, its variance, σ_y^2, can be calculated by squaring $y(t)$, integrating over time, and dividing by the total time. Performing these same operations on the terms on the right side of the equation, one obtains $.5(A_1^2 + A_2^2)$, or .5 times the sum of the squared amplitudes in the Fourier domain:

$$\sigma_y^2 = .5(A_1^2 + A_2^2) \qquad (4)$$

Therefore, the variance in the time domain is proportional to the total power calculated in the frequency domain. In general, a function $y(t)$ might have a continuous spectrum in the Fourier domain, and its total power would be proportional to the integral of the square of the amplitude across the frequency spectrum (e.g., see Papoulis, 1962, for a formal derivation).

MEASURING THE NOISE INJECTED INTO A CONTROL LOOP

The previous discussion concerns how the magnitude of the remnant is measured. A second important feature of the remnant is the shape of its spectrum in the frequency domain. In dealing with this issue, researchers have typically treated the remnant as a noise injected into the control loop by the person. Such noise could arise from perceptual, motor, or other cognitive processes. The measurements of tracking performance are often insufficient to distinguish among these possibilities, so the remnant is typically treated as a motor noise added onto the output of the linear describing function or a perceptual noise added onto the input to the linear describing function. In order to understand an interesting experimental result found by Levison, Baron, and Kleinman (1969), it is useful to consider the latter approach.

The measurement of remnant magnitude discussed earlier dealt with the closed loop remnant. Namely, if the underlying process contributing to the remnant was perceptual noise, the aforementioned measurements estimated the magnitude of that noise as its effects circulated about the closed loop. More specifically, consider Fig. 19.2. All of the symbols in this figure refer to the Fourier transforms of their respective time signals. R_e is the noise associated with the person's perception of the error signal, H is the person's linear describing function, and V is the linear transfer function of the

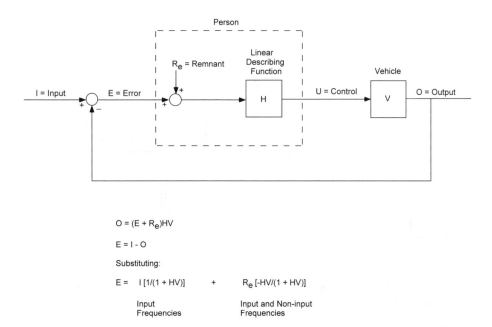

$$O = (E + R_e)HV$$

$$E = I - O$$

Substituting:

$$E = \quad I\,[1/(1 + HV)] \qquad + \qquad R_e\,[-HV/(1 + HV)]$$

Input
Frequencies

Input and Non-input
Frequencies

FIG. 19.2. If perceptual noise associated with observing the error signal is injected into the control loop, it will contribute to the measured error signal. From "A model for human controller remnant" by W. H. Levison, S. Baron, and D. L. Kleinman, 1969, *IEEE Transactions on Man–Machine Systems, MMS-10*, pp. 101–108. Copyright 1969 by IEEE. Adapted by permission.

vehicle being controlled, which is assumed to have linear dynamics. Writing two equations for output, O, and error, E, and performing some algebra (Fig. 19.2), one can see that the closed-loop error signal consists of a component due to the input signal and a component due to the injected noise, R_e. If the perceptual noise hypothesis is correct, then these components correspond to the variances discussed earlier.

Although both components of error will be present at input frequencies, only the remnant will be present at noninput frequencies (assuming the input is a sum of sine waves having integral numbers of cycles over the measurement period; see chap. 12). The power spectrum of the error signal at noninput frequencies therefore represents $|R_e|^2|HV/(1 + HV)|^2$, where the vertical lines indicate the amplitude of the complex variable (see chap. 13). To calculate $|R_e|^2$ alone, divide the measured error power at each noninput frequency by $|HV/(1 + HV)|^2$. This operation translates a closed-loop measurement of remnant into an equivalent amount of injected noise power at the input to the linear describing function, H. Although, strictly speaking, H is only measured at the input frequencies, if it is assumed that it provides a good estimate of the linear describing function for neighboring frequencies as well, then the calculation can be conducted. The result is an estimate of the injected noise power in the neighborhood of each of the input frequencies.

Examples of empirically measured power spectra, ϕ_{Rem}, that correspond to an injected noise source such as $|R_e|^2$ are shown in Fig. 19.3. For this example, the person saw a display of the error signal, the vehicle dynamics were a velocity control system, $K/j\omega$, and the input signal variance was either small (2.6 degrees^2) or large (23 degrees^2). The larger input signal produced a larger error signal. However, when the remnant power spectrum is divided by the total error variance, the two spectra are almost identical. In other words, the injected perceptual noise is proportional to the magnitude of the error variance.

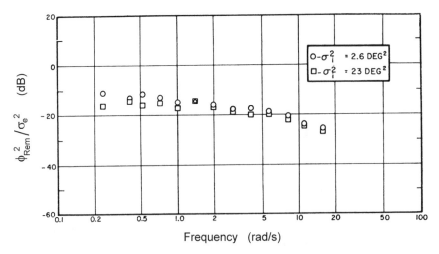

FIG. 19.3.   The estimated injected remnant power in the neighborhood of each input frequency divided by the overall variance of the error signal for a velocity control system with large and small input signals. From "A model for human controller remnant" by W. H. Levison, S. Baron, and D. L. Kleinman, 1969, *IEEE Transactions on Man–Machine Systems, MMS-10*, pp. 101–108. Copyright 1969 by IEEE. Adapted by permission.

A second important feature of these data is the shape of the spectrum. ϕ_{Rem} is approximately constant across frequencies at the low end of the frequency spectrum. A flat spectrum is often referred to as *white noise*. The implication of such a spectrum in the time domain is that the noise is uncorrelated from measurement instant to measurement instant. In other words, the noise is randomly varying from sample to sample. If the spectrum was flat over its entire range, then the interpretation would be that the person injects a perceptual noise on the displayed error signal that randomly varies from instant to instant, and the magnitude of the noise power is proportional to the variance of the error signal.

However, the data exhibit a more complicated structure. Over the entire measurement spectrum, the remnant power resembles a first-order lag. Namely, the spectrum is flat at low frequencies and drops off at approximately 20 db/decade at high frequencies. The term "20 db/decade" means the remnant power drops by a factor of 100 (or equivalently the remnant amplitude drops by a factor of 10) when the measurement frequency increases by a factor of 10 (see chap. 13). In other words, the remnant amplitude injected on the error signal appears to be inversely related to frequency at the high frequency end of the measurement spectrum. One interpretation of this effect is that the person is observing both the error magnitude and the error velocity and is injecting perceptual noise associated with each of these variables. To understand this interpretation, suppose that the person was only observing error rate and was effectively injecting perceptual noise onto that signal (Fig. 19.4, top). The person is considered to estimate the rate of change of the error signal (E multiplied by $j\omega$ in the Fourier transform domain), and noise, $R_{\dot{e}}$, is added to this signal. The remainder of the person's linear describing function is represented as $H/j\omega$, so the total linear describing function for the person, $(j\omega) \times (H/j\omega)$, is simply H. The closed-loop contribution to the measured error signal can be calculated in the same manner as in Fig. 19.2. The result (Fig. 19.4, top) looks very similar to that in Fig. 19.2, with the exception that there is an additional factor of $1/j\omega$ multiplying the remnant term. In other words, noise $R_{\dot{e}}$ added to the error velocity signal will have the same effects as noise $R_{\dot{e}}/j\omega$ added to the error signal (Fig. 19.4, bottom). This inference can be seen by comparing Figs. 19.2 and 19.4. If the amplitude of $R_{\dot{e}}$ is constant across all frequencies (i.e., the perceptual noise on error rate is white), then the injected remnant measured at the error signal will be inversely proportional to frequency.

Suppose that a person observes both error and error velocity, and respectively injects perceptual noises with flat (white) spectra, R_e and $R_{\dot{e}}$, onto these signals. If the person additively combines error and error rate as part of the linear describing function (i.e., the person generates lead; Fig. 19.5), then the resulting remnant spectrum, ϕ_{Rem}, will have the shape shown in Fig. 19.3 for a velocity control system. Namely, the low frequency remnant will be dominated by noise on error, and the high frequency remnant will be dominated by noise on error velocity. The algebra needed to demonstrate this result follows the same style used in Fig. 19.2 and 19.4 (see Levison et al., 1969). In Fig. 19.5, the total linear describing function for the person is still H; however, it is partitioned into a lead, $K_e + j\omega K_{\dot{e}}$, and H divided by the lead. The product of the lead and H divided by the lead is simply H. More intuitively, a lead acts like a gain at low frequencies and a differentiator at high frequencies. Therefore, the remnant pattern at low and high frequencies should correspond to the respective models of remnant in Figs. 19.2 and 19.4.

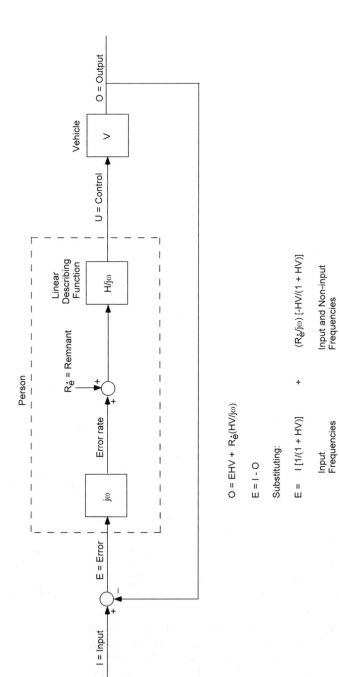

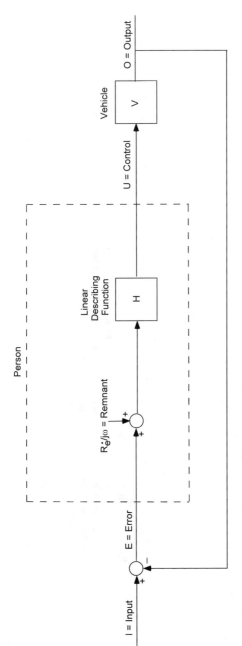

FIG. 19.4. (Top) If perceptual noise, $R_{\dot{e}}$ associated with observing the error velocity is injected into the control loop, then it will contribute to the measured error signal with a multiplicative factor inversely proportional to frequency. (Bottom) An equivalent result would occur if perceptual noise $R_{\dot{e}}/j\omega$ were associated with observing error magnitude.

229

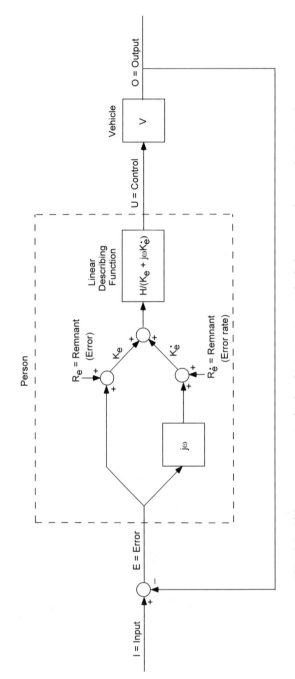

FIG. 19.5.   If the person generates lead and injects noise onto both the error and error velocity signals, then both sources of remnant will contribute to the measured error signal. From "A model for human controller remnant" by W. H. Levison, S. Baron, and D. L. Kleinman, 1969, *IEEE Transactions on Man–Machine Systems, MMS-10*, pp. 101–108. Copyright 1969 by IEEE. Adapted by permission.

The observed remnant pattern with a velocity control is consistent with the presence of a lead (Fig. 19.3); however, the person's overall describing function, H, typically resembles a simple gain for this control system (Fig. 14.4). The remnant pattern suggests that this mathematically simple form of H is generated by a lead and a lag, two processes that roughly cancel each other's effects in the frequency domain (Levison et al., 1969). The lead seems to reflect visual processing (given a primary visual display), and the lag may reflect strategic aspects of control similar to people's behavior with a position control (Fig. 14.6).

These results indicate that the mathematical structure of a person's overall describing function, H, may not reflect the complexity of the underlying behavioral processes (see also chap. 25). The perceptual noises can be thought of as disturbance signals that are independent of each other and of the input signal (Levison et al., 1969). Their circulation through the closed-loop from different points of entry can reveal additional details of behavioral organization (see chap. 22 for additional studies with multiple signals).

Next consider other vehicle dynamics. If the vehicle dynamics are a position control system, the person would be expected to primarily respond to the error signal and H will resemble a lag (Fig. 14.3). As can be seen in Fig. 19.6, the error remnant spectrum is approximately flat across the entire range of measurement frequencies for these dynamics, the pattern expected for error alone. On the other hand, if the vehicle dynamics are a more challenging acceleration control system, then it would be expected that the person would primarily respond to the error velocity signal, and H will resemble a lead with a low break frequency (Fig. 14.5). As can be seen in Fig. 19.7, the remnant power drops off inversely with frequency over nearly the entire range of measurement frequencies, the pattern expected for error velocity. There is only a small flat region at low frequencies. These trends are consistent with expectations based on the crossover model (chap. 14). However, it should be noted that the break frequency (chap. 13) in the remnant spectrum does not typically equal the estimated

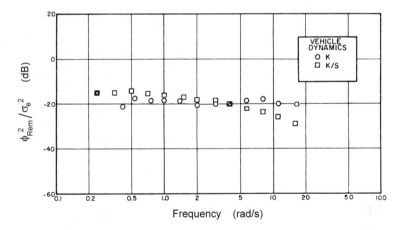

FIG. 19.6.   Remnant spectra for a position control system and a velocity control system. From "A model for human controller remnant" by W. H. Levison, S. Baron, and D. L. Kleinman, 1969, *IEEE Transactions on Man–Machine Systems, MMS-10*, pp. 101–108. Copyright 1969 by IEEE. Adapted by permission.

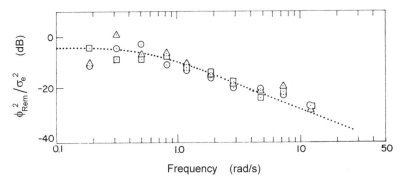

FIG. 19.7.   Remnant spectra for an acceleration control system for three trials by the same participant. The dotted curve represents a first-order lag. From "Corroborative data on normalization of human operator remnant" by H. R. Jex and R. E. Magdaleno, 1969, *IEEE Transactions on Man–Machine Systems, MMS-10*, pp. 137–140. Copyright 1969 by IEEE. Adapted by permission.

break frequency in the linear describing function for the same performance (Jex & Magdaleno, 1969; Levison et al., 1969). This difference presumably reflects the contributions of other, non-visual processes to H.

Another interesting comparison for remnant spectra involves foveal and parafoveal vision. The fovea is in the center of the retina, and it has much greater spatial resolution than other regions of the retina. When a tracking display is off to the side from where a person is looking (parafoveal), the remnant power is greater for both error magnitude and error velocity. As estimated via an optimal control model (chap. 17), the rise in the remnant for error magnitude is particularly pronounced, and there is some evidence of a sizable constant component of remnant power that is not proportional to the error variance (Baron & Levison, 1977; Levison, 1971). This result is consistent with the eye's relatively greater sensitivity to motion versus position for parafoveal viewing in comparison with foveal viewing (see also Jex & Magdaleno, 1969). When a person is required to scan several displays to perform simultaneous tracking tasks, for at least some of the time the view of a particular display will be parafoveal, and the remnant will increase (e.g., Jex & Magdaleno, 1969). However, even in the absence of visual scanning, if a person is required to perform simultaneous tasks or monitor multiple signals within a task, the remnant associated with observing each signal will often increase (e.g., Levison, 1979; Levison, Elkind, & Ward, 1971; Wickens, 1976). Such increases in remnant have been interpreted as being inversely related to attention allocation (see Levison, 1979, for a review of modeling).

WEBER'S LAW AND REMNANT

Weber's Law is a relation most typically associated with research on perception. It states that variability associated with perceiving a stimulus is proportional to the magnitude of that stimulus:

$$\sigma_s = kS \tag{5}$$

σ_s is the standard deviation of the perceptual noise associated with the process of perceiving a stimulus of magnitude S, and k is a constant. This relation is equivalent to

$$\sigma_s^2 = k^2 S^2 \tag{6}$$

In other words, the variance of the perceptual noise is proportional to the square of the stimulus magnitude. A more general form of this relation (Getty, 1975) is

$$\sigma_s^2 = k^2 S^2 + c^2 \tag{7}$$

where the constant c^2 represents a component of the perceptual noise that does not change with the stimulus magnitude.

As noted earlier, the perceptual noise associated with a tracking task has been found to have an error component and an error velocity component. Each component has a power level that is approximately uniform across the frequency domain at the point in the control loop at which it is injected. The amplitude of this power level can therefore be modeled as a constant, and it is typically proportional to the variance of the signal being observed. For perception of error,

$$|R_e|^2 = k_{R_e}^2 \sigma_e^2 \tag{8}$$

However, as already noted, the total power of a signal in the frequency domain is proportional to its variance in the time domain. Given that $|R_e|^2$ is constant across the range of the spectrum typically observed via sampling at fixed time intervals (see chap. 12), its integral across the spectrum will be proportional to $|R_e|^2$. In other words, the variance of the remnant in the time domain, $\sigma_{R_e}^2$, is proportional to its uniform power level in the frequency domain, $|R_e|^2$. Substituting into the previous equation yields:

$$\sigma_{R_e}^2 = k^2 \sigma_e^2 \tag{9}$$

In other words, the variance of the perceptual noise associated with perceiving the error signal is proportional to the variance of the error signal itself. A similar relation would describe the perception of error velocity. Typical values for k^2 are about .03 for both error and error velocity when a performer foveally views a single compensatory display (Levison, 1989).

The previous relation is very similar to Weber's Law (see Jex & Magdaleno, 1969) in that the variance of the error signal, σ_e^2, represents the average squared deviation of the error from its mean. If it is assumed that the mean error is approximately zero, then the variance of the error is a measure of the averaged squared magnitude of the error. In the case of parafoveal observation, it is empirically found to be important to add an additional component of perceptual noise not related to the magnitude of the error signal (Levison, 1971):

$$\sigma^2_{R_e} = k^2 \sigma^2_e + c^2 \qquad (10)$$

Although the modeling of remnant has typically emphasized perceptual noise, it is likely that there are components of this variability due to cognitive anticipation processes and movement processes (Levison et al., 1969; Jex & Magdaleno, 1969, for a brief review). In the optimal control model (chap. 17) a separate source of movement variability is included, although its magnitude is considerably less than the perceptual noise (Levison, 1989).

WEBER'S LAW AND DISCRETE MOVEMENTS

Given that remnant follows a relation similar to Weber's Law, it is of interest to consider whether similar relations arise for other movement control tasks. One such example is a task investigated by Schmidt, Zelaznik, Hawkins, Frank, and Quinn (1979). Their task required a person to move a stylus to a target line in a short specified amount of time. Thus, there was a target distance as well as a target time. The target time was very brief, ranging from 140 to 200 ms. The primary dependent measure was spatial endpoint variability. Schmidt et al. (1979) found that the standard deviation of movement endpoints was linearly related to the average velocity of the movement (target distance divided by average movement time). Schmidt et al. argued that persons performing this task use a stereotypic pattern of force as a function of time, and vary this pattern by raising or lowering the overall level of force and by increasing or decreasing the duration of the pattern. Variability in the force parameter and variability in the time parameter was assumed to result in the variability of the spatial endpoints. Schmidt et al. (1979) collected data on two auxiliary tasks to argue that both the force and time parameters followed Weber's Law. In one task, participants attempted to produce commanded levels of force by pressing against a stationary surface for a very brief period. The commanded force level and the produced force were displayed to the performer on an oscilloscope screen. The standard deviation of the produced force was found to be proportional to the force magnitude. In a second task, participants moved a lever back and forth over various distances in time with a metronome. Across trials, the metronome was set to different periods of 500 ms or less. The standard deviation of the duration of the force pattern in this task was found to be proportional to movement time. For both tasks, the performers were instructed to avoid making corrections to the initial movement, and the data for both tasks approximated Weber's Law. Larger forces were proportionally more variable than smaller forces, and longer durations were proportionally more variable than shorter durations.

Building on these efforts, Meyer, Smith, and Wright (1982) mathematically demonstrated that if Weber's Law held for both the force magnitude and duration parameters of a force–time pattern with a particular shape (i.e., a motor program), then the linear relation between movement endpoint variability and average velocity found by Schmidt et al. (1979) would result. The particular shape resembled a distorted sine wave with symmetric accelerative and decelerative sections, although it was also affected by any possible correlation that might exist between the two underlying sources of variability.

This theory is an interesting example of how variability in underlying control parameters, in this case the magnitude and duration of a force–time pattern, could contribute to observed endpoint variability. There have been a number of criticisms of this model, however. Although Weber's Law may be a good approximation to the relation between force variability and force magnitude over a limited range, the relation is most likely exponential over larger ranges of force (Carlton & Newell, 1993; see also chap. 8). The rate of change of force variability for a given increase in force magnitude is smaller for larger forces than for smaller forces. Second, the assumption that the force–time pattern is uniformly speeded up or slowed down in order to maintain a constant shape has also been found to be inaccurate for rapid aimed hand movements (e.g., Zelaznik, 1993).

WEBER'S LAW AND RHYTHMIC TAPPING

Another movement task that has been carefully studied is rhythmic tapping (see Vorberg & Wing, 1996, for a review). In a very simple version of this task, a person initially taps a finger to the steady beat of a metronome. The metronome is then turned off, and the person is required to continue tapping at the same steady pace. Wing and Kristofferson (1973) studied the pattern of variability of the intertap intervals and found that the durations of immediately neighboring taps were negatively correlated. However, intervals that were not immediate neighbors were uncorrelated. They accounted for this pattern of variability by postulating two underlying processes—a central pattern generator or timekeeper that produces intervals, C_i, approximately equal to the experimentally required duration, and delays, M_i, due to the various processes involved in implementing the physical tapping motion of the limb (Fig. 19.8). The observed intertap interval I_i is equal to $C_i + M_i - M_{i-1}$. Similarly,

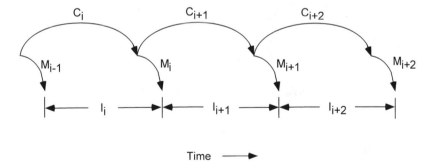

FIG. 19.8. Intervals C_i generated by a central timekeeper and intervals M_i due to delays in implementing the tapping motion. The observed intertap intervals are I_i. From "Response delays and the timing of discrete motor responses" by A. M. Wing and A. B. Kristofferson, 1973, *Perception & Psychophysics, 14,* pp. 5–12. Copyright 1973 by the Psychonomic Society. Adapted by permission.

the subsequent intertap interval I_{i+1} is equal to $C_{i+1} + M_{i+1} - M_i$. The central timing intervals and the movement implementation delays are assumed to vary from tap to tap independently from each other and from all previous central timing intervals and implementation delays. However, note that two immediately neighboring intertap intervals such as I_i and I_{i+1} have one underlying interval in common, M_i. It appears with a positive sign for I_i (i.e., increasing M_i increases I_i) and with a negative sign for I_{i+1} (i.e., increasing M_i decreases I_{i+1}). These relations are a simple consequence of the adjacency of the intervals. Across a long sequence of taps, these relations produce a negative correlation between successive intertap intervals. In contrast, intervals that are not immediate neighbors will be uncorrelated, because they do not share any common sources of variation. The model therefore agrees with the observed pattern of intertap correlations[1] (Wing & Kristofferson, 1973).

If the underlying central timekeeper and the movement delays are really independent processes that contribute to the overall tapping variability, then it should be possible to separately influence them by appropriate experimental manipulations. In fact, if the duration of the required intertap intervals is varied across trials (i.e., the metronome is set to different target intervals from one trial to the next), then the estimated variance of the central timekeeper intervals changes, but the variance of the movement implementation delays remains approximately constant (Wing, 1980). Wing (1980) initially reported that Var(C) increased linearly with the mean interval duration; however, Ivry and Corcos (1993) noted that the linear relation between Var(C) and the square of the mean interval duration was just as strong. Furthermore, the extrapolated intercept was near zero for this latter relation and therefore seemed more meaningful than the negative intercept obtained by Wing (1980). Namely, as the produced interval approached zero, its associated variance also approached zero. The linear relation between central timekeeper variance and the squared mean interval duration is therefore another example of Weber's Law (see also Ivry & Hazeltine, 1995).

In another test of the independence of the two sources of motor variability, Ivry and Keele (1989) tested persons with peripheral nerve damage to one arm. Analyses of tapping performance with the injured arm and the noninjured arm revealed a larger motor variance for the injured arm, but no difference in central timekeeper variance between the arms. Peripheral nerve damage selectively influenced the motor implementation variability. Thus, the experimental evidence for two different sources of intertap variability seems strong.

---

[1]As noted by Wing and Kristofferson (1973), separate estimates of the variances of the C_i's and the M_i's can be calculated from the observable intertap intervals as follows:

$$Var(I) = Var(C) + 2Var(M)$$
$$Cov(I_i, I_{i+1}) = -Var(M)$$

where Var represents variance and Cov represents covariance. Therefore,

$$Var(C) = Var(I) + 2Cov(I_i, I_{i+1})$$
$$Var(M) = -Cov(I_i, I_{i+1})$$

OVERVIEW

Weber's Law, which is most typically associated with perceptual judgments, has been found to describe the variability in a number of movement tasks—closed-loop tracking, very brief aimed movements, and rhythmic tapping. The latter two tasks have been assumed to have open-loop control structures. Each of the three analyses considered two sources of variability—position and velocity perceptual noise in the tracking tasks, force and time variability in the brief movement task, and central timing and movement implementation variability in the tapping task. Whereas the modeling of remnant in tracking tasks has emphasized perceptual noise, the modeling of the brief aimed movements and rhythmic tapping have emphasized motor noise. However, Ivry and Hazeltine (1995) noted that the perception of brief temporal intervals follows a Weber's Law relation that quantitatively mimics that for motor variability in rhythmic tapping, and they considered the possibility of a single underlying noise source. Comprehensive models of performance must consider how overall performance is related to variability in perceptual, motor, and other cognitive processes.

Although the present chapter has concentrated on perceptual noise, motor noise, and central timing noise, the process of adaptation may also contribute in substantial ways to performance variability (e.g., McRuer, Graham, Krendel, & Reisener, 1965). Namely, if a performer were to execute a single strategy with perfect consistency, then the variability of performance might be at a minimum. However, if the performer is constantly adaptively adjusting the values of various parameters in order to search for a best strategy, then the measured performance will be more variable. In tracking tasks, the performer might deliberately introduce small perturbations into the control movements for the purpose of getting more information about the dynamic structure of the system being controlled and to facilitate the adaptive adjustment of performance parameters (e.g., Flach, 1990). Such behavior would be related to the concept of "persistent excitation" in adaptive control theory (e.g., Narendra & Annaswamy, 1989, chap. 6; Slotine & Li, 1991, p. 331). Namely, very simple input patterns may not supply sufficient information for an adaptive control scheme to converge to the appropriate parameter values. The human performer may deal with this problem by introducing deliberate perturbations that are uncorrelated with the input in a tracking task and that would also contribute to the remnant.

REFERENCES

Allen, R. W., & Jex, H. R. (1972). A simple Fourier analysis technique for measuring the dynamic response of manual control systems. *IEEE Transactions on Systems, Man, and Cybernetics, SMC-2*, 630–643.

Baron, S., & Levison, W. H. (1977). Display analysis with the optimal control model of the human operator. *Human Factors, 19*, 437–457.

Carlton, L. G., & Newell, K. M. (1993). Force variability and characteristics of force production. In K. M. Newell & D. M. Corcos (Eds.), *Variability and motor control* (pp. 15–36). Champaign, IL: Human Kinetics.

Elkind, J. I. (1956). *Characteristics of simple manual control systems* (Tech. Rep. No. 111). Lexington, MA: MIT Lincoln Laboratory.

Flach, J. M. (1990). Control with an eye for perception: Precursors to an active psychophysics. *Ecological Psychology, 2,* 83–111.

Getty, D. (1975). Discrimination of short temporal intervals: A comparison of two models. *Perception and Psychophysics, 18,* 1–8.

Ivry, R. B., & Corcos, D. M. (1993). Slicing the variability pie: Component analysis of coordination and motor dysfunction. In K. M. Newell & D. M. Corcos (Eds.), *Variability and motor control* (pp. 415–447). Champaign, IL: Human Kinetics.

Ivry, R. B., & Hazeltine, R. E. (1995). Perception and production of temporal intervals across a range of durations: Evidence for a common timing mechanism. *Journal of Experimental Psychology: Human Perception and Performance, 21,* 3–18.

Ivry, R., & Keele, S. (1989). Timing functions of the cerebellum. *Journal of Cognitive Neuroscience, 1,* 136–152.

Jex, H. R., & Magdaleno, R. E. (1969). Corroborative data on normalization of human operator remnant. *IEEE Transactions on Man–Machine Systems, MMS-10,* 137–140.

Levison, W. H. (1971). *The effects of display gain and signal bandwidth on human controller remnant* (Aerospace Medical Research Laboratory Tech. Rep. No. 70-93). OH: Wright-Patterson Air Force Base.

Levison, W. H. (1979). A model for mental workload in tasks requiring continuous information processing. In N. Moray (Ed.), *Mental workload: Its theory and measurement* (pp. 189–218). New York: Plenum.

Levison, W. H. (1989). The optimal control model for manually controlled systems. In G. R. McMillan, D. Beevis, E. Salas, M. H. Strub, R. Sutton, & L. Van Breda (Eds.), *Applications of human performance models to system design* (pp. 185–198). New York: Plenum.

Levison, W. H., Baron, S., & Kleinman, D. L. (1969). A model for human controller remnant. *IEEE Transactions on Man–Machine Systems, MMS-10,* 101–108.

Levison, W. H., Elkind, J. I., & Ward, J. L. (1971). *Studies of multi-variable manual control systems: A model for task interference* (NASA CR-1746). Washington, DC: NASA.

McRuer, D. T., Graham, D., Krendel, E., & Reisener, W., Jr. (1965). *Human pilot dynamics in compensatory systems* (AFFDL-TR-65-15). OH: Wright-Patterson Air Force Base.

McRuer, D. T., & Krendel, E. S. (1957). *Dynamic response of human operators* (Wright Air Development Center Tech. Rep. No. 56-524; ASTIA Document NR AD-110693).

Meyer, D. E., Smith, J. E. K., & Wright, C. E. (1982). Models for the speed and accuracy of aimed movements. *Psychological Review, 89,* 449–482.

Narendra, K. S., & Annaswamy, A. M. (1989). *Stable adaptive systems.* Englewood Cliffs, NJ: Prentice-Hall.

Papoulis, A. (1962). *The Fourier integral and its applications.* New York: McGraw-Hill.

Schmidt, R. A., Zelaznik, H. N., Hawkins, B., Frank, J. S., & Quinn, J. T. (1979). Motor-output variability: A theory for the accuracy of rapid motor acts. *Psychological Review, 86,* 415–441.

Slotine, J. J. E., & Li, W. (1991). *Applied nonlinear control.* Englewood Cliffs, NJ: Prentice-Hall.

Vorberg, D., & Wing, A. (1996). Modeling variability and dependence in timing. In H. Heuer & S. W. Keele (Eds.), *Handbook of perception and action* (Vol. 2, pp. 181–262). New York: Academic Press.

Wickens, C. D. (1976). The effects of divided attention on information processing in manual tracking. *Journal of Experimental Psychology: Human Perception and Performance, 2,* 1–13.

Wing, A. M. (1980). The long and short of timing in response sequences. In G. E. Stelmach & J. Requin (Eds.), *Tutorials in motor behavior* (pp. 469–486). Amsterdam: North-Holland.

Wing, A. M. (1992). The uncertain motor system: Perspectives on the variability of movement. In D. E. Meyer & S. Kornblum (Eds.), *Attention and Performance XIV: Synergies in experimental psychology, artificial intelligence, and cognitive neuroscience* (pp. 708–744). Cambridge, MA: MIT Press.

Wing, A. M., & Kristofferson, A. B. (1973). Response delays and the timing of discrete motor responses. *Perception & Psychophysics, 14,* 5–12.

Zelaznik, H. N. (1993). Necessary and sufficient conditions for the production of linear speed–accuracy trade-offs in aimed hand movements. In K. M. Newell & D. M. Corcos (Eds.), *Variability and motor control* (pp. 91–115). Champaign, IL: Human Kinetics.

20

Lifting a Glass of Juice

Reaching out to pick up a glass of beer may not seem like a very difficult activity, at least in the early evening. The glass is on the table, you reach, pick it up, carry it to your mouth and drink. No-one stops to ask "How did you do that?", and you would be very surprised if they did. The ease with which we do this makes us think that there is nothing to explain. . . . However, it is just these very simple activities that can often be the hardest to explain.
—Smyth, Collins, Morris, and Levy (1994, p. 97)

Intelligence does not mean dependence on the same rigid structures across task contexts. Intelligence means the ability to adapt, to fit behavior and cognition to the changing context. A smart system seems unlikely to ever do exactly the same thing twice. Rather, a smart system would shift its behavior slightly to fit the nuances of the particular context or would shift radically — jump to an all-new state — if the situation demanded it.
—Thelen and Smith (1995, p. 244)

Chapters 14 and 15 dealt with adaptive aspects of feedback control and considered the method of gradient descent. This chapter considers adaptive aspects of feedforward control in combination with feedback control. Feedforward control involves the generation of movement based on transformations of the time history of a desired trajectory.

One type of situation that involves adaptive control is the lifting of objects that vary in weight. For example, as people gradually drink a glass of orange juice during a meal, the glass may be nearly full at the beginning of the meal and nearly empty near the end of the meal. Nevertheless, they are able to lift the glass to their mouth without spilling the juice despite the variations in its weight. Occasionally, people probably do make mistakes. If individuals think they are picking up a full glass, but in fact they are picking up a nearly empty glass, they may move the glass too forcefully toward their mouth, have to slow down very rapidly in midmovement, and end up spilling some of the liquid. Similarly, if someone who is about to hand an individual a package decides to play a practical joke and tells the person that the package is very light when in fact it is quite heavy, the person may end up dropping it or perhaps pulling a back muscle. The individual was not anticipating a heavy load, and did not prepare their muscles accordingly. Both of these examples indicate that people do adapt their motions to the varying weights of objects that they transport.

The problem of adapting a movement pattern to initially unknown loads has been studied in the context of robotic motion control as well as in terms of human motion control. To establish some intuitions about the ways people might deal with this adaptive task, it is useful to consider how engineers might design adaptive robotic control systems.

Two different approaches as to how adaptation might be accomplished are called direct and indirect adaptive control (e.g., Astrom & Wittenmark, 1989; Narendra & Annaswamy, 1989; Slotine & Li, 1991). *Direct adaptive control* involves adjusting movement parameters to produce a desired movement pattern. *Indirect adaptive control* involves two relatively independent processes—estimating the dynamic properties of the system being controlled and implementing a desired movement pattern. Both approaches involve parameter adjustment as well as movement generation. Analyzing the overall stability of an adaptive system is therefore more complicated than analyzing a system with fixed parameters.

Whereas the adaptive method of gradient descent discussed in the previous chapter can lead to stable convergence to a desired style of movement, there is no certainty that this convergence will always occur. If the initial style of movement of the control system is very different from the desired style, then gradient descent may not always be effective. A general method of adaptive adjustment that does lead to effective convergence to the desired style of movement for all initial conditions is associated with work of the Russian scientist Lyapunov. The method involves continuously evaluating performance in terms of a composite index that includes both deviations from the desired movement trajectory as well as parameter adjustment errors. The composite performance index is called a *Lyapunov function*. If it can be shown that this function is made smaller and smaller over time regardless of the initial conditions, then the convergence to the desired style of performance is assured. In other words, the system is said to be "globally stable."

A SIMPLIFIED MODEL OF THE ARM

To see how a Lyapunov function might be used to design an adaptive control system, first consider a very simple example that corresponds to having a robot flex its elbow to lift a glass of juice (adapted from Craig, 1989, and Slotine & Li, 1991; see also Atkeson, 1989; Lewis, Abdallah, & Dawson, 1993). Assume that the plane of the motion is fixed, so that it only involves one degree of freedom, namely, changing the angle of the elbow over time to lift the glass. Also, for simplicity, consider the arm to be a massive object to which a motor applies some force in order to move it. According to basic physics, we know that a force, F, applied to a mass, m, determines its acceleration, d^2x/dt^2, where x is position:

$$F = m\left(\frac{d^2x}{dt^2}\right) \tag{1}$$

or equivalently

$$\frac{F}{m} = \left(\frac{d^2x}{dt^2}\right) \tag{2}$$

The greater the mass, m, of an object, the larger the force, F, that is needed to produce a given acceleration. This relation applies to unconstrained linear motion (i.e., if the robot had a forearm that was not connected to its upper arm at the elbow). However, the forearm movement for lifting the glass of juice is approximated as a rotation about the elbow. The rotary analog to force is called *torque*, τ.[1] The more torque that is applied about the elbow joint by the motor, the faster the angular acceleration, $d^2\theta/dt^2$, where θ is the angle of the forearm relative to horizontal.

The rotary analog to mass is called the *moment of inertia*, and it takes into account both the mass of the arm and how far that mass is from the pivot point. The farther the mass is from the pivot point, the more torque that is needed to produce a given angular acceleration. In fact, the necessary torque increases as the square of the distance between the mass and the pivot point. For example, if a person had a much longer forearm, it would take greater torque and hence stronger muscular contraction to lift the same glass of juice, even if the person's arm weighed as much as it does now. The exact calculation of the moment of inertia would be performed by multiplying each small bit of mass in the arm, the glass, and the juice by the square of its distance from the elbow and adding them up. For simplicity, the number that would result from such a calculation will be represented as ml^2, as though an equivalent mass m, were located at a single distance l from the elbow, the length of the forearm. The rotary analog of Equation 1 is thus:

$$\tau = ml^2\left(\frac{d^2\theta}{dt^2}\right) \tag{3}$$

or equivalently

---

[1] If a tangential force, F, is applied at a distance, l, from a pivot point (e.g., an elbow), then the resulting torque is Fl. In other words, torque increases with the applied tangential force and with the distance from the pivot point. When you and a friend play on a seesaw, you adjust your distance from the pivot point so that each of you applies a nearly equal magnitude of torque, but in opposite directions. Then small kicks against the ground (additional torques) can be used to reverse the direction of movement of the seesaw and make it oscillate. Torque is measured in units which reflect both force and distance from a pivot point, e.g., Newton-meters.

$$\frac{\tau}{ml^2} = \left(\frac{d^2\theta}{dt^2}\right) \tag{4}$$

This equation represents the proportionality between the applied torque and the angular acceleration of the arm. The greater the moment of inertia, ml^2, the greater the torque, τ, that will have to be applied to produce a given angular acceleration.

Equation 3 ignores the effects of gravity. Gravity applies an additional torque to the arm equal to $mgl\cos\theta$. In other words, the torque due to gravity depends on the angle of the arm. Gravity pulls downward with a force, mg, and the component of this downward force that is tangential to the motion of the arm, $mg\cos\theta$, will result in a torque about the elbow (Fig. 20.1). The torque is equal to the product of the tangential force and the length of the arm. The component of the downward force that is radial or parallel to the elbow, $mg\sin\theta$, will not contribute to the rotation of the arm. The gravitational torque will be greatest when the arm is horizontal ($\cos\theta = 1$) and gravity acts totally tangentially to the motion of the arm. The gravitational torque will be zero when the arm is vertical ($\cos\theta = 0$) and gravity acts totally radially (along the arm). For in-between angles, $mg\cos\theta$ is the fraction of the total gravitational force, mg, that acts tangentially on the arm of length l. This torque can be added to Equation 4 as a torque that sums with the torque, τ, generated by the robot motor:

$$\frac{(-mgl\cos\theta + \tau)}{ml^2} = \left(\frac{d^2\theta}{dt^2}\right) \tag{5}$$

The negative sign indicates that the gravitational torque is downward.

Equation 5 represents the simplified dynamics of a single link arm, and it is pictured in Fig. 20.2. Two torques, one generated by the motor and one due to gravity, sum together, are divided by the effective moment of inertia, ml^2, and are integrated twice to determine the angle of the arm, θ. The sum of the torques determines the angular acceleration or the limb, and two integrations represent the transformation of angular acceleration into angular velocity and of angular velocity into angular position. In other words, the arm is an example of an acceleration control system. More elaborate models might include friction and springiness of the arm, as well as the complicating effects of multiple links (e.g., a hand, forearm, and upper arm). However, this simple model will be sufficient to illustrate some important aspects of lifting a glass of juice.

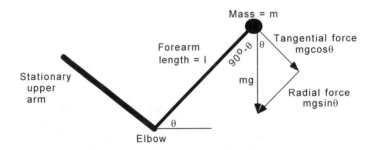

FIG. 20.1.   Gravitational torque acting on the forearm $= -l(mg\cos\theta)$.

A CONTROL STRATEGY WHEN THE MASS
IS KNOWN

Suppose that the equivalent mass of the glass of juice and arm, m, is known. Suppose further that the desired movement trajectory, θ_d, can be approximated as half of a cycle of a sine wave, going from its minimum, $-A$, to its maximum, $+A$:

$$\theta_d = A \sin(\omega t - \pi/2) \qquad 0 < t < \pi/\omega \tag{6}$$

Then a control strategy to implement this desired movement pattern is to make the torque, τ, applied by the robot motor equal to:

$$\tau = mgl\,\cos\theta + ml^2\left(\frac{d^2\theta_d}{dt^2}\right) + ml^2\left[K_1\left(\frac{de}{dt}\right) + K_2 e\right] \tag{7}$$

where $e = \theta_d - \theta$, the instantaneous error or difference between the desired and actual arm angle. K_1 and K_2 are constants. The total torque, τ, consists of the sum of three functions that serve three different purposes.

The first term in Equation 7 represents a feedback control signal that is a nonlinear function of the actual angle of the forearm, θ. This component of τ is intended to cancel the gravitational torque, which can be considered to be a nonlinear disturbance acting on the limb (Fig. 20.2). If the expression for τ shown in Equation 7 is substituted into Equation 5, which represents the effects of both gravity and τ on the limb, the gravitational influence will effectively be eliminated (i.e., the first two terms sum to zero):

$$\left\{-mgl\,\cos\theta + mgl\,\cos\theta + ml^2\left(\frac{d^2\theta_d}{dt^2}\right) + ml^2\left[K_1\left(\frac{de}{dt}\right) + K_2 e\right]\right\} / ml^2 = \frac{d^2\theta}{dt^2} \tag{8}$$

Using feedback control to cancel an unwanted nonlinear influence of some environmental force is a commonly used strategy in many different types of control systems (e.g., Slotine & Li, 1991, chap. 6). Its application to robotic control is referred to as *computed-torque control*.

The second term in Equation 7 represents feedforward operating on the time history of the desired trajectory, θ_d. Because the arm is an acceleration control system (two integrators, Fig. 20.2), the torque component for the feedforward control must be proportional to the acceleration of the desired trajectory. When this torque is integrated twice, the desired position will result. In other words, the feedforward component of torque effectively uses an inverse model of the dynamic response of the arm (e.g., Jordan, 1996; see also chap. 16). The feedforward control is generated by twice differentiating the desired trajectory, θ_d, and then multipling this desired acceleration pattern by ml^2. The dynamic response of the arm inverts these processes by effectively multiplying this torque component by $1/ml^2$, and integrating it twice to generate the actual arm angle, θ. Note that if the initial condition of the arm is quiescent and there are no unexpected perturbations on the arm, then the error, e, will be zero. The only two uncancelled nonzero terms in Equation 8 would be the two acceleration terms, that is, the actual acceleration of the arm, $d^2\theta/dt^2$, would equal the desired acceleration, $d^2\theta_d/dt^2$. The inverse model of the arm used for feedforward control does not include the effects of gravity, because that effect has already been cancelled out. If the gravitational torque had not been separately cancelled out, then the feedforward term would have had to take gravity into account (e.g., add a term $mgl\cos\theta_d$, if the desired arm angle, θ_d, closely approximates the actual arm angle, θ) and/or the error nulling part of the control strategy would have larger errors to diminish.

The third term in Equation 7 represents feedback control to null out any differences between the desired and actual arm angles. Namely, if the initial condition of the arm is not quiescent and/or there are various perturbations, then the actual arm angle will not equal the desired arm angle. The error will be nonzero, and some additional component of torque will be necessary to null out this error. As discussed in previous chapters on the McRuer crossover model (chaps. 14 and 16), the use of lead (i.e., an additive combination of error and error velocity) is an effective error nulling strategy with an acceleration control system. This control strategy is also referred to as proportion plus derivative (PD) control. Moving all the terms in Equation 8 to the left side and noting that $d^2\theta_d/dt^2 - d^2\theta/dt^2 = d^2e/dt^2$, the following is obtained:

$$\frac{d^2e}{dt^2} + K_1\left(\frac{de}{dt}\right) + K_2 e = 0 \tag{9}$$

This equation resembles the prototypical equation for a linear second-order system:

$$\frac{d^2e}{dt^2} + 2\zeta\omega_n\left(\frac{de}{dt}\right) + \omega_n^2 e = 0 \tag{10}$$

where ζ is the damping ratio, and ω_n is the undamped natural frequency (chap. 6). To achieve critical damping (i.e., $\zeta = 1$), set $K_1 = 2\omega_n$ and $K_2 = \omega_n^2$. The larger the value of ω_n, the faster any errors will be nulled out, but at a cost of using larger torques.

In summary, if the effective mass of the arm plus glass of juice is known, then Equation 8 represents a combination of feedback to cancel gravitational effects, feedforward of the desired movement trajectory, and feedback for error nulling. The combination of these three components of torque is one way a robot arm could be controlled (e.g., Craig, 1989; Slotine & Li, 1991). Whether people use a similar strategy is an interesting question.

INDIRECT ADAPTIVE CONTROL
WITH AN INITIALLY UNKNOWN MASS

The previous procedure (Equation 7) for generating torque, τ, assumed that the effective mass of the glass of juice plus the arm, m, was known. This m serves as a gain or scalar multiplier of the torque pattern, in that each of the terms in Equation 7 contains m as a factor. If m is not known, then that parameter in Equation 7 needs to be adap-

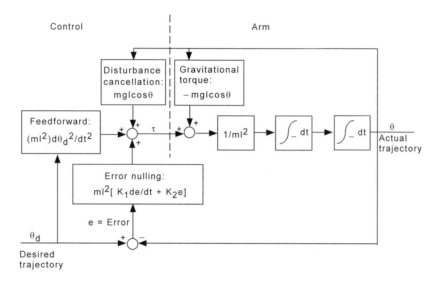

FIG. 20.2.   Control of a single-link robot arm when the effective mass, m, is known.

tively adjusted. There are a number of approaches to adjusting control parameters while the control process is ongoing. A primary distinction among these approaches is that some are called *direct*, and some are called *indirect*. Direct adaptive control uses the tracking error, e, the difference between the desired and actual trajectories, to adjust the control parameter (Fig. 20.3b). In contrast, indirect adaptive control uses a model of the controlled system (e.g., the arm) to predict the dynamic relation between its response and the applied torque. The prediction error (i.e., the difference between the predicted relation and the observed relation) is used to adjust an estimate of the unknown parameter, m_{est} (Narendra & Annaswamy, 1989; Fig. 20.3a). This estimate is then used to implement a control pattern such as Equation 7. There are also hybrid approaches that utilize both the tracking error and the prediction error to adjust the unknown parameter (e.g., Duarte & Narendra, 1989; Slotine & Li, 1989). However, this chapter simply presents examples of direct and indirect adaptive control.

For indirect adaptive control of the single-link robot arm, try to estimate the relation between the torque applied to the arm, τ, and the dynamic response of the arm. It is mathematically convenient to consider τ as a function of θ rather than vice versa (e.g., Slotine & Li, 1989). According to Equation 5:

$$\tau = m\left[l^2\left(\frac{d^2\theta}{dt^2}\right) + gl\cos\theta\right] \qquad (11)$$

Namely, torque is proportional to an additive combination of limb acceleration and the cosine of limb angle. A dynamic model of the arm would represent this same relation, but with m replaced by an estimate of m, namely, m_{est}. The predicted torque, $\hat{\tau}$, would be:

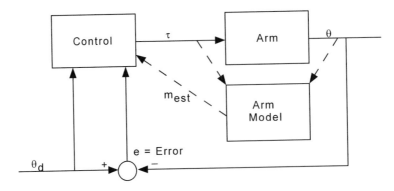

a. Indirect adaptive control

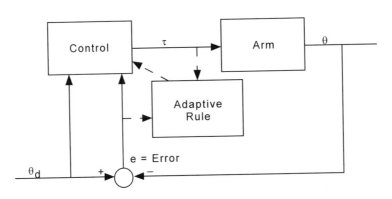

b. Direct adaptive control

FIG. 20.3.   Indirect and direct adaptive control. Solid lines indicate the flow of information for control as in Fig. 20.2. Dashed lines indicate the flow of information for adaptive
adjustment of the control.

$$\hat{\tau} = m_{est}\left[l^2\left(\frac{d^2\theta}{dt^2}\right) + gl\,\cos\theta\right] \tag{12}$$

If τ, $d^2\theta/dt^2$, and θ are all observable,[2] and g and l are known constants, then the numerical value of m_{est} might be chosen to minimize the squared deviations between τ
and $\hat{\tau}$ (i.e., the squared prediction error):

$$\left\{\tau_i - m_{est}\left[l^2\left(\frac{d^2\theta_i}{dt^2}\right) + gl\,\cos\theta_i\right]\right\}^2 \tag{13}$$

---

[2]If the acceleration is not directly observable, but the velocity $d\theta/dt$ is, then it is possible to similarly estimate the mass from a low-pass filtered version of Equation 12 that has $d\theta/dt$ as the highest derivative (see
Slotine & Li, 1991, pp. 360–362).

where the subscript i indicates samples of torque, acceleration, and limb angle taken at times t_i. This problem is equivalent to finding the slope of the best fitting line through a set of noisy data, namely, a linear regression problem, $y_i = m_{est}x_i$, where $y_i = \tau_i$, and $x_i = (l^2 d^2\theta_i / dt^2 + gl\cos\theta_i)$. The accuracy of the estimated mass, m_{est}, will improve as the number of samples increases, if τ, $d^2\theta / dt^2$, and θ vary over time. m_{est} can be calculated recursively in a number of different ways (e.g., Slotine & Li, 1989), successively updating m_{est} on the basis of each new sample (chaps. 18 and 24; see Atkeson, An, & Hollerbach, 1986, for a more advanced treatment of the estimation problem). The estimated mass, m_{est}, can be substituted for m in Equation 7 to implement the indirect adaptive control. Thus, the estimation process and the control process can proceed together; however, the form of the estimation process does not depend on the details of the control process.

DIRECT ADAPTIVE CONTROL WITH AN INITIALLY UNKNOWN MASS

In indirect adaptive control, the processes of estimating the unknown parameter(s) and implementing a control strategy that utilizes that estimate are independent. The independence of these processes provides flexibility to a system designer, who may use any of a variety of different approaches for either process. However, proving the stability of such a system for large initial inaccuracies of the unknown parameters may be difficult to achieve. In contrast, direct adaptive control does not treat these processes independently, but uses the tracking error to adjust the unknown control parameters. Typically, the direct approach minimizes a composite function that includes both tracking error and parameter estimation error. This composite function is called a *Lyapunov function*, after the Russian mathematician who first suggested this style of adaptive control. For the single link robot arm, a possible Lyapunov function, L, is a quadratic function of deviations from the desired trajectory and errors in estimating the arm's moment of inertia. In particular:

$$L = \left(\frac{1}{2}\right)\left[ml^2s^2 + \left(\frac{1}{k}\right)(ml^2 - m_{est}l^2)^2 \right] \tag{14}$$

where $s = de/dt + \omega_n e$. In this case, s can be considered a "quickened" error signal (see chaps. 9 and 26), a measure of anticipated deviations from the desired trajectory. The second additive term in L is the difference between the actual moment of inertia of the arm and the estimate of that quantity. Because l is assumed to be known, the only uncertainty in the estimate of the moment of inertia arises from m_{est}. If m_{est} is updated in a clever manner (Slotine & Li, 1991), then L will gradually decrease over time, and never increase. In fact, if the motion of the arm is sufficiently complex, then L will approach zero, which implies that the actual trajectory will approximate the desired trajectory, and the estimated mass will approximate the actual mass.

In particular, suppose that the torque is generated by the control strategy in Equation 7, but using m_{est} in place of m. Suppose also that K_1 and K_2 have been adjusted for critical damping with a frequency ω_n:

$$\tau = m_{est}l^2\left[\frac{g\cos\theta}{l} + \left(\frac{d^2\theta_d}{dt^2}\right) + 2\omega_n\left(\frac{de}{dt}\right) + \omega_n^2 e\right] \tag{15}$$

If the estimated moment of inertia, $m_{est}l^2$, is updated in the following manner (Slotine & Li, 1991):

$$\frac{d(m_{est}l^2)}{dt} = k\left(\frac{\tau}{m_{est}l^2}\right)s \tag{16}$$

then dL/dt will never be positive. (k is a gain that may be set to achieve efficient adjustment of $m_{est}l^2$.) To see how this inference follows, first calculate dL/dt:

$$\frac{dL}{dt} = ml^2 s\left(\frac{ds}{dt}\right) + \frac{1}{k}(ml^2 - m_{est}l^2)\left(\frac{-dm_{est}l^2}{dt}\right) \tag{17}$$

Substituting from Equation 16 and simplifying:

$$\frac{dL}{dt} = ml^2 s\left(\frac{ds}{dt}\right) + s\left(\tau - \frac{m\tau}{m_{est}}\right) \tag{18}$$

Substituting for the τ on the left from the relation representing the arm dynamics (Equation 5, or equivalently Equation 11), and substituting for the τ on the right from the control strategy (Equation 15), and simplifying, the following is obtained:

$$\frac{dL}{dt} = ml^2 s\left(\frac{ds}{dt}\right) - ml^2 s\left[\frac{d^2 e}{dt^2} + 2\omega_n\left(\frac{de}{dt}\right) + \omega_n^2 e\right]$$

$$\frac{dL}{dt} = ml^2 s\left(\frac{ds}{dt}\right) - ml^2 s\left[\frac{ds}{dt} + \omega_n s\right]$$

$$\frac{dL}{dt} = -ml^2 \omega_n s^2 \tag{19}$$

Because m, l, and ω_n are all positive, and s^2 is greater than or equal to zero, dL/dt will either be negative or zero, but not positive. Given a sufficiently variable trajectory, L will approach zero, meaning that m_{est} will approach the actual mass m, and the trajectory error, $\theta_d - \theta$, will get very small.[3] Note that the clever choice for updating the estimated moment of inertia (Equation 16) strongly depended on the particular control strategy, unlike the previous indirect adaptive control. However, it is also possible to use Lyapunov functions with certain examples of indirect adaptive control (e.g., Narendra & Annaswamy, 1989).

---

[3]In general, if the trajectory has a very simple structure, it is possible for the trajectory error to approach zero without the parameter estimation error approaching zero (e.g., Narendra & Annaswamy, 1989; Slotine & Li, 1991).

ADAPTIVE CONTROL IN HUMAN PERFORMANCE

The previous examples of adaptive control are computationally complex, and it is uncertain whether human movement closely parallels these possible approaches to movement control. Important aspects of human movement control may involve prediction of the consequences of motor commands to overcome internal time delays (e.g., Bushan & Shadmehr, 1999), and the motor commands may be more complex than simple torques (e.g., Giszter, Mussa-Ivaldi, & Bizzi, 1993). Nevertheless, the qualitative properties of robotic models may be useful for thinking about human motion control (Hollerbach, 1982). For example, in Fig. 20.2 the torque, τ, is generated from three distinct transformations: transformations on the desired trajectory, θ_d, which represent an inverse model of the arm dynamics; transformations on the actual trajectory, θ, which are a model of gravitational dynamics used to cancel such effects; and transformations on the difference between the actual and desired trajectories, e, to null out small deviations. These distinctions are useful in interpreting experiments on human adaptive control of movement.

People seem to be very good at executing movements under different external loads if they have some estimate of the load. For example, Laquaniti, Soechting, and Terzuolo (1982) had people point to different targets using multijoint arm movements. The movements started with the forearm held out horizontally, and participants knew whether they were moving a 2.5 kg weight or a light piece of styrofoam on any given trial. Crossplots of the horizontal and vertical positions of the wrist and of the shoulder and elbow angles revealed approximately the same trajectory with either the heavy or the light load. These data suggest that the desired trajectory to a target (i.e., the multijoint equivalent of θ_d in Fig. 20.2) does not change as the load is varied. In the simplified one-dimensional control problem in Fig. 20.2, changing the mass of the object being moved would require a change in the parameter m in the torque generation process, which would correspond to a change in gain. This is by no means a trivial change, as suggested by the aforementioned adaptive control techniques, and it would be more involved for a multilink arm, e.g., Atkeson et al., 1986. However, this type of change may be simpler than one involving other alterations in the torque generation process.

For a more elaborate example, in an experiment involving multisegment horizontal arm movements, Shadmehr and Mussa-Ivaldi (1994) required human participants to grasp a two-joint robot arm and move its endpoint to various stationary targets. As noted by many previous investigators, the human participants' movement trajectories to each of the targets was in a straight line in the two-dimensional movement space. The experimenters interpreted this straight line trajectory as the desired movement trajectory (analogous to θ_d in Fig. 20.2). They then introduced a velocity dependent force on the participants' movements by means of small motors coupled to the joints of the movable robot arm. Movement in certain directions produced forces that retarded movement, similar to moving through water; movement in other directions produced forces that aided the movement. When these additional forces were first introduced, the participants' movements to the targets exhibited marked degrees of curvature that varied depending on the direction of the movement. However, as they practiced moving in this unusual force field over a series of nearly 1,000 targets, participants' movement trajectories gradually approached straight line trajectories once again.

The investigators' interpretation of this change was that participants gradually learned how the pattern of perturbing forces were related to their movements, and generated torques to cancel these additional forces in a manner roughly analogous to the way the gravitation forces are canceled in Fig. 20.2.[4] An alternate interpretation of the return to straight line trajectories is that participants simply stiffened their arms to make them less sensitive to perturbations. This strategy would be analogous to increasing K_2 (and possibly K_1) in the error-nulling aspect of the control strategy in Fig. 20.2. The investigators noted, however, that removing the unusual perturbing forces by turning off the motors attached to the robot arm resulted in a re-introduction of curved trajectories. The curvatures were now in the opposite directions from when the unusual forces were first introduced. If participants were simply increasing the stiffness of their arms, then this aftereffect of removing the unusual forces would not occur. It would occur if the participants had learned to generate an additional component of torque to cancel the unusual forces. This additional torque would not be needed once the motors were turned off, and would produce the oppositely directed curvatures.

An additional issue that arises in multijoint movement (and not in the single joint motion in Fig. 20.2) is whether the movement control occurs in a coordinate system based in environmental coordinates or in joint coordinates. For example, in moving to an externally specified target, a person must perceive the target position in space and somehow translate that information into appropriate torque commands to the various joints. The intermediate steps in that process are a subject of active investigation. Whether the desired trajectory is internally represented in environmental and/or joint coordinates and to what level of detail is a topic of debate (e.g., Hollerbach, 1990). Shadmehr and Mussa-Ivaldi (1994) argued that the learning of force cancellation occurs in joint coordinates. Evidence for this latter conclusion is their participants' superior transfer to novel targets when the perturbing forces generated by the motors depend on joint velocities rather than on the velocity of the manipulator endpoint in environmental coordinates.

In summary, this chapter has considered formal models of torque generation that involve feedforward, feedback cancellation of environmental disturbances, and error nulling. Both direct and indirect approaches were considered for adaptively adjusting the torque in response to unknown changes in the mass of the object moved. It was assumed that the dynamic model of the arm was known, and the only uncertainty was in the value of the mass. In practice, any dynamic model may only be a reasonable approximation to the actual physical system being controlled. There is considerable interest in designing adaptive algorithms that are "robust" (i.e., still perform reasonably well) in the presence of certain classes of unmodeled dynamics and/or perturbations (e.g., Ioannou & Sun, 1996). Such efforts may eventually provide additional insight into human performance. A more difficult problem is to begin

---

[4]It could also be hypothesized that participants learned a new relation between the desired trajectory and the required torque for feedforward control. The more elaborate arm movement model of Bushan and Shadmehr (1999) includes such a mechanism.

with a totally unknown dynamic system (e.g., a bicycle, a car, or an airplane) and to learn both the dynamic structure and the parameters in order to generate appropriate control. Approaches to this problem raise issues analogous to the distinction between direct and indirect adaptive control (e.g., Jordan, 1996).

REFERENCES

Astrom, K. J., & Wittenmark, B. (1989). *Adaptive control*. Reading, MA: Addison-Wesley.

Atkeson, C. G. (1989). Learning arm kinematics and dynamics. *Annual Review of Neuroscience, 12*, 157–183.

Atkeson, C. G., An, C. H., & Hollerbach, J. M. (1986). Estimation of inertial parameters of manipulator loads and links. *International Journal of Robotics Research, 5*, 101–119.

Bushan, N., & Shadmehr, R. (1999). Computational nature of human adaptive control during learning of reaching movements in force fields. *Biological Cybernetics, 81*, 39–60.

Craig, J. J. (1989). *Introduction to robotics*. Reading, MA: Addison-Wesley.

Duarte, M. A., & Narendra, K. S. (1989). Combined direct and indirect adaptive control of plants with relative degree greater than one. *Automatica, 25*, 3.

Giszter, S. F., Mussa-Ivaldi, F. A., & Bizzi, E. (1993). Convergent force fields organized in the frog's spinal cord. *Journal of Neuroscience, 13*, 467–491.

Hollerbach, J. M. (1982). Computers, brains and the control of movement. *Trends in Neurosciences, 5*, 189–192.

Hollerbach, J. M. (1990). Planning of arm movements. In D. N. Osherson, S. M. Kosslyn, & J. M. Hollerbach (Eds.), *Visual cognition and action: An invitation to cognitive science* (Vol. 2, pp. 183–211). Cambridge, MA: MIT Press.

Ioannou, P. A., & Sun, J. (1996). *Robust adaptive control*. Upper Saddle River, NJ: Prentice-Hall.

Jordan, M. I. (1996). Computational aspects of motor control and motor learning. In H. Heuer & S. W. Keele (Eds.), *Handbook of motor control* (pp. 71–120). San Diego, CA: Academic Press.

Laquaniti, F., Soechting, J. F., & Terzuolo, C. A. (1982). Some factors pertinent to the organization and control of arm movements. *Brain Research, 252*, 394–397.

Lewis, F. L., Abdallah, C. T., & Dawson, D. M. (1993). *Control of robot manipulators*. New York: MacMillan.

Narendra, K. S., & Annaswamy, A. M. (1989). *Stable adaptive systems*. Englewood Cliffs, NJ: Prentice-Hall.

Shadmehr, R., & Mussa-Ivaldi, F. A. (1994). Adaptive representation of dynamics during learning of a motor task. *Journal of Neuroscience, 14*, 3208–3224.

Slotine, J. J. E., & Li, W. (1989). Composite adaptive control of robot manipulators. *Automatica, 25*, 4.

Slotine, J. J. E., & Li, W. (1991). *Applied nonlinear control*. Englewood Cliffs, NJ: Prentice-Hall.

Smyth, M. M., Collins, A. F., Morris, P. E., & Levy, P. (1994). *Cognition in action*. Hove, UK: Lawrence Erlbaum Associates.

Thelen, E., & Smith, L. B. (1995). *A dynamic systems approach to the development of cognition and action*. Cambridge, MA: MIT Press.

Sine Wave Tracking
Is Predictably Attractive

> *There are a lot of good reasons why rhythmical movements are a good place to start. Rhythmical behaviors are ubiquitous in biological systems. Creatures walk, fly, feed, swim, breathe, make love, and so forth. Rhythmical oscillations are archetypes of time-dependent behavior in nature, just as prevalent in the inanimate world as they are in living organisms. Although they may be quite complicated, we have the deep impression that the principles underlying them should possess a deep simplicity. Ordering or regularity in time is important also for technological devices, including computers.*
>
> —Kelso (1995, pp. 46–47)

Given that the sum of a large number of sine waves is used in Fourier analysis to approximate human tracking of random signals, an additional question might concern how well humans manually track single sine waves. The short answer is "very well" if the frequency is not too high. However, the pattern of responses across single sine waves of different frequencies does not resemble the patterns seen for random input tracking. For a fixed linear system, a Fourier analysis of its response to a random input signal could be performed, and then, from the transfer function, the amplitude and phase lag for single sine waves could be predicted. They would be the same for the two types of inputs. That is not so for the human tracker. For moderate to high frequencies, the response to single sine waves is much better in amplitude ratio (closer to 1) and phase lag (closer to 0 degrees) than would be predicted from the describing functions for random appearing input signals (Krendel & McRuer, 1968; Stark, 1966). The human tracker is more adaptive than a fixed linear system and takes advantage of the predictability of the sinusoidal input pattern. This chapter discusses different mathematical models of how the human might accomplish this task.

As noted in chapter 11, sine waves correspond to temporal patterns characterized by three parameters: amplitude, frequency, and phase. How well does the human tracker do at matching these pattern parameters? For manual tracking with an explicit display of the sinusoidal input signal, a well-practiced performer tends to match the frequency and maintain relatively small phase lags or leads up to about 2 Hz (e.g., Leist, Freund, & Cohen, 1987; Pew, Duffendack, & Fensch, 1967; Stark, 1966). Beyond about 2 Hz, the eyes stop actively tracking the sinusoidal input signal and remain relatively stationary, and it becomes difficult to stay in phase with the target (Leist et al., 1987). Frequency matching errors may also develop (e.g., Pew et al., 1967).

Amplitude matching can also become difficult at this frequency for certain limbs. In the case of hand and wrist movements, maximal movement amplitude begins to decrease around 2 Hz due to temporal overlap of agonist–antagonist muscle activations that begin to cancel each other (Hefter, Reiners, & Freund, 1983; Leist et al., 1987). However, provided that maximal movement amplitudes are not required, amplitude matching can persist at frequencies beyond those for which phase matching is not possible (Leist et al., 1987; Magdaleno, Jex, & Johnson, 1969). This type of behavior has been called *range control* by Freund (1986).

OPEN-LOOP LEAD

Consider a number of possible control strategies that a person might use to perform amplitude, frequency, and phase matching of a single sinusoid. Given that people exhibit small phase lags out to about 2 Hz, they cannot be simply relying on closed-loop feedback control as was used in compensatory tracking of random inputs. To see this point, simply refer back to the describing functions in chapter 14, and note the large phase lags at frequencies of 1 Hz.

One alternative would be to rely on feedforward (open-loop) control (e.g., Magdaleno et al., 1969; Pew et al., 1967; Ware, 1972).[1] To a first-order approximation, the human controller using a position control system might be approximated as an inherent time delay of about 200 ms and an open-loop lead that attempts to cancel that delay (Ware, 1972). As indicated in chapters 7 and 14, a lead consists of an additive combination of position and velocity information (referred to as "speed anticipation" by Poulton, 1952). This is a linear model, so its response frequency will be the same as the input frequency. If the weightings on position and velocity are adjusted appropriately, then the phase lag can be zero, and the response amplitude can match the input amplitude for a single sine wave.

To appreciate this point, consider Fig. 21.1, in which the input sine wave, $A\sin(\omega t)$, is represented as a rotating vector with amplitude, A, in a complex space. The horizontal axis is the real axis, and the vertical axis is the imaginary axis (refer to chap. 13). The projection of the rotating vector onto the real axis is the observable signal, a sinusoidal pattern. If the person's response is also sinusoidal in form, it can be represented as a second rotating vector. However, the human tracker has an inherent time

---

[1]Although feedforward control would be easiest to implement with a pursuit display that permitted direct view of the input signal, Pew et al. (1967) found that with extended practice sine wave tracking with pursuit displays (input and output) and compensatory displays (error only) were roughly equal. As noted by Krendel and McRuer (1968), the pursuit mode of response can be executed with even a compensatory display if the performer uses the error signal and their own response to estimate the input signal.

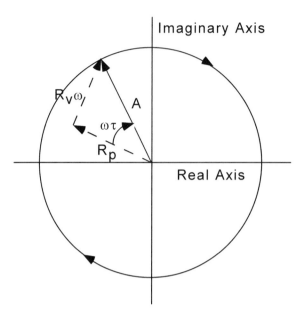

FIG. 21.1.   An open-loop lead strategy for sinusoidal tracking. The input sinusoid has amplitude A. The sinusoidal response matches the input with an additive combination of components with amplitudes R_p and $R_v\omega$, respectively proportional to delayed input position and velocity.

delay, so the response vector would be delayed (i.e., it would have a phase lag corresponding to the time delay, τ, times the input frequency, ω). If a component of the human tracker's response is proportional to the input position, this component, $R_p\sin(\omega t - \omega\tau)$, can be represented as a vector with amplitude R_p and phase lag $\omega\tau$ relative to the input vector. If another component of the human tracker's response is proportional to the input velocity, $R_v\omega\cos(\omega t - \omega\tau)$, then it can be represented as another vector with amplitude $R_v\omega$. In that the cosine function leads the sine function by 90°, the $R_v\omega$ vector is perpendicular to the R_p vector. In the open-loop lead model, these two components are added to produce the tracker's total response. The two vectors are added geometrically by placing the beginning of the $R_v\omega$ vector at the end of the R_p vector. From this geometric construction, it is apparent that R_p must equal $A\cos(\omega\tau)$, and $R_v\omega$ must equal $A\sin(\omega\tau)$ in order for the response to match the phase and amplitude of the input (see Ware, 1972, for an alternate derivation). Thus, the appropriate weightings of delayed input position and velocity depend on the sinusoidal frequency and on the person's effective time delay.

For very slow frequencies (i.e., small ω) the component of the person's response that is proportional to velocity will be small because the phase lag, $\omega\tau$, will be small. Both the phase lag and the response component proportional to velocity will increase in magnitude as the sine wave frequency increases. For example, when the person's effective time delay corresponds to one-eighth cycle of the sine wave (i.e., $\omega\tau = 360°/8 = 45°$), then the R_p vector will trail the input vector by 45°, and the R_p and $R_v\omega$ vectors will be of equal length. Note that the strategy of using a lead to overcome the person's time delay is a general one that could rely on instantaneous estimates of the input signal's position and velocity and therefore could be used as well in pursuit tracking of random appearing, relatively unpredictable inputs (e.g., Elkind, 1956; Ware, 1972) and in pursuit tracking of simple temporal patterns with different shapes (e.g., Jagacinski & Hah, 1988). However, in these cases, the input position and velocity would not have the simple sinusoidal form indicated in Fig. 21.1, and the lead could only approximately compensate for the time delay. For the sine wave, the lead is theoretically capable of perfectly overcoming a time delay.

Researchers believe that the aforementioned strategy or some similar strategy is used for low frequency sine waves (e.g., Magdaleno et al., 1969; Pew et al., 1967; Poulton, 1952; Ware, 1972). At high frequencies, limitations in estimating the input velocity might reduce the effectiveness of both amplitude and phase matching. There is strong evidence that at high frequencies an alternate strategy is used. One common observation is that at very high frequencies (about 2 Hz or higher), people produce frequency errors (e.g., Ellson & Gray, 1948; Noble, Fitts, & Warren, 1955; Pew et al., 1967). Namely, the frequency of their sinusoidal motion tends to vary away from the frequency of the input signal and produce beat frequencies in the error pattern (see Fig. 12.6). Very marked frequency errors also occur in the sinusoidal tracking of young children at frequencies below 1 Hz (Mounoud, Viviani, Hauert, & Guyon, 1985). These types of errors do not correspond to the open-loop lead model, because the model's response is a linear transformation of the input (i.e., a weighted sum of delayed position and velocity). The response frequency should therefore always match the input frequency according to this model. If human performers exhibit frequency-matching errors, then they cannot be relying on open-loop lead or other linear transformations of the sinusoidal target (e.g., see Noble et al., 1955; Mounod et al., 1985).

ADAPTIVELY ADJUSTING AN INTERNALLY GENERATED PATTERN

Another control strategy for performing sine wave tracking would be to generate a sinusoidal pattern and then adaptively adjust the amplitude, frequency, and phase based on observations of the input (e.g., Krendel & McRuer, 1968; Magdaleno et al., 1969; Pew et al., 1967; Poulton, 1952). In other words, the shape of the response is generated from some internal pattern generator rather than simply being a linear transformation of the input. The parameters of this ongoing pattern are then adaptively adjusted to match the input. If at very high frequencies this adjustment process began to break down, then the commonly observed frequency errors would result.

As a heuristic exercise in thinking about this problem, consider how the phase of the response pattern might be adjusted to match the input pattern. Once again, let the input pattern be $A\sin(\omega t)$. Let the response pattern be $A_r\sin(\omega_r t + \theta_r)$. For simplicity, assume that the two amplitudes match, $A_r = A = 1$. However, the response frequency, ω_r, may not exactly match the input frequency, ω. Also, the initial response phase, θ_r, may not match the input phase of zero. Negative values of θ_r correspond to phase lag, and positive values correspond to phase lead.

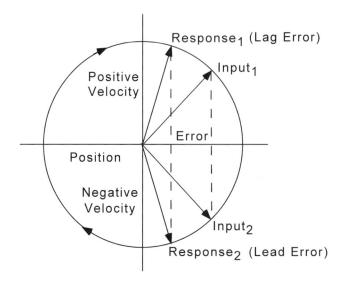

FIG. 21.2.   The importance of the sign of the response velocity for distinguishing lag errors from lead errors and determining the required direction of response phase adjustment. The vertical axis is velocity, and the horizontal axis is position (i.e., the phase plane).

A simple control strategy with some similarity to the crossover model would be to adjust the phase in proportion to the magnitude of the tracking error. Namely:

$$\text{error}(t) = \sin(\omega t) - \sin(\omega_r t + \theta_r) \tag{1}$$

$$d(\theta_r)/dt = K[\text{error}(t - \tau)][\text{sign(response velocity)}] \tag{2}$$

Equation 2 indicates that the response phase is adjusted at a rate that is proportional to the tracking error. K is the proportionality constant. The larger the error, the more quickly θ_r is adjusted. A time delay, τ, is included as usual to indicate that the adjustments are not made instantaneously. A value of $\tau = .15$ s is used in the following exercise.

The last term in Equation 2 is the sign of the response velocity. This term is included to get the appropriate polarity of the adjustment over most of the oscillation.[2] Namely, if the error is positive as in Fig. 21.2 (position of Input_1 – position of $\text{Response}_1 > 0$), and the person is moving in a direction that will make the response position more positive (response velocity > 0), then the response is lagging the input. θ_r should be increased to advance the response phase, as will occur by Equation 2. However, the same positive error could result from the response leading the input (position of Input_2 – position of $\text{Response}_2 > 0$), in which case θ_r should be decreased to reduce the lead error. Given that Response_2 has a negative velocity, multiplication

---

[2]This term effectively makes the parameter adjustment be in the direction that decreases the instantaneous squared error (e.g., see Astrom & Wittenmark, 1989).

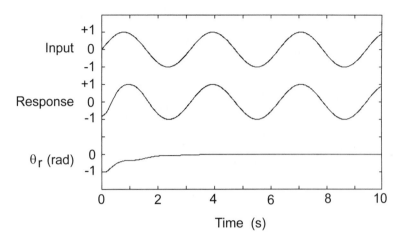

FIG. 21.3.   Nulling out a phase difference between a sinusoidal input and response.

by the sign of the response velocity in Equation 2 will permit the correct polarity for this adjustment.

Although the form of Equation 2 has some elements in common with the crossover model, an important difference is that it is nonlinear. Namely, there is a nonlinear relation between the error and θ_r. Multiplication of one variable by another (e.g., the error multiplied by the sign of the response velocity) is a nonlinear operation. Also, the sign function itself is nonlinear. Adaptive controllers are generally nonlinear (Astrom & Wittenmark, 1989). This point is raised because nonlinear systems typically exhibit a range of behaviors that are more complex than those exhibited by linear systems (e.g., Thompson & Stewart, 1986).

In order to examine the variety of behaviors exhibited by this model, suppose first that the response frequency matches the input frequency, $\omega = \omega_r = 1$. However, suppose that the initial response phase is $\theta_r = -1$ rad, which does not match the input. If K is set at some low value (e.g., $K = 1.5$), then the phase discrepancy is eliminated over a few cycles of the motion (Fig. 21.3).

Next consider what happens if the response frequency does not match the input frequency. Suppose that $\omega_r = 2$ r/s and remains constant, $\theta_r = 0$ rad initially, and ω gradually increases from 2 to 7 rad/s (Fig. 21.4). The difference between the input frequency and the response frequency creates a phase lag that the adaptive control strategy attempts to null out (Equations 1 and 2). However, this control strategy is not capable of achieving a perfect match to the input under these circumstances. It basically makes up for the frequency difference by having an increasing phase lag. This phase lag makes the error larger and $d\theta_r/dt$ higher (Equation 2), which effectively raises the frequency of the response (i.e., the total response phase, $\omega_r t + \theta_r$, increases more quickly). However, if the phase lag were to go to zero, then the basis for this effective increase in response frequency would be lost.

As the input frequency continues to increase, there comes a point where the phase lag becomes so great that the frequency matching is lost. To demonstrate this loss of frequency matching, the ratio of the input frequency to the response frequency has

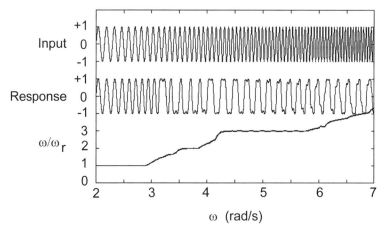

FIG. 21.4.   The inability of an error-based phase adjustment (Equation 2) alone to match the sinusoidal response to an input sinusoid with gradually increasing frequency.

been graphed in Fig. 21.4.[3] Note that the frequency ratio remains equal to 1.0 even as the input frequency passes 2.5 rad/s. However, just prior to reaching 3 rad/s, the frequency ratio starts to increase, which implies that the simple phase matching strategy can no longer keep up.

As the input frequency continues to increase, the frequency ratio might be expected to continuously increase. However, nonlinear systems are often surprising. The response no longer looks sinusoidal, its frequency slows down, and there are several regions in the graph of the frequency ratio versus input frequency that are approximately flat. These flat regions are largest in the neighborhood of simple integer ratios such as 2/1 and 3/1. However, there are also smaller flat regions discernible around 5/3 = 1.67 and 10/3 = 3.33. Such regions are sometimes referred to as *mode locking* regions in the vocabulary of nonlinear systems theory. Such regions are common in systems in which two oscillators are coupled together by weak forces, such as the phase coupling in the present example (e.g., Bak, 1986). Different models of mode locking have also been explored as possible descriptions of multilimb coordination involving oscillatory movement patterns (e.g., Peper, Beek, & van Wieringen, 1991; Treffner & Turvey, 1993).

Although the model behavior generated by Equations 1 and 2 is interesting, an important question concerns whether it corresponds to what people actually do. Given the task of synchronizing their movements with a sinusoid of increasing frequency without regard for amplitude matching, adults have maintained very good frequency and phase matching up to a frequency of 2.7 Hz (Wimmers, Beek, & van Wieringen, 1992). Over this range, the performers did not exhibit systematic frequency drifting or regions of stable ratios other than 1 to 1. Therefore, the adaptive scheme in Equations 1 and 2 is not a good model of this behavior.

---

[3]To generate the frequency ratio at the bottom of Fig. 21.4, the rate of change of the input frequency was much slower than in the simulation for the upper two plots, that is, the input frequency took longer to increase from 2 rad/s to 7 rad/s in order to provide a more accurate estimate of the frequency ratios achievable by Equations 1 and 2.

Better model performance is achieved if the rate of response phase adjustment, $d(\theta_r)/dt$, is proportional to the phase difference between the input and response, $\omega t - (\omega_r t + \theta_r)$. The response frequency then matches the input frequency as the input frequency is gradually increased from 2 rad/s to 7 rad/s. However, there is a linearly increasing phase lag that gets large at the higher frequencies unless there is assumed to be an implausibly large value for K. This behavior is analogous to the way a simple lag (chap. 4) falls farther and farther behind a constantly accelerating input (e.g., see Wilts, 1960, chap. 3). In this analogy, phase corresponds to position, frequency corresponds to velocity, and rate of change of frequency corresponds to acceleration.

Although these models represent only two possible ways of adjusting phase, they suggest that phase adjustment alone is insufficient for following a sine wave with increasing frequency. People are more adaptable and seem to be able to adjust the frequency of their response as well as their phase.[4] However, as already noted, at very high frequencies people's frequency matching breaks down.

Additional equations similar to Equations 1 and 2 could be postulated to model frequency adjustment and amplitude adjustment. The exact form of such equations would depend on whether the person is assumed simply to rely on instantaneous error and/or error velocity for adaptive parameter adjustment, or whether relative phase, frequency, and amplitude are assumed to be perceived and used to implement such adaptive behavior. Empirical tests of such equations need to be performed. However, there is evidence that amplitude and frequency matching can be maintained even when phase matching is lost at high frequencies (e.g., Leist et al., 1987; Magdaleno et al., 1969). Such behavior suggests that the adaptive matching process is multidimensional.

Although an internal sinusoidal pattern generator can be postulated as a unit of behavioral organization, various dynamic models also can be hypothesized that would generate sinusoidal patterns. For example, a linear combination of a mass, spring, and damper (chap. 6) has an impulse response consisting of damped sinusoidal motion. If such a system is driven by a sinusoidal input, then a sinusoidal output results. Liao and Jagacinski (2000) used such a model to describe sinusoidal tracking. As in a number of other second-order dynamic models of motion, the performer is assumed to be controlling the damping and/or springiness of the system, which in turn influence the amplitude, frequency, and phase of the motion. (In the simple sinusoidal pattern generator model described earlier, it was assumed that the phase was controlled more directly; Equations 1 and 2). In the open-loop lead model described earlier, it was assumed that the limb position was controlled more directly.) Although sometimes researchers identify the oscillatory dynamics in such mass-spring–damper models with the peripheral characteristics of a massive limb controlled by opposing, springy muscles and associated peripheral reflexes (e.g., Feldman, 1986; Latash, 1992), the dynamic model can capture a combination of central and peripheral aspects of perceptual-motor coupling in a more holistic fashion (e.g., Viviani & Mounoud, 1990). Complex aspects of such models can include nonlinear damping and/or spring characteristics, which are beyond the scope of this chapter.

---

[4] For related discussions of these issues in the domain of rhythmic models of attention, see Large and Jones (1999) and McAuley (1994).

OSCILLATORY BEHAVIOR AND ATTRACTORS

Another indication that amplitude matching may be independent of frequency or phase matching comes from the observation that experimental participants seem to be readily able to concentrate on frequency and phase matching if they are given instructions not to be concerned about amplitude matching. For example, Wimmers et al. (1992) asked experimental participants to track a horizontally moving sinusoidal target signal with side to side movements of their lower arm. The visual display of the target signal did not include any additional display of arm movement or of tracking error, and participants were able to choose the amplitude of their arm movements without having to match the target amplitude. On some trials, participants were instructed to maintain 0° phase lag between their arm movements and the target (in-phase); on other trials, they were instructed to maintain 180° phase lag between their arm movements and the target (anti-phase).

As the frequency of the target sinusoid was gradually increased from 1.5 Hz to 2.7 Hz within a single trial, participants were able to maintain the in-phase relation between the target and their arm movements. However, the participants were unable to maintain the anti-phase pattern over this frequency range. When the target frequency reached the neighborhood of 2 Hz, the standard deviation of the phase lag increased, and there was a transition from the anti-phase to the in-phase mode of tracking.[5] Additionally, the time that it took participants to recover from brief perturbing torque pulses rapidly increased in the anti-phase mode just before the transition frequency.

Transitions from anti-phase to in-phase movement patterns have been previously observed in simultaneous movements of multiple limbs (e.g., Haken, Kelso, & Bunz, 1985). Additionally, it was observed that if the target frequency was reduced after an anti-phase to in-phase transition, the in-phase mode persisted (i.e., the reverse transition did not occur). As in the Wimmers et al. (1992) experiment, just before the mode transition there was an increase in the time to recover from perturbations, which led to an increase in phase variability. This phenomenon has been termed *critical slowing* (e.g., Kelso & Scholz, 1987).

The formal modeling of this behavior has relied on the concept of an *attractor*. An attractor is a locus in state space to which a system will return if it is perturbed (e.g., Crutchfield, Farmer, Packard, & Shaw, 1986; Kelso, Ding, & Schoner, 1993). A *state space* is a space in which the dimensions correspond to a set of variables that, if specified along with any external forces, uniquely determine the future behavior of the system. The two-dimensional phase plane is an example of a state space for simple systems that only require the specification of position and velocity to determine their subsequent behavior. More complicated systems require higher dimensional state spaces to determine their subsequent behavior. The behavior of a system over time forms a trajectory in such a space without any branch points (i.e., the trajectory never acts as though it has reached a fork in the road at which it sometimes turns right and sometimes turns left). If the trajectory did have a branch point, then additional infor-

---

[5]The performance was classified as being in the in-phase or anti-phase mode if the phase lag was within ±45° of the required values of 0° or 180°, respectively.

mation would be required to determine which branch it would take. However, a state space, by definition, includes sufficient information to avoid such indeterminacy. Therefore, system trajectories in a state space do not exhibit branch points (e.g., Ogata, 1967).

The shape of a set of trajectories in state space resulting from various perturbations of a system may reveal certain invariant properties. For example, if the system always returns to a single point in the state space after a perturbation, the behavior is said to be characterized by a "point attractor" (Fig. 21.5, A). Paralleling the descriptions of bimanual coordination (Haken et al., 1985), in the Wimmers et al. (1992) experiment the temporal variations in the phase difference (lag or lead) between the target sinusoid and the horizontal arm movement can be described in terms of two competing point attractors—a strong attractor corresponding to 0° phase difference and a weaker attractor corresponding to 180° phase difference. The human performer behaves so as to nullify perturbations away from these two point values, one for in-phase movement and one for anti-phase movement. If it is assumed that both attractors become weaker as the frequency of motion increases (so that the phase difference is less and less tightly regulated), then the transition from anti-phase tracking to in-phase tracking can be attributed to the stronger in-phase point attractor dominating the weaker anti-phase point attractor. Namely, even though the performer is trying to exhibit anti-phase movement at high frequencies, the tendency for in-phase tracking is simultaneously present, and at a high enough frequency a perturbation can cause the stronger in-phase tendency to begin to dominate the performer's behavior.

Point attractors are sometimes represented as *valleys*, or *potential functions*, with the system's relative phase being analogous to a ball that is drawn toward the bottom of the valley (Fig. 21.6). The bottom of the deeper valley corresponds to 0° phase, the

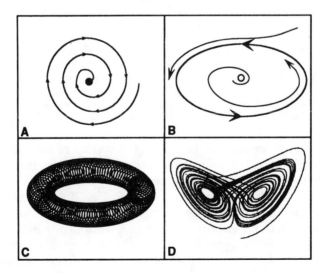

FIG. 21.5.   Examples of a point attractor (*A*), limit cycle attractor (*B*), quasiperiodic attractor (*C*), and chaotic attractor (*D*). From "Chaos, self-organization, and psychology" by S. Barton, 1994, *American Psychologist, 49,* pp. 5–14. Copyright 1994 by APA. Reprinted by permission.

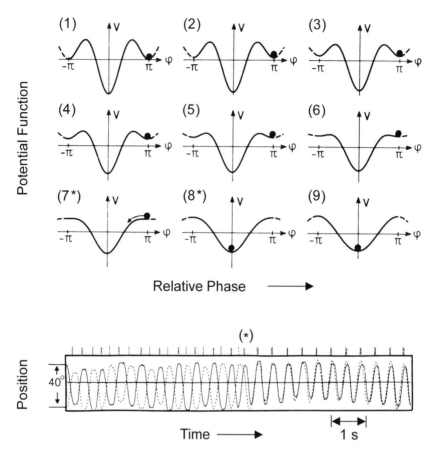

FIG. 21.6. (Top) Competing point attractors for in-phase and anti-phase motion that weaken as the movement frequency in increased. (Bottom) The positions of two hands over time. The asterisk is located at a transition from anti-phase to in-phase motion, which is also depicted in (7*) and (8*). From "A theoretical model of phase transitions in human hand movements" by H. Haken, J. A. S. Kelso, and H. Bunz, 1985, *Biological Cybernetics, 51*, p. 348, fig. 1 and p. 350, fig. 5. Copyright 1985 by Springer-Verlag. Adapted by permission.

point attractor for in-phase motion. Similarly, the bottom of the shallower valleys corresponds to 180° phase, the point attractor for anti-phase motion. Depictions of the weakening competing attractors as the frequency increases and a mode transition from anti-phase to in-phase behavior are shown in Fig. 21.6 (Haken et al., 1985).

Another behavior that can be approximated as a point attractor is the *critical tracking task* (chap. 15). In the critical tracking task without any external input (Jex, McDonnell, & Phatak, 1966), perturbations due to the person's own unsteadiness are amplified by unstable dynamics. The person acts to bring the cursor back to the center of the screen and overcome this noise amplification. The overall system acts as a stable point attractor. However, the time constant of the unstable dynamics becomes shorter, the person lowers their gain, and the strength of the point attractor becomes weaker. Eventually the person loses control of the system, which can be described as the point attractor becoming a point repellor (i.e., the overall system becomes unstable).

A second common type of attractor is a *limit cycle attractor* (e.g., Crutchfield et al., 1986; Kelso et al., 1993; Fig. 21.5, B). A repeating oscillatory motion will exhibit a closed orbit in a state space, and this closed orbit is a limit cycle attractor. For example, a second-order damped linear system driven by a sinusoidal input will exhibit a limit cycle. Perturbations away from the limit cycle will be nulled out. A person performing spontaneous oscillatory motions (Kelso, Holt, Rubin, & Kugler, 1981), sinusoidal synchronization without an amplitude constraint (Wimmers et al., 1992), or sinusoidal tracking (e.g., Liao & Jagacinski, 2000) can also be described as a limit cycle attractor. Although the transition from anti-phase to in-phase noted previously can be described in terms of point attractors for relative phase, the motion of the person's arm in such tasks can also be described as a limit cycle oscillator (Haken et al., 1985; Wimmers et al., 1992). Short perturbations of the oscillatory movement, such as experimentally applied torque pulses, are nulled out, and the motion returns to its cyclic state space orbit (Kelso et al., 1981; Wimmers et al., 1992).

The McRuer crossover model approximates people's performance in the critical tracking task, but also in many other compensatory tracking tasks. In such tasks, the person plus control system basically act as an attractor trying to null out perturbations away from the input signal. The shape of the input signal therefore determines the form of the attractor. If the input signal consists of a sum of sine waves that are not harmonically related, as could be the case in generating a relatively unpredictable input signal, the input signal will map out a high dimensional torus in a state space that represents each of the sine waves. For example, if there were only two non-harmonically related sinusoids, then their combined trajectory would look like a three-dimensional doughnut or torus in such a state space (Fig. 21.5, C; see also Abraham & Shaw, 1982). One oscillation would correspond to going around the doughnut from inside to outside and back again. The other oscillation would correspond to going around the doughnut in an orthogonal direction (i.e., mimicking the O shape). The combination of these two movements would cover the surface of the doughnut with a nonintersecting trajectory if the two sinusoids were not harmonically related (i.e., their frequency ratio was an irrational number). An attractor exhibiting this shape is called "quasiperiodic" (e.g., Crutchfield et al., 1986). The sum of a large number of nonharmonically related sinusoids would result in a higher dimensional torus. The person trying to track such a signal would be approximated by the crossover model and would effectively act as a quasiperiodic attractor, although the higher frequency components of the person's response would typically be amplified or attenuated and delayed relative to the input pattern (see Fig. 15.4). A quasi-periodic attractor can also result when a nonlinear oscillator receives a small amplitude sinusoidal input (e.g., Thompson & Stewart, 1986).

A fourth type of attractor that has enjoyed much popularity, but has yet to make an impact on manual control, is a chaotic attractor (Fig. 21.5, D). Such attractors exhibit instability that amplifies small perturbations as well as an overall stability to form an attractor. The instability combines with a successive "folding over" of the orbit in state space to form a fractal pattern. Examples of chaotic attractors include certain patterns of behavior exhibited by dripping faucets (Crutchfield et al., 1986;

Shaw, 1984; see chap. 8), beating hearts (Garfinkel, Spano, Ditto, & Weiss, 1992), and even some motions of celestial objects in the solar system (Wisdom, 1987), examples that previously might have been incorrectly thought of as limit cycle attractors.

It should be noted that the various types of attractors have been described in terms of topological features. Very roughly speaking, topology is a mathematical language for describing qualitative differences in shapes. A point, a closed cycle, a quasi-periodic torus, and a chaotic attractor are all topologically distinct. A good deal of theoretical work has related these types of descriptions to differential equation descriptions of nonlinear systems (e.g., Thompson & Stewart, 1986). Additionally, there has been a good deal of theoretical work describing changes in system performance. Changes that alter the qualitative nature of the attractor are considered to be more fundamental changes than those that do not. The more fundamental changes, called *bifurcations*, are often not reversible (e.g., Hockett & Holmes, 1987; Jackson, 1991). For example, the nonreversible loss of stability in a critical tracking task would be a more fundamental change than a reversible decrease in gain of someone performing a compensatory tracking task. In the former case, a point attractor becomes a repellor. In the latter case, the qualitative nature of the attractor is not altered, even though the style of tracking may become more sluggish.

Although it is possible to characterize manual control behavior in terms of the language of attractors, an important question is whether anything is gained by this type of description. For the purposes of illustration, some of the previous examples of attractors have been based on behaviors that are well approximated by nearly linear models, e.g., the crossover model. However, the advantages of describing behavior in terms of attractors become most apparent in control tasks involving strongly nonlinear systems. Such systems may exhibit surprisingly diverse qualitative changes in behavior (i.e., bifurcations) for small changes in system parameters. Examples of such changes have been found in multi-limb coordination (e.g., see Kelso, 1995), but a review of this area is beyond the scope of this chapter. In the context of challenging nonlinear control problems, the qualitative, topological modeling that is entailed in the concept of attractors is leading toward more qualitative control theoretic models and design techniques (e.g., Colonius & Kliemann, 2000; Zhao, 1994; see chap. 27). Additionally, measures for characterizing system performance may be more qualitative as well. For example, for many complex systems it is difficult to know a priori what are a sufficient set of different variables to construct a state space. One development in nonlinear systems theory is a proof that for a nonlinear system it is possible to construct a state space from a single measured variable (Takens, 1980). The dimensions of the state space are the measured variable and successively delayed copies of that same variable (Crutchfield et al., 1986; see also Parker & Chua, 1989). The multidimensional spatial representation created by cross-plotting these delayed copies of the same variable captures the qualitative aspects of the underlying attractors. Such techniques may prove useful in analyzing control systems in terms of their underlying attractors.

It is important to emphasize that although much of present control theory deals with linear models, those models become part of a larger nonlinear system once the process of adaptively adjusting the parameters is explicitly considered (e.g., Astrom & Wittenmark, 1989). Although nonlinear systems theory has introduced the concept of "control parameters" that strongly influence performance (e.g., the required frequency of an oscillatory movement; Haken et al., 1985), explicit process models of how such parameters are adjusted are often lacking. Models of such processes will most likely be adaptive control structures. Theories of adaptive control may provide a bridge between classical control theory and nonlinear systems theory (see also Pressing, 1998).

OTHER SINUSOIDAL TASKS

This chapter has concentrated on the tracking of single, highly predictable sine waves. Previous chapters have dealt with tracking signals composed of a sum of a large number of sine waves, which are relatively unpredictable to the human performer (e.g., Elkind, 1956; McRuer & Krendel, 1957; Sheridan & Ferrell, 1972). In between these two extremes are signals that have intermediate degrees of complexity and predictability. Magdaleno et al. (1969) categorized a number of these intermediate cases in terms of the dimensions of waveform shape complexity and waveform time variations. Shape complexity corresponds to the number and pattern of predominant frequency components in a repeating pattern. For example, a sum of two or three harmonically related sinusoids would be more complex than a single sinusoid. Time variations could arise from sine waves whose parameters are temporally modulated (e.g., Jagacinski, Greenberg, Liao, & Wang, 1993) or by adding narrowband or broadband noise to a repeating pattern.

An interesting combination of two sine waves is to have one sine wave in the horizontal dimension and another sine wave in the vertical dimension. If these two sine waves have the same frequency, but differ in phase, then the two-dimensional pattern is typically an ellipse. If the phase difference is one-quarter cycle and the two amplitudes are equal, then the two-dimensional pattern is a circle. When adults are asked to draw ellipses, the time course of their movements is well approximated by the combination of a vertical and horizontal sinusoid (Laquaniti, Terzuolo, &Viviani, 1983; Soechting, Lacquaniti, & Terzuolo, 1986; Viviani & Schneider, 1991). This finding is particularly interesting because the experimental task constrains the shape of the two-dimensional path, but allows the performer to choose the time course for traversing that path.

If the experimental task is to track an input that follows an elliptical pattern, then both the path and the time course by which it is traversed are experimentally constrained. Adults perform better if the time course of the input is generated from the sum of a horizontal and vertical sine wave. If another pattern of velocity is experimentally required to perform the elliptical shape, then participants do more poorly and increasingly so at higher movement frequencies (Viviani & Mounoud, 1990). These results suggest that a preferred spatiotemporal patterning exists that resembles the sum of a horizontal and vertical sinusoid. The underlying movement generating mechanisms might involve horizontal and vertical sinusoidal pattern generators (e.g., Laquaniti et al., 1983), pursuit control involving two-dimensional lead and an exploitation of resonances in the perceptual-motor coupling (Viviani & Mounoud,

1990), and/or possibly a smoothness constraint such as minimizing the rate of change of acceleration or "jerk" (Flash & Hogan, 1985; Wann, Nimmo-Smith, & Wing, 1988). In any case, Viviani and Mounoud (1990) noted that these results indicate a type of spatiotemporal compatibility between spontaneous drawing motions and tracking (see also, Viviani, 1990).

When people perform one-dimensional tracking of the sum of two sine waves with a pursuit display, they are able to anticipate both frequencies very well if the two frequencies are low (e.g., .1 and .3 Hz). However, if one frequency is low and the other is high (e.g., .1 and .6 Hz), then they typically anticipate the high frequency sine wave with an effective time delay well below 100 ms, whereas the effective delay for the low frequency sine wave is above 200 ms (e.g., Yamashita, 1989). Such a result is counter to the typical pattern of the crossover model for random appearing inputs. Although experiments with extended practice remain to be conducted, these data suggest that it is difficult to perceive and/or generate an internal pattern corresponding to the sum of two sine waves of very different frequencies.

The previous examples are illustrative of some of the constraints on uni-manual tracking of patterns of intermediate complexity. This general area is much larger and less fully explored than the tracking of very simple patterns, such as the single sine wave, and the tracking of very complex patterns, such as the random-appearing sum of many sine waves. This domain of inquiry has the potential to reveal many additional insights into the nature of movement control.

REFERENCES

Abraham, R. H., & Shaw, C. D. (1982). *Dynamics – The geometry of behavior* (Part 1: Periodic behavior). Santa Cruz, CA: Aerial Press.

Astrom, K. J., & Wittenmark, B. (1989). *Adaptive control*. Reading, MA: Addison-Wesley.

Bak, P. (1986). The Devil's staircase. *Physics Today, 39*, 38–45.

Barton, S. (1994). Chaos, self-organization, and psychology. *American Psychologist, 49*, 5–14.

Colonius, F., & Kliemann, W. (2000). *The dynamics of control*. Boston: Birkhauser.

Crutchfield, J. P., Farmer, J. D., Packard, N. H., & Shaw, R. S. (1986). Chaos. *Scientific American, 254*, 46–57.

Elkind, J. I. (1956). *Characteristics of simple manual control systems* (Tech. Rep. No. 111). Lexington, MA: MIT Lincoln Laboratory.

Ellson, D. G., & Gray, F. (1948). Frequency responses of human operators following a sine wave input. USAF, Air Materiel Command, Memo. Rep. No. MCR-EXD-694-2N, Wright-Patterson AFB, Ohio. Cited in Noble, Fitts, & Warren (1955).

Feldman, A. G. (1986). Once more on the equilibrium-point hypothesis (λ model) for motor control. *Journal of Motor Behavior, 18*, 17–54.

Flash, T., & Hogan, N. (1985). The coordination of arm movements: An experimentally confirmed mathematical model. *Journal of Neuroscience, 5*, 1688–1703.

Freund, H. J. (1986). Time control of hand movements. *Progress in Brain Research, 64*, 287–294.

Garfinkel, A., Spano, M. L., Ditto, W. L., & Weiss, J. N. (1992). Controlling cardiac chaos. *Science, 257*, 1230–1235.

Haken, H., Kelso, J. A. S., & Bunz, H. (1985). A theoretical model of phase transitions in human hand movements. *Biological Cybernetics, 51*, 347–356.

Hefter, H., Reiners, K., & Freund, H. J. (1983). Mechanisms underlying the limitation of speed and amplitude of rapid movements. *Neuroscience Letters*, Supplement *14*, S158.

Hockett, K., & Holmes, P. (1987). Nonlinear oscillators, iterated maps, symbolic dynamics, and knotted orbits. *Proceedings of the IEEE, 75*, 1071–1080.

Jackson, E. A. (1991). *Perspectives of nonlinear dynamics* (Vol. 1). New York: Cambridge University Press.

Jagacinski, R. J., Greenberg, N., Liao, M., & Wang, J. (1993). Manual performance of a repeated pattern by older and younger adults with supplementary auditory cues. *Psychology and Aging, 8*, 429–439.

Jagacinski, R. J., & Hah, S. (1988). Progression-regression effects in tracking repeated patterns. *Journal of Experimental Psychology: Human Perception and Performance, 14*, 77–88.

Jex, H. R., McDonnell, J. D., & Phatak, A. V. (1966). A "critical" tracking task for manual control research. *IEEE Transactions on Human Factors in Electronics, HFE-7*, 138–145.

Kelso, J. A. S. (1995). *Dynamic patterns.* Cambridge, MA: MIT Press.

Kelso, J. A. S., Ding, M., & Schoner, G. (1993). Dynamic pattern formation: A primer. In L. B. Smith & E. Thelen (Eds.), *A dynamic systems approach to development: Applications* (pp. 13–50). Cambridge, MA: MIT Press.

Kelso, J. A. S., Holt, K. G., Rubin, P., & Kugler, P. N. (1981). Patterns of human interlimb coordination emerge from the properties of non-linear, limit cycle oscillatory processes: Theory and data. *Journal of Motor Behavior, 13*, 226–261.

Kelso, J. A. S., & Scholz, J. P. (1987). Cooperative phenomena in biological motion. In H. Haken (Ed.), *Complex systems: Operational principles in neurobiology, physical systems and computers* (pp. 124–149). Berlin: Springer-Verlag.

Krendel, E. S., & McRuer, D. T. (1968). Psychological and physiological skill development: A control engineering model. In *Proceedings of the Fourth Annual NASA-University Conference on Manual Control* (NASA SP 192). University of Michigan, Ann Arbor, MI.

Laquaniti, R., Terzuolo, C., & Viviani, P. (1983). The law relating the kinematic and figural aspects of drawing movements. *Acta Psychologica, 54*, 115–130.

Large, E. W., & Jones, M. R. (1999). The dynamics of attending: How we track time varying events. *Psychological Review, 106*, 119–159.

Latash, M. L. (1992). Virtual trajectories, joint stiffness, and changes in the limb natural frequency during single-joint oscillatory movements. *Neuroscience, 49*, 209–220.

Liao, M., & Jagacinski, R. J. (2000). A dynamical systems approach to manual tracking performance. *Journal of Motor Behavior, 32*, 361–378.

Leist, A., Freund, H. J., & Cohen, B. (1987). Comparative characteristics of predictive eye–hand tracking. *Human Neurobiology, 6*, 19–26.

Magdaleno, R. E., Jex, H. R., & Johnson, W. A. (1969). Tracking quasi-predictable displays. In *Proceedings of the Fifth Annual NASA-University Conference on Manual Control* (NASA SP-215). MIT, Cambridge, MA.

McAuley, J. D. (1994). Time as phase: A dynamic model of time perception. In A. Ram & K. Eiselt (Eds.), *Proceedings of the 16th Annual Conference of the Cognitive Science Society* (pp. 607–612). Hillsdale, NJ: Lawrence Erlbaum Associates.

McRuer, D. T., & Krendel, E. S. (1957). *Dynamic response of human operators* (WADC Tech. Rep. No. 56-524). Wright-Patterson Air Force Base, Ohio.

Mounod, P., Viviani, P., Hauert, C. A., & Guyon, J. (1985). Development of visuomanual tracking in 5- to 9-year-old boys. *Journal of Experimental Child Psychology, 40*, 115–132.

Noble, M., Fitts, P. M., & Warren, C. E. (1955). The frequency response of skilled subjects in a pursuit tracking task. *Journal of Experimental Psychology, 49*, 249–256.

Ogata, K. (1967). *State space analysis of control systems.* Englewood Cliffs, NJ: Prentice-Hall.

Parker, T. S., & Chua, L. O. (1989). *Practical numerical algorithms for chaotic systems.* New York: Springer-Verlag.

Peper, C. E., Beek, P. J., & van Wieringen, P. C. W. (1991). Bifurcations in polyrhythmic tapping: In search of Farey principles. In J. Requin & G. E. Stelmach (Eds.), *Tutorials in motor neuroscience* (pp. 413–431). Dordrecht: Kluwer.

Pew, R. W., Duffendack, J. C., & Fensch, L. K. (1967). Sine-wave tracking revisited. *IEEE Transactions on Human Factors in Electronics, HFE-8*, 130–134.

Poulton, E. C. (1952). The basis of perceptual anticipation in tracking. *British Journal of Psychology, 43*, 295–302.

Pressing, J. (1998). Referential behavior theory: A framework for multiple perspectives on motor control. In J. P. Piek (Ed.), *Motor behavior and human skill: A multi-disciplinary perspective* (pp. 357–384). Champaign, IL: Human Kinetics.

Shaw, R. (1984). *The dripping faucet as a model chaotic system.* Santa Cruz, CA: Aerial.

Sheridan, T. B., & Ferrell, W. R. (1972). *Man–machine systems.* Cambridge, MA: MIT Press.

Soechting, J. F., Lacquaniti, F., & Terzuolo, C. A. (1986). Coordination of arm movements in three-dimensional space: Sensorimotor mapping during drawing movement. *Neuroscience, 17,* 295–311.

Stark, L. (1966). Neurological feedback control systems. In F. Alt (Ed.), *Advances in bioengineering and instrumentation* (pp. 291–359). New York: Plenum.

Takens, F. (1980). Detecting strange attractors in turbulence. *Lecture notes in mathematics* (pp. 366–381). Berlin: Springer-Verlag.

Thompson, J. M. T., & Stewart, H. B. (1986). *Nonlinear dynamics and chaos.* Chichester, England: Wiley.

Treffner, P. J., & Turvey, M. T. (1993). Resonance constraints on polyrhythmic movement. *Journal of Experimental Psychology: Human Perception and Performance, 19,* 1221–1237.

Viviani, P. (1990). Common factors in the control of free and constrained movements. In M. Jeannerod (Ed.), *Attention and Performance XIII: Motor representation and control* (pp. 345–373). Hillsdale, NJ: Lawrence Erlbaum Associates.

Viviani, P., & Mounod, P. (1990). Perceptuomotor compatibility in pursuit tracking of human two-dimensional movements. *Journal of Motor Behavior, 22,* 407–433.

Viviani, P., & Schneider, R. (1991). A developmental study of the relationship between geometry and kinematics in drawing movements. *Journal of Experimental Psychology: Human Perception and Performance, 17,* 198–218.

Wann, J., Nimmo-Smith, I., & Wing, A. M. (1988). Relation between velocity and curvature of movement: Equivalence and divergence between a power law and minimum-jerk model. *Journal of Experimental Psychology: Human Perception and Performance, 14,* 622–637.

Ware, J. R. (1972). An input adaptive, pursuit tracking model of the human operator. In *Proceedings of the Seventh Annual Conference on Manual Control* (NASA SP-281, pp. 33–45). University of Southern California, Los Angeles, CA.

Wilts, C. H. (1960). *Principles of feedback control.* Reading, MA: Addison-Wesley.

Wimmers, R. H., Beek, P. J., & van Wieringen, P. C. W. (1992). Phase transitions in rhythmic tracking movements: A case of unilateral coupling. *Human Movement Science, 11,* 217–226.

Wisdom, J. (1987). Urey Prize Lecture: Chaotic dynamics in the solar system. *Icarus, 72,* 241–275.

Yamashita, T. (1989). Precognitive behavior in tracking of targets with 2 sine waves. *Japanese Psychological Research, 31,* 20–28.

Zhao, F. (1994). Extracting and representing qualitative behaviors of complex systems in phase space. *Artificial Intelligence, 69,* 51–92.

Going with the Flow:
An Optical Basis for the
Control of Locomotion

> The center of the flow pattern during forward movement of the animal is the direction of movement. More exactly, the part of the structure of the array from which the flow radiates corresponds to that part of the solid environment toward which he is moving. If the direction of movement changes, the center of flow shifts across the array, that is, the flow becomes centered on another element of the array corresponding to another part of the solid environment. The animal can thus, as we would say, "see where he is going." . . . To aim locomotion at an object is to keep the center of flow of the optic array as close as possible to the form which the object projects.
>
> —Gibson (1958/1982, p. 155)

How do drivers know when to initiate braking as they approach a stopped line of traffic? What is a "safe" distance of separation when following cars on the highway? When making a left-hand turn across traffic, how large does the "gap" have to be to permit a safe passage? How do pilots know when to begin descent for a safe and gentle touchdown? How do pilots know whether they have adequate airspeed to avoid a wing stall?

The perceptual component of the optimal control model generally assumes that the human operator is an "ideal observer (perceiver)" or Kalman filter with inputs that are functions of the "state variables" of the process being controlled. In order to address the aforementioned questions related to vehicular control, the vehicle dynamics (i.e., the controlled process) would typically be modeled as an inertial system where the state variables are the position and velocity for each axis of motion. Thus, optimal control of braking would require the operator to know both the distance to the collision boundary and the approach velocity. At higher velocities it would be necessary to initiate braking at greater distances from the obstacle than it would for lower velocities. How does the driver judge distance and velocity? The speedometer is one potential source of information about velocity. Do drivers typically look at the speedometer in order to decide when to initiate braking? Most drivers do not. Most drivers are attending to information available in the head-up view through the windshield of the car (and the rear and side view mirrors). As the vehicle moves through the world, the light reflected from surfaces of the ground and objects (e.g., other vehicles, traffic signs, buildings, trees, etc.) produces a stream of optical texture across the windshield and mirrors, or more generally, across the field of view. This *optical flow field* can be an important source of feedback that closes the loop to allow control of locomotion. In fact, it might be argued that the flow field is an essential source of information, because the speedometer only tells the speed of the vehicle, not the relative speed with respect to that of other vehicles. The speedometer is indicative of range rate only in the case that the collision object is stationary. This chapter discusses how structure in the optical flow field can specify the state of the locomotion process and how the degree of specificity can be an important constraint on the control process.

SPEED OF LOCOMOTION

First consider how information in an optical flow field specifies the speed of self-motion. Because of the inertial properties of vehicles, the speed of motion is an important component of the vehicle state. Whereas people may have difficulty judging absolute speed of motion, they generally have a good sense of their speed with respect to road and traffic conditions (without having to sample the speedometer). For example, they consistently adjust speed appropriately to merge into a gap in highway traffic and they can generally adjust speed appropriately to road conditions (e.g., to handle a tight curve). But there are situations where people misjudge their speed with regard to vehicular control. And, somewhat ironically, the situations where control fails often provide key insights into how the person–vehicle control system works. It is important to keep in mind, however, that the goal is to understand how the control system works. Perceptual research has had a tendency to fixate on illusions and perceptual errors as the critical phenomena and has sometimes had difficulty explaining how perception functions in the dynamic context of action (e.g., skilled driving).

Edge Rate

A common situation where drivers tend to misjudge speed is when they exit a highway after long periods of driving at highway speeds. Under these conditions, drivers often underestimate their speed and are surprised when they realize they are attempting to navigate a curve at an unsafe speed—they become aware of the error when their wheels squeal or when the car skids into the turn. Usually, this misjudgment can be corrected before a collision occurs, but in some situations (e.g., traffic circles in Great Britain) this error can be a contributing factor to accidents.

One potential source of information about speed is the local texture rate (edge or discontinuity rate). For example, a driver might count the rate at which stripes (painted at equal intervals on the highway) pass underneath the car. The number of stripes per second would be higher at higher speeds than at lower speeds. However, it is unlikely that drivers would explicitly count the stripes. Rather, an impression of speed is created by the rate at which the texture elements (e.g., stripes on the highway) flow across the field of view. This impression might have a generally monotonic relation to the number of stripes per second. The greater the count of stripes, the greater the impression of speed.

However, a common feature of the human perceptual system is adaptation or habituation. That is, prolonged exposure to a constant stimulus causes the system to become less sensitive to that stimulus. For example, the skin adapts to constant temperature or to constant pressure. Water that feels hot when a hand is first immersed in it will feel cooler after prolonged exposure (even though the temperature has not changed). This can result in a negative aftereffect. That is, if the hand was moved from the hot water to water at room temperature, then the water would feel cooler than if the hand had not been first exposed to the warmer water. A compelling demonstration of this effect is often used in introductory psychology courses. A student places one hand in hot water and another hand in cold water. After time is allowed for adaptation, both hands are placed in the same bucket with water at room temperature. The same water will feel cool to one hand (the one adapted to the hot water) and warm to the other hand. The visual system appears to adapt to constant motion in a similar way. This affects sensitivity to speed and can result in negative after effects (a nonmoving stimulus can appear to be moving in the opposite direction from the motion of the adapting stimulus). Thus, after driving at constant highway speeds (i.e., constant texture rate) for an extended period, the impression of speed may habituate. That is, even though the texture rate is unchanged, the impression of speed may be attenuated (in the same way that the warm water feels cooler after long exposure).

Denton (1976, 1977, 1980) hypothesized that adaptation may be a contributing factor for the misjudgment of speed that leads to inadequate braking when exiting highways. To counteract the effects of speed adaptation, Denton manipulated the distances between edges in the textured surface. For example, he painted yellow stripes at the approaches to traffic circles so that the distance between stripes was gradually reduced (using an exponential function) as the traffic circle was approached. Thus, moving over the stripes at a constant speed would produce an accelerating edge rate. If the impression of speed is dependent on edge rate, then this manipulation might counteract the effects of speed adaptation. Denton (1980) called this technique *control of speed by illusion* (COSBI). He found that this manipulation had a significant effect on judgments of speed in a driving simulator. Further, field tests showed that COSBI led to significant reductions (approximately 20%) in the mean approach speed to an intersection. At the Newbridge Roundabout situated on the dual-lane Glasgow-Edinburgh M8, the application of COSBI resulted in a reduction from 14 accidents reported in the 12 months prior to installation of COSBI to only 2 accidents reported in the 16 months following installation (Denton, 1980, p. 402). As a result of the application of COSBI at 37 sites throughout the United Kingdom, accident rates were reduced at 29 of the sites. This amounted to an overall reduction of accidents by about two thirds.

Denton's work provides evidence that perception depends on the geometric structure of optic flow. Denton was able to change control performance by manipulating the flow geometry. In environments where textures are spaced at regular intervals (or even stochastically regular) average edge rate will be directly proportional to velocity (see also Larish & Flach, 1990). Thus, edge rate reliably specifies speed. However, this reliable source can be biased by either manipulation of the geometric structure (as demonstrated by Denton's work) or by changes in the receptivity of the perceptual system (e.g., speed adaptation). The evidence for speed adaptation and the success of Denton's intervention suggests that, when driving automobiles, the human observer tunes to edge rate as a generally reliable source of information about speed. This seems to lead to generally successful vehicular control, but can be a contributing factor to accidents in those situations where environmental and perceptual factors conspire to undermine the reliability of this information source.

GLOBAL OPTICAL FLOW RATE

A mathematical analysis of motion perspective is presented in terms of the optical flow-pattern reflected from a surface to an eye. It is shown that the variables of this flow-pattern are specific not only to the "depth" of the surface but also to the movement of an O [observer]. Assuming that these variables are stimuli for perception they can determine not only the experience of a stable tridimensional world, but provide a basis for the judgments required for the control of locomotion in that world. (Gibson, Olum, & Rosenblatt, 1955, p. 385)

Denton's work suggests one way to measure flow. Basically, the strategy is to implicitly count the number of edges (texture elements or discontinuities) that pass by a particular optical locus (e.g., the edge of your windshield). However, the flow rate can also be measured as the angular change associated with a particular texture element. What is the amount of angular change per unit of time? This is the strategy that Gibson et al. (1955) adopted in order to build a mathematical model of the flow field. Figure 22.1 illustrates the optical geometry. The optical position for a point P is specified by its angular position δ. The optical flow rate for the point P is then the rate of change of δ:

$$d\delta/dt = \left(V/s\right)(\sin \delta \cos \delta + \sin^2 \delta \cos \theta \cot \beta) \tag{1}$$

As shown in the equation, three factors combine multiplicatively to determine the flow rate of a texture element. The first factor is the speed at which the observer is moving (V). This is a global factor that scales the flow rate for all points in the optic array. The faster the observer moves, the faster is the global optical flow rate.

The second factor affecting flow rate is the distance to the surface texture measured along the glide path (OG) or, in the case of level flight, measured along the normal line to the surface (OL). For level flight ($\beta = 0$), the optical flow equation reduces to:

$$d\delta/dt = \left(V/h\right)(\sin^2 \delta \cos \theta) \tag{2}$$

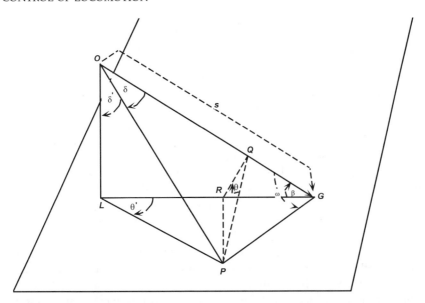

FIG. 22.1. A geometric analysis of optical flow is shown (after Gibson, Olum, & Rosenblatt, 1955). P represents an arbitrary point on a surface. O is the observer's position relative to the surface. The observer is moving along the path OG. The coordinates of the optical flow field are the angle (δ) and the rate of change of angle associated with each point on the surface that reflects light to the observer. From "Parallax and perspective during aircraft landings" by J. J. Gibson, P. Olum, & F. Rosenblatt, 1955, *American Journal of Psychology, 68*, 372–385. Copyright 1955 by the Board of Trustees of the University of Illinois. Adapted and used with permission of the University of Illinois Press.

where h is the eyeheight of the observer (*OL* in Fig. 22.1). The distance to the surface (h or s — distance along the glidepath *OG*) is also a global factor that scales the speed for all points in the optical array. The greater the distance from the observer to the textured surface, the slower the global flow rate. Two pilots flying at the same ground speed but at different altitudes would see different flow rates. The flow rate would be higher for the pilot that was closer to the ground.

The third factor in the flow equations is the term containing the trigonometric functions. This term accounts for local properties of flow. That is, points on the same ground surface will have different flow rates depending on the optical positions within the flow field. Texture elements near the aim point will flow slowly. Texture elements closer to the observer will flow more quickly (e.g., fence posts at the edge of the highway) and elements toward the horizon will flow more slowly (e.g., distant mountains). For the control of speed, the global (or distributional) properties of the flow seem to be most important.

For an observer whose distance from the ground is fixed ($\beta = 0$), global optical flow rate (V/h; h is the length of vertical projection *OL* in Fig. 22.1) provides reliable feedback about the speed of locomotion. However, if altitude varies, then global optical flow rate can be ambiguous with respect to speed of locomotion. For example, when the first 747 aircraft were introduced, there was a sudden increase in damage to the landing gear resulting from pilots attempting to turn the aircraft while taxiing at too great a speed. The pilots of these aircraft had been trained in 707 and 727 aircraft.

Safe turn speeds for the 747 were the same as for the 707 and 727, but the pilots were consistently taxiing at nearly double the safe speed when in the 747. How can this be explained? Owen and Warren (1982) postulated that the geometry of global optical flow rate may have been a contributing factor. The cockpit for the 747 is on top of the fuselage. The result is that the effective altitude (or eyeheight) of the pilot in the 747 is about twice as high as in the 707 and 727. Thus, the hypothesis is that pilots were judging their taxi speed by the global optical flow rate. When taxiing in the 747, the flow rate would match a safe flow rate (relative to the experience in the 727 and 707) when the vehicle was actually moving at approximately double the safe speed.

The ambiguity created by global optical flow rate has been hypothetically linked to a very dangerous situation connected with low altitude flight. The U.S. military loses a significant number of aircraft due to a problem labeled controlled flight into terrain (CFIT; Haber, 1987). CFIT refers to situations where an aircraft flies into the ground with no obvious mechanical failure, no dangerous weather or meteorological conditions, and apparently no medical problems. The cause seems to be pilot control error. The hypothetical explanation is as follows. The first assumption is that pilots utilize the global optical flow rate as a primary indicant of airspeed. As already noted, this is a smart strategy, as long as altitude is roughly constant. Airspeed is a critical determinant of lift. If airspeed becomes too low, then the amount of lift will be less than the pull of gravity, and the aircraft will essentially begin to fall out of the sky. The minimum airspeed necessary to keep the aircraft in flight is referred to as the stall speed. A wing stall is a situation where the lift generated by the wings is less than the force of gravity.

Now imagine a maneuver (e.g., a sharp turn) where the aircraft simultaneously loses altitude and air speed. As is discussed in later sections of this chapter, there are numerous optical changes, in addition to optical flow rate, associated with the loss of altitude (e.g., the expansion of objects on the ground). It is assumed that pilots easily detect loss of altitude. However, global optical flow rate seems to be fundamental to the perception of airspeed. Here is where the danger lies. The loss of altitude causes an increase in global optical flow rate. This increase may compensate or mask any decreases due to loss of air speed—leading a pilot to misjudge airspeed. Thus, the aircraft may reach a stall speed without the pilot being aware. In other words, it is possible that when the pilot tries to initiate a climb to avoid collision with the ground, there is inadequate power (i.e., airspeed) and the aircraft stalls at an altitude where recovery is not possible (or perhaps the aircraft does not stall but the pilot runs out of room—that is, the pilot cannot generate enough lift in time to avoid collision with the ground).

Consider some examples of the impact of changes in altitude and airspeed on the global optical flow rate. First, compare the effect due to a loss of 100 m in altitude as a function of the initial cruising altitude (1 or .5 nautical miles) for a speed of 200 knots:

1 nautical mile = 1852 m
1 knot = 1 nautical mile/hour or .5144 m/s

From an altitude of 1 nautical mile:

(200 knots × .5144 m/s)/1852 m = .0555 eyeheights/s
(200 knots × .5144 m/s)/1752 m = .0587 eyeheights/s

Thus, there is a 5.8% increase in global optical flow rate due to the 100 m loss in altitude. From an altitude of .5 nautical miles:

$$(200 \text{ knots} \times .5144 \text{ m/s})/ \ 926 \text{ m} = .1111 \text{ eyeheights/s}$$
$$(200 \text{ knots} \times .5144 \text{ m/s})/ \ 826 \text{ m} = .1246 \text{ eyeheights/s}$$

Thus, there is a 12.2% increase in global optical flow rate due to the 100 m loss in altitude. Note that the same 100 m loss of altitude has an increased impact on global optical flow rate at lower altitudes. The effect of altitude loss on optical flow is a function of the fractional change in altitude.

Suppose that a loss of 50 knots in air speed was combined with a loss of 200 m in altitude (initial speed = 200 knots; initial altitude = .5 nautical miles). What would be the net impact on global optical flow rate?

$$(150 \text{ knots} \times .5144 \text{ m/s})/726 \text{ m} = .1063 \text{ eyeheights/s}$$

Compare this with the flow rate for 200 knots at .5 nautical miles. In this case, a 25% loss in airspeed resulted in only a 4% loss in global optical flow rate due to the compensatory effect due to altitude loss. Note the result if there had been only loss of airspeed with no associated loss of altitude:

$$(150 \text{ knots} \times .5144 \text{ m/s})/926 \text{ m} = .0833 \text{ eyeheights/s}$$

Again compare this with the flow rate for 200 knots at .5 nautical miles. In this case, a 25% loss in airspeed results in a 25% loss in global optical flow rate. This example illustrates the potential for pilots who utilize flow rate as an indication of airspeed to underestimate loss of airspeed when there is an associated loss in altitude.

In controlling locomotion, the structure of optical flow fields may be a critical source of information that allows the operator to close the loop. To the degree that the structure is specific to the relevant state variables, stable control is to be expected. On the other hand, if the structure is not specific to the relevant state variables, then instabilities are expected. In the case of global optical flow, the specificity is bounded. That is, under conditions where altitude is constant, global optical flow rate reliably specifies changes in speed. However, in situations where a significant change of altitude accompanies changes in speed, global optical flow rate is no longer a reliable source of information about speed.

COLLISION CONTROL

CFIT is a special example of the more general problem of collision control. As people navigate through the world, they must steer so as to avoid colliding with obstacles. For driving a car, Gibson and Crooks (1982) introduced the construct of the "field of safe travel," which they defined as "the field of possible paths which the car may take unimpeded" (p. 120). The depth of this field is bounded by "frontal obstacles such as other vehicles, policemen, and stop lights" and the width of the field is bounded by "obstacles like curbs, ditches, soft shoulders, walls, parked cars, pedestrians, or white lines" (pp. 121–122). Within the safe field of travel, there is a "minimum stopping zone" that is "dependent on the speed of the car—and . . . on the condition of the road surface and brakes" (p. 123). The question to be considered within this section is the degree to which the field of safe travel is specified by structure in the optical flow field.

In their mathematical analysis of optical flow Gibson et al. (1955) noted that although global optical flow may be ambiguous with respect to independent judgments about speed and altitude change (as discussed earlier), the time to collision is well specified: "Ground-speed and altitude are not, however, independently determined by the optical information. A more rapid flow-pattern may indicate either an increase in speed or a decrease in altitude. Length of time before touching down, however, seems to be given by the optical information in a univocal manner" (p. 382).

Thus, the ratio of velocity to distance that specifies global optical flow rate specifies the instantaneous time-to-contact (assuming that speed is constant). In addition to the flow rate, there is a specific optical pattern associated with collision. That is, a potential collision obstacle will expand symmetrically in the field of view. Thus, Gibson (1958/1982) provided the following formula for soft collisions: "Contact without collision is achieved by so moving as to cancel the centrifugal flow of the optic array at the moment when the contour of the object or the texture of the surface reaches that angular magnification at which contact is made" (p. 156).

The pattern of symmetrical expansion associated with a potential collision is called *looming* (Gibson, Schiff, & Caviness, 1962; Schiff, 1965). Lee (1976) noted that the looming pattern itself could provide information about time-to-contact with the looming object. Figure 22.2 illustrates a simple collision event. In this simple case (see Tresilian, 1991, for discussion of more complicated optical situations), an object of a fixed radius (r) travels along the line of sight toward an observer at a constant speed ($\dot{x}$). The angle (θ) subtended by the object at the observer can be described as a function of the radius of the object (r) and the distance between the object and the observer (x):

$$\theta = \tan^{-1}\left(\frac{r}{x}\right) \tag{3}$$

Making an approximation that holds for small angles:

$$\tan(\theta) \approx \theta \tag{4}$$

the equation becomes:

$$\theta \approx \frac{r}{x} \tag{5}$$

and differentiating both sides of the equation with respect to time, yields:

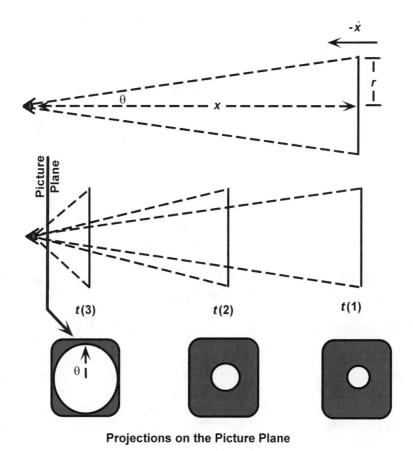

Projections on the Picture Plane

FIG. 22.2.   Illustration of the geometry of inverse fractional expansion rate (Tau).

$$\dot{\theta} \approx \frac{-\dot{x}r}{x^2} \tag{6}$$

Dividing Equation 5 by Equation 6 yields the following:

$$\frac{\theta}{\dot{\theta}} = \left(\frac{x^2}{-\dot{x}r}\right) \times \left(\frac{r}{x}\right) = \frac{x}{-\dot{x}} \tag{7}$$

Assuming constant speed:

$$\frac{\theta}{\dot{\theta}} \approx T_{EYE} \tag{8}$$

where T_{EYE} represents time to collision with the eye, or "Tau."

Recently, the variable Tau has taken on two different definitions. Tau is sometimes used to indicate the instantaneous time-to-contact. In this case, it refers to the ratio of the range to the rate of change of range with respect to the collision object:

$$\frac{\text{range (m)}}{\text{range rate (m/s)}} = \text{time-to-collision (s)} \tag{9}$$

Alternatively, Tau is used to indicate the optical specification of time-to-contact or the inverse fractional expansion rate. That is the ratio of the visual angle to the visual expansion rate:

$$\frac{\text{optical angle (deg)}}{\text{optical expansion rate (deg/s)}} = \frac{1}{\text{fractional expansion rate (s}^{-1})}$$

$$\approx \text{time-to-collision (s)} \tag{10}$$

Here what is most interesting is the inverse fractional expansion rate, as a potential optical specification of the time-to-collision. The term *inverse fractional expansion rate* is used to avoid the confusions associated with the Tau variable.

Inverse fractional expansion rate is an example of a higher order optical invariant. From a control theoretic perspective, this signifies that two state variables (position and velocity) are integrated within a single measurement. Typically, position and velocity have been assumed to be observed independently (e.g., see chaps. 14, 16, and 17). Then they must be combined (weighted) by a control algorithm to compute the appropriate or the optimal response to the demands of an inertial system. However, inverse fractional expansion rate is a perceptual variable through which the integration may happen within the flow field. In other words, the variables are combined outside of the control system (outside the head in the case of a human controller). Now the person can observe a single variable (fractional expansion rate) and base a response on that single variable. Even though only one variable is being observed, the person is still "taking into account both position and velocity," meeting the demands of the inertial control problem. Thus, inverse fractional expansion rate is a higher order optical primitive relative to the lower order primitives (e.g., angle and angular rate or position and velocity).

The idea of a higher order optical invariant has important implications for performance. If people are responding to the higher order variable, then they should be indifferent to changes in the underlying variables that leave the higher order variable the same. In this case, the response to the same inverse fractional expansion rate should be invariant (at approximately a constant time-to-contact) regardless of the particular lower order factors that contribute to the ratio. Examples of two lower order factors that combine to determine inverse fractional expansion rate are speed and the size of the collision object.

Analyses of a wide range of natural situations where humans and animals must control collisions (e.g., plummeting gannets: Lee & Reddish, 1981; table tennis: Bootsma & Van Wieringen, 1990; volleyball: Lee, Young, Reddish, Lough, & Clayton, 1983; and car drivers: Lee, 1976) show that, consistent with the idea of a higher order optical invariant, humans and other animals are capable of responding at the correct time over a range of speeds and with objects of varying sizes. However, numerous controlled laboratory studies have found that there are consistent effects on human performance of both the speed of closure and the size of objects. Studies that have varied the speed of closure find that participants tend to respond early, and they consistently respond earlier to slower events than to faster events (Li & Laurent, 1995; McLeod & Ross, 1983; Schiff, Oldak, & Shah, 1992; Sidaway, Fairweather, Sekiya, & McNitt-Gray, 1996). Studies that have varied the size of objects find that participants respond earlier to larger objects than to smaller objects (Caird & Hancock, 1994; DeLucia, 1991; DeLucia & Warren, 1994; Van der Kamp, Savelsbergh, & Smeets, 1997).

Several alternatives to the inverse fractional expansion rate hypothesis have been suggested based on the performance patterns. For example, the tendency for subjects to respond early to slower approaches and to larger objects is consistent with the hypothesis that people are using simply optical expansion rate as the optical information for controlling collisions (Smith, Flach, Dittman, & Stanard, 2001). Smith et al. (2001) found evidence of an attunement process. Early in practice, with a discrete collision control task, participants responded as if they were using an optical boundary based on expansion rate alone. However, with practice, participants did better than would be expected based on expansion rate alone. Performance late in practice suggested that the optical boundary reflected a combination of optical angle and expansion rate.

Figure 22.3 shows an optical state space. As seen in earlier chapters, the state space for a second-order control system would typically have the dimensions of position and velocity. However, in Fig. 22.3, these dimensions have been replaced by their optical correlates (optical angle and expansion rate). Distinctive patterns in this space are associated with different optical criteria. A fractional expansion rate criterion

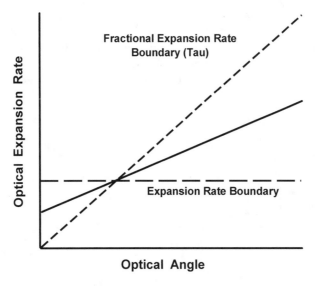

FIG. 22.3.   An illustration of the optical state space for collisions. The lines reflect different optical boundaries that might be used to control collisions.

(Tau) would be a diagonal line (slope > 0) with zero intercept. The inverse of the slope would be the time-to-contact. An expansion rate criterion would be a horizontal line (slope = 0). A criterion that reflected a weighted combination of angle and expansion rate would be a diagonal line (slope > 0) with a nonzero intercept. If the person was using one of these strategies in a discrete control task, then the performance should be distributed along the appropriate optical boundary in the state space.

The optical state space can be used to represent more than the optical boundaries. Action regions and collision events can also be traced in this space. Figure 22.4 illustrates a discrete collision task used by Smith et al. (2001). Balls approached in a head-on collision path, and the participant's task was to release a virtual pendulum so that it would make contact with the ball at the "ideal hitting position" that would send the ball directly back from where it came (a line drive). There was a time window around the ideal hitting position in which the pendulum would make contact with the ball and deflect it down (a foul ball). If the pendulum was released outside this time window (either too early or too late), then the ball would be missed altogether (a strike). Figure 22.5 shows the hit zone as a white region within the optical state space. If the pendulum is released at optical coordinates that fall within this zone, then the pendulum will make contact with the ball. The solid line near the middle of the hit zone represents the coordinates of an ideal hit. The curves extending from the lower left corner of the graph toward the upper right corner illustrate the optical trajectories for balls moving at different constant speeds. The steepest (leftmost) curve is a trajectory for the fastest ball (20 m/s), and the shallowest curve is a trajectory for the slowest ball (5 m/s). The optical state space allows the event trajectories for a given trial type (ball of a particular speed) to be visualized relative to the conjoint effects of the action constraints and the information constraints.

The points in Fig. 22.5 show the optical coordinates at which the pendulum was actually released (these are means averaged over four participants, although the general pattern was consistent for each individual). The open symbols are for performance early in practice. Note that the solid black line fit to these data is in close correspondence to a constant expansion rate boundary (shown as the dashed line). The filled symbols are for performance after several practice sessions. Note that the solid line fit to these data has a nonzero slope and a nonzero intercept. This suggests

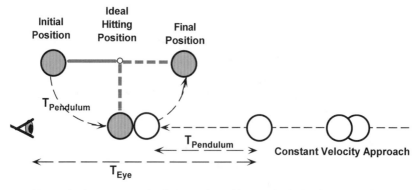

FIG. 22.4.   The discrete control collision task used by Smith, Flach, Dittman, and Stanard (2001).

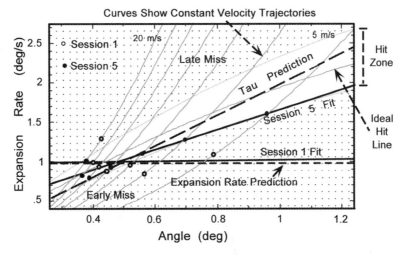

FIG. 22.5.   An optical state space showing event trajectories and performance data as a function of the conjoint constraints on action (white region shows boundaries for hitting zone) and information (dashed lines show optical boundaries).

that the participants were using an optical boundary involving both angle and expansion rate, but they do not appear to be using fractional expansion rate (shown as dashed-line). Perhaps with further practice, participants would learn to use the fractional expansion rate (performance seems to be moving in that direction).

The key point for this section is to illustrate how state space representations can be adapted to situations were the "measured variables" are not traditional Newtonian measures of motion (position and velocity). Rather, the measured variables reflect constraints within optical flow fields (optical angles and angular rates). When the Newtonian variables are converted to optical correlates, the resulting optical state space can be useful for thinking about and evaluating alternative hypotheses about the optical basis for control. The state space can also be useful for visualizing the conjoint relations between action and information constraints (as in Fig. 22.5). This can help shed light on the degree to which different optical variables are able to specify the relevant action constraints, and thus whether the control loop can be closed through the optical array.

ALTITUDE CONTROL

As an aircraft approaches the ground, the angular extent of objects on the ground expands in the optical flow field. For example, the runway will "loom" as it is approached during a normal landing. The horizon will remain stationary in the visual field. The far edge of the runway will move up in the field of view (getting closer to the horizon), the near edge of the runway will move down in the field of view (getting farther from the horizon), and the sides of the runway will "splay out." This expansion pattern reflects the laws of visual perspective. Figure 22.6 illustrates the

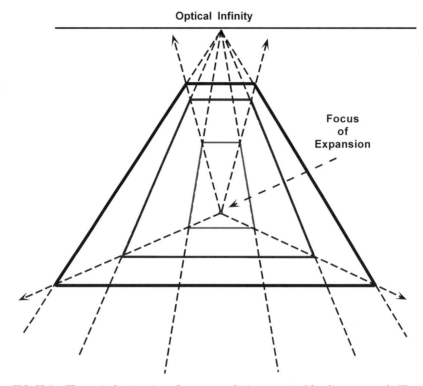

FIG. 22.6.   The optical expansion of a runway during a normal landing approach. The solid lines show the edges of the runway. The dashed lines illustrate the optical constraints associated with optical infinity and the focus of expansion.

looming of the runway. The solid lines show the runway shape as viewed from three different distances along the glide path (closer views are wider). The dashed lines show the optical constraints: lines parallel to the direction of viewing project to optical infinity, and the entire pattern expands symmetrically around the aim point of the movement.

The overall expansion illustrated with the looming runway in Fig. 22.6 can be decomposed into two more elemental flow components: splay and depression angle. These two components may independently contribute to the control of altitude. Warren (1982) identified optical splay angle as a potential source of information for altitude control. Edges on a flat ground plain that are parallel to the direction of motion (and gaze) will appear to converge at the vanishing point on the horizon. The angle of convergence (splay angle) will be a function of the lateral distance of the edges from the line of motion (e.g., in the case of the runway in Fig. 22.6, this reflects the width of the runway). It will also be a function of the altitude of the observer. Figure 22.6 shows that when the plane is high above the runway, the splay angle is small and the angle increases as the plane gets lower.

The splay angle can be computed using the following equation:

$$S = \tan^{-1}\left(\frac{Y_g}{h}\right) \qquad (11)$$

where Y_g is equal to the lateral distance to the edge and h is the eyeheight (altitude of the observer).

The rate of change in splay angle with respect to change in observer position is specified by the following equation:

$$\dot{S} = \left(\frac{-\dot{h}}{h}\right)\cos S \sin S + \left(\frac{\dot{Y}_g}{h}\right)\cos^2 S \tag{12}$$

The first term in this sum indexes change in splay as a function of change in altitude ($\dot{h}$). The negative sign indicates that as altitude decreases, splay angle increases, and vice versa. Note that change in splay due to change in altitude is inversely related to altitude. That is, the same change in altitude will have a smaller impact on splay angle if the initial altitude is higher. As noted by Warren (1988), " 'sensitivity' of the display [optical splay rate] varies inversely with altitude; the lower the altitude, the more change in visual effect for equivalent altitude change commands. At very low altitudes this optical activity is dramatic and even 'optically violent' " (p. A121). The ratio of change in altitude to altitude is a global factor (it effects every parallel edge in the field of view) and, for this reason, has been termed *global perspectival splay rate* (Wolpert, 1987). The sine and cosine functions in the first term of the splay rate equation index the dependence of splay rate on the optical position in the field of view. Edges with zero splay (perpendicular to the horizon) and ±90° splay angle (the horizon) will have a splay rate of zero. From these minima, the absolute change in splay angle for a given change in altitude will increase to a maximum at an initial splay angle of ±45°.

The second term in the splay rate equation indexes change in splay that results from lateral (side-to-side) motion of the observer. That is, as the observer moves closer to an edge, the splay angle gets smaller, and as the observer moves farther from an edge, the splay angle gets larger. Note that the change in splay angle due to lateral motion is also an inverse function of eyeheight (i.e., altitude). The higher the observer is, the smaller the change in splay angle will be for a given lateral motion. The cosine-squared term indicates the impact of position. The change in splay due to lateral motion is greatest for an edge directly below the observer (0°), and decreases to a minimum at the horizon (90°).

To test for the relative contribution of splay angle for control of altitude, Wolpert, Owen, and Warren (1983) measured observers' ability to detect loss of altitude in a flight simulator (simulating flight at a constant forward speed) using three types of ground texture. The three ground textures are shown in Fig. 22.7. The *Grid* texture contains the full optical expansion; the *Splay* texture contains only the splay component of the optical expansion; and the *Depression* texture has eliminated the splay component of optical expansion. The results showed that detection of altitude loss was best with the Splay texture. It was nominally worse with Grid texture and it was significantly worse with the Depression texture. These results were interpreted as evidence that splay angle was a critical component of the overall expansion pattern for judging change in altitude.

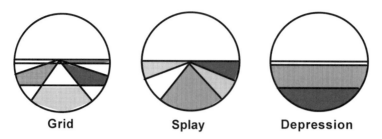

Grid   **Splay**   **Depression**

FIG. 22.7.   Three types of optical texture: Grid, Splay, and Depression.

Johnson, Bennett, O'Donnell, and Phatak (1988; Johnson & Phatak, 1990) also examined control of altitude using similar textures to those used in the Wolpert et al.
(1983) study. However, there were two important differences between the two studies. Wolpert et al. used a psychophysical judgment task (their observers were passive).
Johnson et al., however, used an active regulation task. That is, participants in the Johnson et al. study were required to continuously track a constant altitude (to null out a
quasi-random disturbance). The second difference was the flight platform that was
simulated. Wolpert et al. simulated a fixed wing aircraft cruising at a constant forward
velocity. Johnson et al. simulated a rotary wing aircraft (helicopter) in a hover. In apparent contradiction to the Wolpert et al. study, Johnson et al. found that performance
with Depression texture (no splay) was superior. Performance was nearly equivalent
with the Grid texture, and was significantly worse with Splay texture.

How can the contradictory results of the Wolpert et al. and Johnson et al. studies
be explained? Does it reflect the difference between active and passive observers? Or
does it reflect differences in the flight context: constant velocity cruise versus hover?
First consider the optics of the Depression texture that Johnson et al. found superior.
The optical depression angle for an edge parallel to the horizon can be computed using the following equation:

$$\delta = \tan^{-1}\left(\frac{x_g}{h}\right) \tag{13}$$

where x_g is the distance from the observer to the edge and h is the eyeheight (altitude) of the observer.

An edge directly below an observer has a depression angle of 0°, and the horizon has
a depression angle of 90°. The rate of change of depression angle is:

$$\dot{\delta} = \left(\frac{-\dot{h}}{h}\right)\cos\delta\sin\delta + \left(\frac{\dot{x}_g}{h}\right)\cos^2\delta \tag{14}$$

where $-\dot{h}$ is the rate of change of eyeheight (altitude) and $\dot{x}_g$ is the fore-aft velocity.

Note that the first terms in the equations for splay and depression angle are similar. The global rate of change of both optical elements are a function of fractional
change in altitude. Also, the peak rate of change is at 45° in both cases. This is because

these elements are components of the symmetrical expansion associated with approach to an obstacle (in this case the ground). The second terms in the equations for splay and depression angle are different. Splay angle changes with lateral motion, but depression angle changes as a function of fore-aft motion. Note that the global factor in the second term of the depression rate equation is global optical flow rate. Thus, motion of the depression edges is directly proportional to forward velocity and inversely proportional to eyeheight. The depression edge directly below the observer has the greatest velocity, and the edge velocity decreases to zero at the horizon.

Johnson et al. designed their altitude tracking task to take advantage of the differential effects of lateral and fore-aft motion on the different textures. The task was to maintain a constant altitude against a sum-of-sines disturbance to this axis of motion. However, the vehicle also was buffeted by sum-of-sines disturbances on the lateral and fore-aft axes (even though the control only affected altitude). Further, the disturbances on each axis were constructed of sine waves of different frequencies, as illustrated in Fig. 22.8. Thus, frequency was a signature that would allow the individual disturbances to be "traced" through the control system. If the controller responded specifically to the altitude, then a spectral analysis of the control should show peaks at the frequencies associated with the altitude disturbance (as shown in Fig. 22.8). If the controller was responding specifically to the splay component, then there should be peaks in the control spectrum associated with both altitude and lateral disturbances, because changes in both of these quantities affect perceived splay. Similarly, if the controller was responding specifically to the depression component, then there should be peaks in the control spectrum associated with both altitude and fore-aft disturbances. Note that control power associated with either the lateral or fore-aft disturbances is "noise" or "crosstalk" with respect to the altitude control task. Johnson et al. found that Depression texture resulted in overall less tracking error with respect to altitude and greater power at the frequencies associated with the altitude disturbance. They also found a greater amount of crosstalk in the Grid and Depression conditions resulting from the fore-aft disturbance (visible only in the depression component of texture). This pattern suggests that the depression component of the expansion pattern had a strong impact on control.

The use of frequency to differentiate which of multiple possible disturbances a control system is responding to is a very useful strategy. This strategy can be used in any situation where there are multiple signals that might compete for the controller's attention (e.g., different sensory modalities). If the competing signals can be isolated so that they have distinctive but overlapping frequency characteristics, then the frequency content of the control motion can be used to infer the relative impact of the different signals on the control activity. This strategy will work if the control system is quasi-linear. If there are strong nonlinearities in the controller, then the nonlinear characteristics of the controller may have their own distinctive frequency signature.

Now back to the question of why Wolpert et al. found a display with splay only to be superior and Johnson et al. found a display with only depression to be superior for regulating altitude. First, consider the splay and depression texture components of the runway expansion in Fig. 22.6. Note that the relation between altitude and splay (the side edges of the runway) is monotonic, that is, when altitude decreases splay always increases, and vice versa. However, the relation between depression and alti-

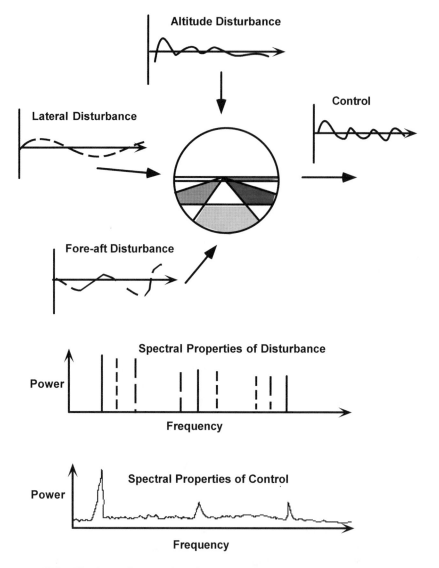

FIG. 22.8.   This figure illustrates how frequency signatures can be used to trace multiple disturbances though a control system. The power spectrum for the control motion suggests that this control system is responding specifically to the altitude disturbance (after Flach, 1990).

tude is not monotonic. The angle for the far edge of the runway increases (approaches the horizon – 90°) as altitude decreases, but the angle for the near edge of the runway decreases (moves under the observer – 0°). This is because the expansion pattern is symmetrical around the aim point. This is representative of the situation used by Wolpert et al. For the Johnson et al. study, the aim point was more directly below the aircraft. In the hover task, the flow of splay and depression textures are both monotonically related to altitude.

A more recent study (Flach, Warren, Garness, Kelly, & Stanard, 1997) found an interaction between optical texture and forward velocity that is consistent with the results from both Wolpert et al.'s and Johnson et al.'s studies. In a hover task, depression texture supports slighly better altitude control. However, for regulation of altitude during a high speed cruise (significant forward velocity), the splay component is far superior to the depression component for regulating altitude. Flach et al. (1997) attributed this interaction to the fact that the global optical flow component of the depression rate equation creates depression flow that is unrelated to the altitude control task. This flow is effectively "noise" that masks the changes in depression associated with altitude. In other words, the observers appeared to have had difficulty disambiguating the changes in depression flow associated with altitude from those associated with motion in the fore-aft direction. This again illustrates the ambiguity in the flow associated with altitude and speed changes.

From the perspective of control, there are two important points for this section. First, when modeling the optical information, it is not only important to appreciate the mapping between flow geometry and the state variables (e.g., altitude), but the mapping to other potential noise sources (e.g., fore-aft and lateral disturbances) must also be considered. One caution about generalizing from the studies in this section is warranted. Altitude control was the experimental task used, but altitude control per se may have little to do with flying. It is likely that pilots are regulating collision with the ground, not altitude. Thus, all axes of motion will generally be controlled in a natural flying task. Although there is an ambiguity associated with changes in speed versus changes in altitude, as Gibson et al. (1955) noted, time-to-collision with the ground is unambiguously specified.

The second important point of this section is the use of frequency to "trace" multiple signals through the control system. This strategy can be used to evaluate the controller's ability to respond based on a joint function of multiple input signals. In which case, the frequency components from all signals should produce peaks in the control spectrum (see Jex, Magdaleno, & Junker, 1978 for another example of this strategy). The strategy can also be used to evaluate the controller's ability to select a specific signal from among multiple inputs. In this case, only the frequency components of the target signal should show up in the control spectrum. This strategy will work as long as the controller is quasi-linear. This is because a sine wave will pass through a linear system without changing frequency. As discussed in previous chapters, changes in the amplitude and phase of the sine wave can be used to make inferences about the characteristics of the controller. Frequency characteristics can also be useful for making inferences about nonlinear systems, however, the logic is more complex. Some nonlinearities can be recognized based on distinctive signatures in the frequency domain (e.g., Berge, Pomeau, & Vidal, 1986).

CONCLUSION

The sensors of engineered control systems are generally designed to measure Newtonian dimensions of space and time. Thus, the state variables are typically specified in terms of position, velocity, acceleration, and so on. However, the eyes and other sensors of biological systems are not necessarily sensitive to these dimensions. For example, the eyes (with their two dimensional surface) do not have a direct measure of distance in three-dimensional space. Eyes do not measure linear distance or velocity directly; rather, they are designed to measure angles and angular rates. On the other hand, the biological control system must satisfy the dynamic constraints reflected in Newton's laws and, in fact, many biological systems are remarkably skillful in responding to the dynamic constraints of motion. They move quickly through cluttered environments without collision, and they can move to intercept other moving objects (e.g., catching a fly ball). This suggests that there is a high degree of specificity between the angles and angular rates measured by the eyes and the control dimensions. The major theme of this chapter is that understanding the degree of specificity among the available forms of information (for vision and other modalities) is a critical aspect of analyzing biological control systems.

Gibson's analysis of optical flow with respect to the problem of controlling locomotion is an important exemplar of how to attack the problem of feedback specificity. The study of the geometry of optic flow has helped in understanding the specificity of structure in the optic array to the action constraints associated with controlling locomotion. This provides important insights to the stability boundaries for the control of locomotion. It also explains how locomotion is controlled and the conditions that may lead to instability (e.g., CFIT). Vision is clearly an important modality for controlling locomotion. However, other modalities also play an important role and a complete theory of the control of locomotion must also consider information provided by other modalities.

This chapter illustrates the tight coupling between perception and action that is true of all control systems. Powers (1973) emphasized the importance of this coupling in the title of his book, *Behavior: The Control of Perception*. In a later book, Powers (1998) wrote:

> Perception plays a central role in controlling. This is why the theory behind this book is called Perceptual Control Theory, or PCT. The emphasis in PCT is not only to understand control from an outside observer's point of view, as in engineering control theory, but to grasp how control appears to the controller—that is, to you and me, who occupy our own copies of this marvelous mechanism and participate in running it. Taking this point of view gives us a new slant on perception. (p. 18)

Gibson's (1958) paper on visually controlled locomotion provides vivid hypotheses about how control might appear to the controller (i.e., the moving animal). The examples in this chapter are a sample of some of the empirical work that has been inspired by Gibson's hypotheses. These examples illustrate how control theoretic methodologies (e.g., the state space representation and frequency response measurement) can inform experimental work to help provide a new slant on perception and, perhaps, to provide a deeper understanding of human behavior.

REFERENCES

Berge, P., Pomeau, Y., & Vidal, C. (1986). *Order within chaos: Towards a deterministic approach to turbulence.* (L. Tuckerman, Trans.) New York: Wiley. (Original work published 1984.)

Bootsma, R. J., & van Wieringen, P. W. C. (1990). Timing an attacking forehand drive in table tennis. *Journal of Experimental Psychology: Human Perception and Performance, 16,* 21–29.

Caird, J. K., & Hancock, P. A. (1994). The perception of arrival time for different oncoming vehicles at an intersection. *Ecological Psychology, 6*(2), 83–109.

DeLucia, P. R. (1991). Pictorial and motion-based information for depth perception. *Journal of Experimental Psychology: Human Perception and Performance, 17,* 738–748.

DeLucia, P. R., & Warren, R. (1994). Pictorial and motion-based depth information during active control of self-motion: Size-arrival effects on collision avoidance. *Journal of Experimental Psychology: Human Perception & Performance, 20,* 783–798.

Denton, G. G. (1976). The influence of adaptation on subjective velocit for an observer in simulated rectilinear motion. *Ergonomics, 19,* 409–430.

Denton G. G. (1977). Visual motion aftereffect induced by simulated rectilinear motion. *Perception, 6,* 711–718.

Denton, G. G. (1980). The influence of visual pattern on perceived speed. *Perception, 9,* 393–402.

Flach, J. M. (1990). Control with an eye for perception: Precursors to an active psychophysics. *Ecological Psychology, 2,* 83–111.

Flach, J. M., Warren, R., Garness, S. A., Kelly, L., & Stanard, T. (1997). Perception and control of altitude: Splay and depression angles. *Journal of Experimental Psychology: Human Perception and Performance, 23,* 1764–1782.

Gibson, J. J. (1958/1982). Visually controlled locomotion and visual orientation in animals. *British Journal of Psychology, 49,* 182–194. Also in E. Reed & R. Jones (Eds.), *Reasons for realism* (pp. 148–163). Hillsdale, NJ: Erlbaum.

Gibson, J. J., & Crooks, L. (1938/1982). A theoretical field-analysis of automobile driving. *American Journal of Psychology, 51,* 453–471. Also in E. Reed & R. Jones (Eds.), *Reasons for realism* (p. 119–136). Hillsdale, NJ: Erlbaum.

Gibson, J. J., Olum, P., & Rosenblatt, F. (1955). Parallax and perspective during aircraft landings. *American Journal of Psychology, 68,* 372–385.

Gibson, J. J., Schiff, W., & Caviness, J. (1962). Persistent fear responses in rhesus monkeys to the optical stimulus of "looming." *Science, 136,* 982–983.

Haber, R. N. (1987). Why low-flying fighter planes crash: Perceptual and attentional factors in collisions with the ground. *Human Factors, 29,* 519–532.

Jex, H. R., Magdaleno, E., & Junker, A. M. (1978). Roll tracking effects of G-vector tilt and various types of motion washout. In *Proceedings of the Fourteenth Annual Conference on Manual Control* (NASA CP-2060). University of Southern California, Los Angeles, CA.

Johnson, W. W., Bennett, C. T., O'Donnell, K., & Phatak, A. V. (1988, June). *Optical variables useful in the active control of altitude.* Paper presented at the 23rd Annual Conference on Manual Control. MIT, Cambridge, MA.

Johnson, W. W., & Phatak, A. V. (1990, August). Modeling the pilot in visually controlled flight. *IEEE Control Systems Magazine,* 24–26.

Larish, J. F., & Flach, J. M. (1990). Sources of information useful for perception of speed of rectilinear self-motion. *Journal of Experimental Psychology: Human Perception and Performance, 16,* 295–302.

Lee, D. N. (1976). A theory of visual control of braking based on information about time-to-collision. *Perception, 5,* 437–459.

Lee, D. N., & Reddish, P. E. (1981). Plummetting gannets: A paradigm of ecological optics. *Nature, 93,* 293–294.

Lee, D. N., Young, D. S., Reddish, P. E., Lough, S., & Clayton, T. M. H. (1983). Visual timing in hitting an accelerating ball. *Quarterly Journal of Experimental Psychology, 35A,* 333–346.

Li, F.-X., & Laurent, M. (1995). Occlusion rate of ball texture as a source of velocity information. *Perceptual and Motor Skills, 81,* 871–880.

McLeod, R. W., & Ross, H. E. (1983). Optic flow and cognitive factors in time-to-collision estimates. *Perception, 12,* 417–423.

Owen, D., & Warren, R. (1982). *Perceptually relevant metrics for the margin of safety: A consideration of global optical flow and density variables.* Paper presented at the Conference on Vision as a Factor in Aircraft Mishaps. USAF School of Aerospace Medicine, San Antonio, TX.

Powers, W. T. (1973). *Behavior: The control of perception.* New York: Aldine.

Powers, W. T. (1998). *Making sense of behavior: The meaning of control.* New Canaan, CT: Benchmark.

Schiff, W. (1965). Perception of impending collision: A study of visually directed avoidant behavior. *Psychological Monographs, 79* (604).

Schiff, W., Oldak, R., & Shah, V. (1992). Aging persons' estimates of vehicular motion. *Psychology and Aging, 6,* 60–66.

Sidaway, B., Fairweather, M., Sekiya, H., & McNitt-Gray, J. (1996). Time-to-collision estimation in a simulated driving task. *Human Factors, 38*(1), 101–113.

Smith, M. R. H., Flach, J. M., Dittman, S. M., & Stanard, T. W. (2001). Alternative optical bases for controlling collisions. *Journal of Experimental Psychology: Human Perception and Performance, 27,* 395–410.

Tresilian, J. R. (1991). Empirical and theoretical issues in the perception of time-to-contact. *Journal of Experimental Psychology: Human Perception and Performance, 17,* 865–876.

Van der Kamp, J., Savelsbergh, G., & Smeets, J. (1997). Multiple information sources guiding the timing of interceptive actions. *Human Movement Science, 16,* 787–822.

Warren, R. (1982). *Optical transformations during movement: Review of the optical concomitants of egomotion* (Final Tech. Rep. for Grant No. AFOSR-81-0108). Columbus, OH: Ohio State University, Department of Psychology, Aviation Psychology Laboratory.

Warren, R. (1988). Visual perception in high-speed, low altitude flight. *Aviation, Space, and Environmental Medicine, 59*(11, Suppl.), A116–A124.

Wolpert, L. (1987). *Field of view versus retinal region in the perception of self-motion.* Unpublished doctoral dissertation, Ohio State University, Columbus, OH.

Wolpert, L., Owen, D., & Warren, R. (1983). Eye-height-scaled versus ground-texture-unit-scaled metrics for the detection of loss in altitude. In *Proceedings of the 3rd Symposium on Aviation Psychology* (pp. 475–481). Columbus, OH: Ohio State University.

Fuzzy Approaches to Vehicular Control

> *To the control engineer quantitative languages supporting arithmetic are the natural ones. To support the translation of the vaguer, non-numeric statements that might be made about a control strategy we needed a semi-quantitative calculus. Zadeh's (1973) fuzzy logic seemed to provide a means of expressing the linguistic rules in such a form that they might be combined into a coherent control strategy.*
>
> —Mamdani and Assilian (1975, p. 2)

Fuzzy control refers to the use of imprecise linguistic categories to describe situations and their corresponding control actions. For example, as noted by McNeill and Thro (1994), people typically talk about temperature adjustment in their homes in fuzzy terms. "When it starts to get too hot in here, open the window a little bit," is a fuzzy command. It can be contrasted with a more precise command such as, "When the temperature reaches 82 degrees in here, open the window 1.75 inches." Normally, no one would ever give such a precise instruction; however, much of control theory is in the style of the more precise instruction. Fuzzy control attempts to implement control theory in a style that is closer to the way people speak. Others might go farther and say that fuzzy control is closer to the way people both speak and act. As such, fuzzy control might provide a mechanism by which verbally expressed expertise can be used to design control systems.

MEASUREMENT CAN BE SETSY

The formal theory of how to use imprecise verbal categories to design control systems stems from the work of Zadeh (1965), who introduced the concept of a fuzzy set. Before considering a fuzzy set, however, first consider its more commonly used opposite, a *crisp set*. A set is a collection of objects, or situations, or actions, or other entities that belong to a particular category. The set of all U.S. citizens is a particular category of people. Any individual is either a member of that category or not. Membership is all or none—full membership or no membership. Such a set is called a crisp set. Similarly, if temperature is being measured to the nearest degree, then the set of temperatures ranging from 79.5000000 . . . degrees to 80.4999999 . . . degrees will be classified as 80 degrees. This range of temperatures is another crisp set, that is, any temperature either is in or is not in this set. Measurement of temperature to the nearest degree partitions the temperature continuum into a nonoverlapping collection of crisp sets, each set having a range of one degree. Any particular temperature will have full membership in one of these sets, and zero membership in the other sets (Fig. 23.1, top).

In contrast, a *fuzzy set* permits memberships that are in between none to full. For example, consider the categories that individuals might use to describe the temperature in their living room, as depicted in Fig. 23.1 (bottom) (Driankov, Hellendoorn, & Reinfrank, 1993). Here membership ranges from 0 to 1. Zero means no membership, 1 means full membership, and fractions between 0 and 1 indicate degrees of membership. The temperature of 21°C (70°F) has full membership in the category "comfortable" because it is very typical of that category. It is the individual's preferred temperature. The temperatures 20°C and 22°C have weaker membership in the category "comfortable"; however, they also have somewhat stronger memberships in the categories "cool" and "warm," respectively. In other words, 20°C is somewhat comfort-

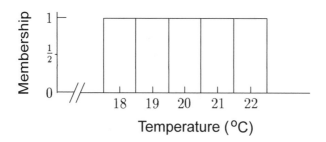

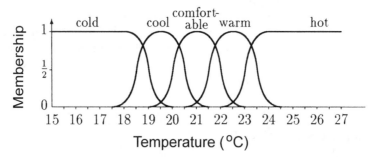

FIG. 23.1. Measurement in terms of (top) crisp sets one degree wide or (bottom) fuzzy sets of variable width and overlap. Bottom drawing from "An introduction to fuzzy control" by D. Driankov, H. Hellendoorn, and M. Reinfrank, 1993, p. 48, fig. 2.5. New York: Springer-Verlag. Copyright 1993 by Springer-Verlag. Adapted by permission.

able, but it is also somewhat cool. Thus, although there is a sharp boundary between the crisp measurement categories of 20°C and 21°C, there is no such sharp delineation between comfortable temperatures and warm temperatures. Similarly, 19°C is too far from the desired temperature to have any membership in the category "comfortable." However, it has strong membership in the category "cool," and somewhat weaker membership in the category "cold." It is also worth noting that the shape of these membership functions and the degree to which they overlap might vary considerably from individual to individual, and hence provide a source of lively family debate on cold winter nights.

FUZZY AND

When fuzzy sets are used to describe a control strategy, they are typically part of a rule that specifies, "If this situation arises, take this action." The description of the situation typically involves three implicit stages: (a) the subjective estimation or crisp measurement of several different quantities; (b) determining the membership of each of the observations or measurements in a relevant fuzzy set; and (c) an "AND-ing" of these memberships to determine how well the combination of observations corresponds to the situation.

For example, consider the fuzzy advice for buying ripe limes at the grocery story: "If a lime looks somewhat yellow and it feels soft, then buy it." The test for ripeness involves the "AND-ing" of two fuzzy sets, "somewhat yellow" and "soft." If either of these two properties is missing (i.e., if the lime looks green rather than somewhat yellow, or if it feels hard rather than soft), then it is not ripe, and consumers should not buy it. Individuals' observation of the color of the lime will have some membership $(0 \leq m_1 \leq 1)$ in the fuzzy set "somewhat yellow," and the feel of the lime will have some membership $(0 \leq m_2 \leq 1)$ in the fuzzy set "soft." If both memberships equal 1, then consumers should clearly buy the lime. If either or both of the memberships equal 0, then they should clearly not buy it. However, a more difficult question is what action to take when the memberships are between 0 and 1.

Although there are many possibilities for combining m_1 and m_2 to determine the degree to which the combination of properties "somewhat yellow and soft" is satisfied (e.g., Hirota, 1989/1994), the two most commonly considered combination rules are: the minimum of m_1 and m_2, and the multiplication $m_1 \times m_2$. These are two possible definitions of a fuzzy "AND-ing" of memberships. Both of these combination rules satisfy the usual logical definition of AND when memberships are 1 and 0, corresponding to true and false. Namely, if either condition is false, then the AND-ing of the conditions is false:

m_1	m_2	minimum of m_1, m_2	$m_1 \times m_1$	
0	0	0	0	(false)
1	0	0	0	(false)
0	1	0	0	(false)
1	1	1	1	(true)

There are several ways to choose between these two alternative combination rules. If fuzzy rules are intended to capture human expertise, then it might be appropriate to mimic human intuitive combination. One approach to choosing among the two AND rules is therefore to measure people's subjective estimates of the membership of elementary statements (e.g., "This lime is somewhat yellow.") and also their subjective estimates of the AND-ing of elementary statements (e.g., "This lime is somewhat yellow and soft."), and empirically determine which combination rule best approximates the relation among their subjective estimates. The two rules make qualitatively different predictions. The minimum rule implies that if m_1 is smaller than m_2, then it does not matter whether m_2 is much larger than m_1 or only a little larger than m_1. The AND-ing of these memberships only depends on the smaller of the two, namely m_1. In contrast, the multiplication rule implies that variations in m_2 when it is larger than m_1 will have some effect on the AND-ing of memberships (see Hirota, 1989/1994; Williams, 1986). The psychological measurement of people's subjective estimates of membership has yielded mixed results, some favoring the multiplication rule (e.g., Oden, 1977) and some favoring the minimum rule (e.g., Thohle, Zimmermann, & Zysno, 1979).

Another approach is to use these different rules for ANDing in implementing various fuzzy control systems, and empirically test which rule results in the best performance. Terano, Asai, and Sugeno (1989/1994) indicated that the most favorable results have generally been found with the minimum rule. In the design of control systems, this rule is the most commonly used form of fuzzy AND. In addition to fuzzy ANDs, there are fuzzy analogs to the operations of OR and NOT as well (e.g., see Hirota, 1989/1994; Oden, 1977). However, in this chapter the fuzzy AND will be sufficient to illustrate two interesting control problems.

FUZZY RULES FOR PARKING A CAR

As already noted, a fuzzy control strategy consists of a set of rules that contains fuzzy sets. As in all control strategies, there are associations between observations and actions. However, fuzzy categories are used in determining which action is most appropriate given any observation. For example, consider a fuzzy control system designed by Sugeno and Murakami (1985) to park a model car as one possible solution to a fuzzy control problem introduced by Zadeh (1968). Maneuvering a car into a garage or a parking space is not an easy task for many people. Therefore, it is an interesting exercise to design a computerized control system to perform this task. Sugeno and Murakami tried to create a system that would mimic human behavior.

The state of the car was described in terms of its two-dimensional position (x, y) relative to perpendicular boundary walls and its heading angle (ψ) (Fig. 23.2). Three sets of rules were used: One set determined whether the car went slowly forward, slowly backward, or stopped; a second set of rules determined the front tire angle, which can be assumed to be proportional to the steering wheel angle, while going forward; and a third set of rules determined the front tire angle while going backward. To appreciate the style of the control rules, it is useful to concentrate on the

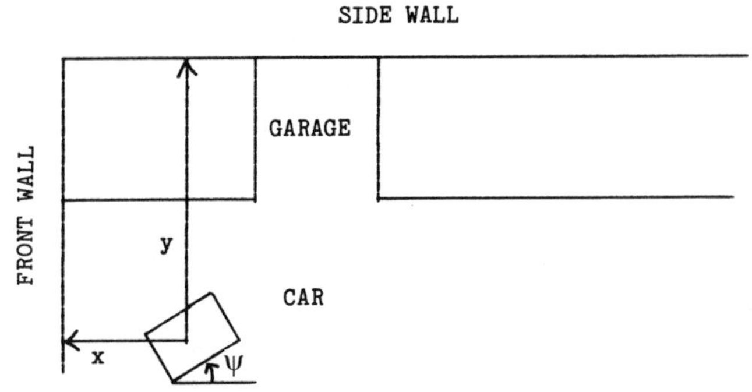

SIDE WALL

FIG. 23.2.   The position of a car is described by its distance from two perpendicular walls (x and y) and by its heading angle (ψ). From "An experimental study on fuzzy parking control using a model car" by M. Sugeno and K. Murakami, 1985, in *Industrial applications of fuzzy control*, edited by M. Sugeno (pp. 125–138). Amsterdam: Elsevier Science Publishers B. V. Adapted by permission.

rules for steering while backing up. The x and y variables were each described by fuzzy sets corresponding to low, medium, and high values, and ψ was described by fuzzy sets corresponding to low and high. Examples of rules were:

Rule 4: If x is low AND y is medium AND ψ is high, then
front tire angle = $2310 - 10.9x - 20.2y + 22.6\psi$ (1)

Rule 7: If x is medium AND y is low AND ψ is low, then
front tire angle = $3000 + 7.74x + 18.1y + 5.40\psi$ (2)

There were a total of 16 such control rules for steering while backing up, and at any point in time, several of these rules might apply to various degrees. The multiplicative version of the fuzzy AND function was used to determine how well a particular set of measurements (x, y, ψ) matched each rule. Namely, for Rule 4 the overall membership for (x is low AND y is medium AND ψ is high) was equal to product of the three individual memberships for (x is low), (y is medium), and (ψ is high). The front tire angle that was actually used to steer the car at any instant was a weighted average of the recommended front tire angles of each of the various rules. The higher the membership of (x, y, ψ) in the fuzzy categories for a given rule, the more strongly it was weighted.

Front tire angle =
[(overall membership of (x, y, ψ) in the categories of Rule 4) ×
   (front tire angle recommended by Rule 4)
 + (overall membership of (x, y, ψ) in the categories of Rule 7) ×
   (front tire angle recommended by Rule 7)
 + . . . etc.] / sum of the overall memberships (3)

This front tire angle was used for a short period over which the car's position (x, y) and heading angle (ψ) would change. Then the rules would be evaluated again to determine what front tire angle should be used next.

Sugeno and Murakami (1985) determined their set of rules by observing a person control an actual model car, and then tried to mimic that behavior. Figures 23.3 and 23.4 show example trajectories by a human controller and by the fuzzy control rules. The fuzzy control rules are not as efficient as the human controller, and sometimes they fail to enter the garage. An important question concerns ways to modify the

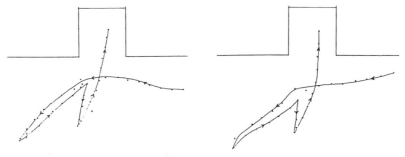

FIG. 23.3. Several car trajectories when a person parks a model car in a garage. The arrows indicate the direction of movement. From "An experimental study on fuzzy parking control using a model car" by M. Sugeno and K. Murakami, 1985, in *Industrial applications of fuzzy control*, edited by M. Sugeno (pp. 125–138). Amsterdam: Elsevier Science Publishers B. V. Reprinted by permission.

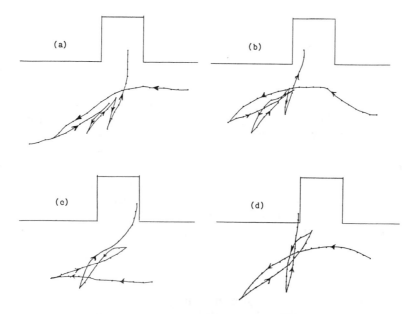

FIG. 23.4. Several car trajectories when a set of fuzzy rules attempts to park a model car. From "An experimental study on fuzzy parking control using a model car" by M. Sugeno and K. Murakami, 1985, in *Industrial applications of fuzzy control*, edited by M. Sugeno (pp. 125–138). Amsterdam: Elsevier Science Publishers B. V. Reprinted by permission.

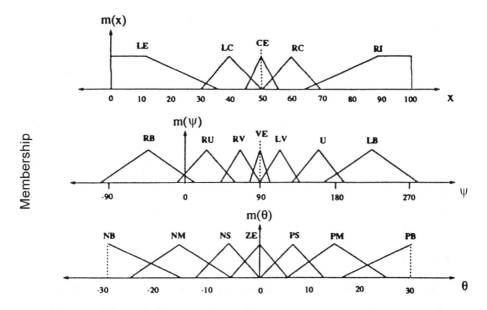

FIG. 23.5.   Fuzzy membership functions for horizontal position (x), heading angle (ψ), and front tire angle (θ) for backing up a vehicle. From "Adaptive fuzzy systems for backing up a truck-and-trailer" by S. Kong and B. Kosko, 1992, *IEEE Transactions on Neural Networks, 3*, pp. 211–223. Copyright 1992 by IEEE. Adapted by permission.

fuzzy rules to make them more efficient and/or make them more closely resemble the human controller.

The task of designing a control system to park a vehicle has been one of continuing interest since Sugeno and Murakami's (1985) early effort (see also chap. 24). A later effort by Kong and Kosko (1992) illustrates a somewhat different way to implement the fuzzy rules. In this example, the task is to back up to a loading dock so that the horizontal position, x, is at the center of the loading dock, and the heading angle, ψ, is vertical. It was assumed that the initial distance from the loading dock was sufficiently large so that vertical position, y, could be ignored in implementing the steering strategy. Horizontal position (x), heading angle (ψ), and front tire angle (θ), which can be assumed to be proportional to the steering wheel angle, were described by various numbers of overlapping fuzzy sets, as shown in Fig. 23.5. The fuzzy rules had the form:

> If the heading angle (ψ) is right of vertical (RV)
> AND the horizontal position (x) is centered (CE),
> then the front tire angle (θ) is a small positive value (PS).   (4)

Kong and Kosko (1992) used the minimum rule to calculate the fuzzy AND. Namely, given a particular observation (x, ψ), the degree to which the "If" part of the previous rule is satisfied is the minimum of ψ's membership in fuzzy set RV and x's membership in fuzzy set CE. Unlike the Sugeno and Murakami (1985) example, the "then" part of the rule is also a fuzzy set (e.g., small positive) rather than a particular front tire angle. In other words, another fuzzy set assigns membership values over a range of recommended front tire angles. However, just as the degree to which the "If" portion of the rule was satisfied determined the weighting of the recommended front tire angle in the Sugeno and Murakami (1985) control system, it should similarly influence how strong a membership is assigned when the recommended front tire angle is a fuzzy set. The memberships of the front tire fuzzy sets were allowed to be no larger than the overall membership of the "If" part of the rule. In Fig. 23.6, the overall membership of the "If" part of the top rule is determined by x's membership in CE, because it is the minimum of the two individual memberships. The front tire membership function is not allowed to exceed this value, and so the shape of the resulting membership function is a trapezoid (shaded region, top right). The front tire membership function for the lower rule in Fig. 23.6 is calculated in a similar manner. However, because the "If" part of the lower rule is not satisfied as strongly as the top rule, the memberships of its recommended front tire angles are limited to a greater degree than in the top rule.

The final step is to take the fuzzy recommendations of all of the rules, and combine them in some way to determine an actual front tire angle. Kong and Kosko (1992) combined the two fuzzy sets of recommended front tire angles and calculated the "centroid," indicated by the arrow in the bottom of Fig. 23.6. Like the Sugeno and Murakami (1985) example, this centroid is a weighted mean. Each front tire angle is weighted by it maximal membership across the set of rules. This process of calculating a particular front tire angle from the recommendation of a number of fuzzy sets is called *defuzzification*. The front tire angle is implemented, the car moves for a short while so that its position (x) and heading (ψ) change, and then the whole process is

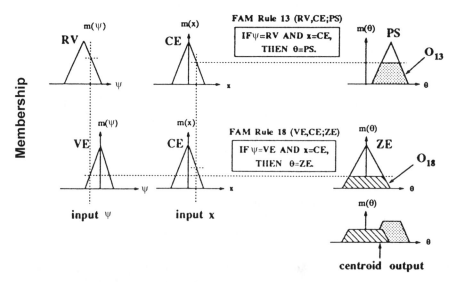

FIG. 23.6.   The combination of two fuzzy control rules to determine the particular front tire angle at one instant in time. From "Adaptive fuzzy systems for backing up a truck-and-trailer" by S. Kong and B. Kosko, 1992, *IEEE Transactions on Neural Networks, 3*, pp. 211–223. Copyright 1992 by IEEE. Adapted by permission.

repeated. New rules will come to dominate the steering behavior as the values of x and ψ change.

This style of control can be contrasted with bang-bang time optimal control (chap. 7), which implements a single extreme control value for each state of a dynamic system (e.g., for each location in the phase plane). Time optimal control exhibits abrupt changes from one extreme control value to another as a system passes through different regions of the phase plane. In contrast, the centroid defuzzification effectively interpolates among a number of competing control values in a smooth, gradual manner.

The calculation procedures used by Kosko and Kong (1992) are typical of many fuzzy control systems (e.g., see Langari & Berenji, 1992; Terano, Asai, & Sugeno, 1989/1994). They can be considered a generalization of the procedures used by Sugeno and Murakami (1985) in which the conditions for action ("If" part of the rule) and the recommended action ("then" part of the rule) are both composed of fuzzy sets. Kong and Kosko (1992) tested their fuzzy controller on a linearized simulation of a vehicle (a truck cab) and had different task constraints from Sugeno and Murakami (1985), so comparison of their results is difficult.

STOPPING A SUBWAY TRAIN

Riding on a subway is not always comfortable for many reasons such as noise, crowds, and jerky starting and stopping. The latter is not always a problem, however, especially if the subway operator is highly experienced. Yasunobu and Miyamoto (1985) attempted to mimic the control patterns of an experienced operator with a fuzzy control system. Their efforts resulted in an automatic control system that was implemented in Japan in 1987 (Yasunobu, 1989/1994). The full control system accelerates from a stop, maintains speed just below the legal limit, and then decelerates to a stop at the next desired location. The style of this control system can be illustrated by how it implements stopping at a desired location.

The subway braking control has only nine different settings. The fuzzy control system contains a nonlinear model of the subway that runs much faster than real time. Every .1 seconds, the present values of subway position and velocity along with the loading due to number of passengers and the upward or downward grade of the tracks are input into the model, and the response to each of the possible control settings is predicted 3 seconds into the future. The predicted speed and position and the associated amount of change in the braking control setting are evaluated in terms of fuzzy sets. For example, one rule might be:

If increasing the braking control by one setting will result in
the *running time* between stations being very good
AND
the predicted *ride comfort* will be good
AND
the predicted *stopping accuracy* will be good,
then increase the braking control by one setting. (Yasunobu, 1989/1994)   (5)

In order to keep the *running time* between stations sufficiently short, braking should not begin before the subway passes a marker that is positioned some distance before the station. High degrees of membership in the fuzzy set of very good running times will occur if braking begins after passing this marker.

The *ride comfort* is evaluated in terms of the time since the last change in the braking control and the magnitude of that change. The longer the interval between successive braking changes and the smaller the changes, the smoother the overall ride will be, and the higher the membership in the fuzzy set of good ride comfort. This evaluation criterion tries to limit the number and size of braking changes.

The *stopping accuracy* is determined from the predicted stopping position output from the fast-time model of the subway. The output of the model is a numerical prediction. However, given that this model is imprecise, the stopping position prediction is taken to be a fuzzy set of possible stopping positions. The prediction of the fast-time model is at the center of this fuzzy set, and has the highest membership function (Fig. 23.7, top). The fuzzy sets corresponding to predicted stopping positions and positions with good accuracy can be ANDed together to determine the degree to which each position is both predicted AND has good accuracy (is close to the desired stopping position) (Fig. 23.7, bottom). The minimum rule is used for this ANDing. The maximum of the resulting memberships can then be considered the fuzzy evaluation that the stopping distance will be good.

In thinking about this evaluation procedure, it is useful to note that if only a single crisp position had been predicted with membership equal to one, then these procedures result in the usual fuzzy evaluation of goodness of stopping accuracy. Namely, the minimum of the memberships of a crisp predicted position and the goodness of stopping accuracy is the membership of goodness of stopping accuracy at the predicted position. The more elaborate evaluation procedure is necessary when an observation or prediction is fuzzy, and the category for evaluating it is also fuzzy. The combination of these fuzzy sets is an example of "fuzzy inference" (e.g., see Hirota, 1989/1994), inferring what control action is appropriate given a fuzzy observation and fuzzy conditions for acting.

Once a membership has been assigned to each of the three evaluation criteria of the control rule (i.e., running time, ride comfort, and stopping accuracy), then the overall evaluation of this rule can be evaluated as the ANDing of these three memberships (i.e., the minimum of these three). Similar evaluations can be conducted for the other possible braking control settings, and the setting with the highest evaluation is executed. One tenth of a second later, the position and velocity of the subway will have changed, and the evaluation process is repeated all over again.

It is interesting to note that unlike classical control theory, the design of this fuzzy control system did not explicitly represent conditions for stable operation. How then does one know that this overall system runs stably? In addition to many computer simulations, over 10,000 tests were conducted with actual subway trains in order to assure that instability would not arise and also to fine-tune the system. The stopping error of the system rarely exceeds 30 cm (1 ft) (Yasunobu, 1989/1994).

Theoretical efforts to bring stability considerations more explicitly into the design of fuzzy control systems are being pursued (e.g., Driankov et al., 1993; Langari & Berenji, 1992). There have also been efforts to make fuzzy control systems adaptive or

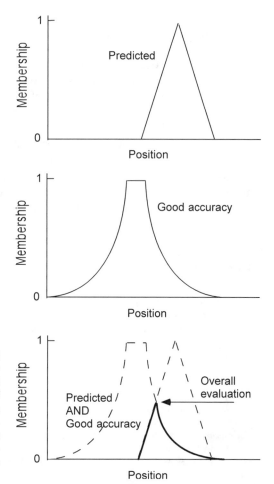

FIG. 23.7. The goodness of predicted stopping accuracy can be determined by first ANDing the fuzzy set of predicted stopping positions with the fuzzy set of good accuracy (heavy lines in the bottom graph). The maximum degree of membership in the resulting fuzzy set is the overall evaluation of good stopping accuracy (based on Yasunobu & Miyamoto, 1985).

self-tuning, often by combining them with neural networks. This topic is touched on in chapter 24.

WHERE DO THE FUZZY RULES COME FROM?

Fuzzy control systems can be considered to be a type of *expert system*. Namely, they attempt to take linguistically expressed knowledge of human experts and represent it as a set of If–then rules. The number of rules in fuzzy control systems is typically on the order of from 30 to 100, which is at least a factor of 10 less than the typical number of rules found in nonfuzzy expert systems (e.g., Hirota, 1989/1994).

Linguistic expression of expert knowledge is not always complete or accurate, however, especially when that knowledge concerns perceptual-motor skills. Sugeno and Murakami (1985) noted that although verbal reports of expert knowledge may be a useful starting point, such reports may need to be supplemented by observing the operator's performance and empirically deriving a set of rules to describe the observed behavior. They additionally suggested that it may be easier to model the human operator's interactions with a system than to model the system to be controlled and subsequently design a set of control rules. This latter approach can be pursued, however, when human experts are not available, or when their level of performance is not of sufficient quality. The model of the system to be controlled may then be a fuzzy one, or fuzzy control may be used as the top level of a hierarchical control system (e.g., Langari & Berenji, 1992).

This chapter has provided a very brief introduction to fuzzy control. For more detailed discussions, the reader is referred to the references at the end of this chapter. Additionally, the collection of readings by Dubois, Prade, and Yager (1993) provides a historically based overview and bibliography ranging forward from the earliest application of fuzzy control to a steam engine by Mamdani and Assilian (1975).

REFERENCES

Dubois, D., Prade, H., & Yager, R. R. (Eds.). (1993). *Readings in fuzzy sets for intelligent systems.* San Mateo, CA: Morgan Kaufmann.

Driankov, D., Hellendoorn, H., & Reinfrank, M. (1993). *An introduction to fuzzy control.* New York: Springer-Verlag.

Hirota, K. (1994). Fuzzy set theory for applications. In T. Terano, K. Asai, & M. Sugeno (Eds.), *Applied fuzzy systems* (C. Aschmann, Trans.) (pp. 9–50). Cambridge, MA: AP Professional. (Original work published 1989)

Kong, S., & Kosko, B. (1992). Adaptive fuzzy systems for backing up a truck-and-trailer. *IEEE Transactions on Neural Networks, 3,* 211–223.

Langari, R., & Berenji, H. R. (1992). Fuzzy logic in control engineering. In D. A. White & D. A. Sofge (Eds.), *Handbook of intelligent control: Neural, fuzzy, and adaptive approaches* (pp. 93–140). New York: Van Nostrand Reinhold.

Mamdani, E. H., & Assilian, S. (1975). An experiment in linguistic synthesis with a fuzzy logic controller. *International Journal of Man–Machine Studies, 7,* 1–13.

McNeill, F. M., & Thro, E. (1994). *Fuzzy logic: A practical approach.* Cambridge, MA: AP Professional.

Oden, G. C. (1977). Integration of fuzzy logical information. *Journal of Experimental Psychology: Human Perception and Performance, 3,* 565–575.

Sugeno, M., & Murakami, K. (1985). An experimental study on fuzzy parking control using a model car. In M. Sugeno (Ed.), *Industrial applications of fuzzy control* (pp. 125–138). Amsterdam: Elsevier Science.

Terano, T., Asai, K., & Sugeno, M. (Eds.). (1994). *Applied fuzzy systems.* (C. Aschmann, Trans.). Cambridge, MA: AP Professional. (Original work published 1989)

Thohle, U., Zimmermann, J. J., & Zysno, P. (1979). On the suitability of minimum and product operators for the intersection of fuzzy sets. *Fuzzy Sets and Systems, 2,* 167–180.

Williams, R. J. (1986). The logic of activation functions. In D. E. Rumelhart & J. L. McClelland (Eds.), *Parallel distributed processing* (pp. 423–443). Cambridge, MA: MIT Press.

Yasunobu, S. (1994). Automatic train operation. In T. Terano, K. Asai, & M. Sugeno (Eds.), *Applied fuzzy systems* (C. Aschmann, Trans.) (pp. 115–128). Cambridge, MA: AP Professional. (Original work published 1989)

Yasunobu, S., & Miyamoto, S. (1985). Automatic train operation system by predictive fuzzy control. In M. Sugeno (Ed.), *Industrial applications of fuzzy control* (pp. 1–18). Amsterdam: Elsevier Science.

Zadeh, L. A. (1965). Fuzzy sets. *Information and Control, 8,* 338–353.

Zadeh, L. A. (1968). Fuzzy algorithms. *Information and Control, 12,* 94–102.

Zadeh, L. A. (1973). Outline of a new approach to the analysis of complex systems and decision processes. *IEEE Transactions on Systems, Man and Cybernetics, SMC-3,* 28–44.

Learning to Control Difficult Systems: Neural Nets

A learning control system *is one that has the ability to improve its performance in the future, based on experiential information it has gained in the past, through closed-loop interactions with the plant and its environment. . . . In the context of control, learning can be viewed as the automatic incremental synthesis of multivariable functional mappings and, moreover, . . . connectionist systems provide a useful framework for realizing such mappings. . . . Learning is required when these mappings cannot be determined completely in advance because of a priori uncertainty (e.g., modeling error).*
— Baker and Farrell (1992, pp. 36–38)

The various feedforward and feedback control techniques discussed in the previous chapters have assumed considerable knowledge about the dynamics of the system being controlled. In the case of complex machines or living systems, such knowledge is often absent. One tool for controlling such systems is to use control systems that gradually learn about the unknown dynamic structure. Multilayer feedforward neural networks and adaptive fuzzy sets are two structures that can accomplish such learning (for a brief introduction to others, see Astrom & McAvoy, 1992). This chapter provides a brief introduction to these methods and discusses their use in learning to back up a truck and trailer.

NEURAL NETWORKS AS A MATHEMATICAL APPROXIMATION TECHNIQUE

Before considering neural networks, it is helpful to consider first the simpler problem of trying to approximate a set of two-dimensional data consisting of pairs of points, (x_i, y_i), with a straight line (e.g., Jordan, 1996):

$$y'_i = ax_i + b \tag{1}$$

a and b are constants, and y'_i is an estimate of y_i. The goal is to choose values for a and b that will minimize the squared discrepancies or errors between the predicted values, y'_i, and the actual data points, y_i, summed across all the data points. In other words, the goal is to minimize $\sum E_i^2 = \sum (y_i - y'_i)^2$. This problem is typically referred to as *linear regression*, and the equations for calculating a and b based on observing an entire set of (x_i, y_i) pairs can be found in many statistics texts.

Suppose, however, that rather than using these equations, the problem is approached by gradually adjusting a and b after observing each individual (x_i, y_i) pair. Namely, randomly sample a data point (x_i, y_i) from the set of available data, observe the discrepancy, $E_i = (y_i - y'_i) = (y_i - ax_i - b)$, and then adjust a and b accordingly. Then, randomly sample another data point and repeat the process of adjustment. Given some arbitrary initial values, a_o and b_o, these will gradually be adjusted to more nearly optimal values. The crucial question is how one adjusts a and b after observing E_i.

One approach is to calculate how sensitive E_i^2 is to small perturbations of a and b, and then make adjustments proportional to these sensitivities (e.g., see Hertz, Krogh, & Palmer, 1991; Jordan, 1996). In more mathematical terminology, calculate how much E_i^2 would change from a small increment in a or b. For example, the ratio of the change in E_i^2 to a change in a while treating b as a constant is called the *partial derivative* of E_i^2 with respect to a, $\partial E_i^2/\partial a$. It is possible to adjust a by an amount proportional to this ratio and follow a similar procedure for b. This technique is also called *gradient descent* (e.g., see Hertz et al., 1991). Namely,

$$\partial E_i^2/\partial a = 2E_i \partial E_i/\partial a = 2E_i \partial(y_i - ax_i - b)/\partial a = -2E_i x_i$$
$$\partial E_i^2/\partial b = 2E_i \partial E_i/\partial b = 2E_i \partial(y_i - ax_i - b)/\partial b = -2E_i$$
$$a \rightarrow a - p(\partial E_i^2/\partial a) = a + 2pE_i x_i$$
$$b \rightarrow b - p(\partial E_i^2/\partial b) = b + 2pE_i \tag{2}$$

where p is a small multiplicative constant. The negative signs appear before p in the adjustment rule in order to adjust a and b in directions that will decrease E_i^2. Note that if x_i is small, then a is not changed much, because it could not have contributed much to the discrepancy, E_i (i.e., $y'_i = ax_i + b$ is approximately equal to b, when x_i is small). Conversely, when x_i is large, E_i is very sensitive to the value of a, and a is adjusted by a relatively large amount. In contrast, the sensitivity of E_i to changes in b does not contain x_i as a multiplicative factor. This regression problem is pictured in Fig. 24.1 (top).

If the relation between x_i and y_i is nonlinear—for example, $y_i = \sin(x_i)$—then the linear approximation obtained by adjusting a and b will not be very good. A more elaborate approximation technique might be used. For example, consider Fig. 24.1 (middle). Here $y'_i = f(c_1 h_1 + c_2 h_2) = f[c_1 f(a_1 x_i + b_1) + c_2 f(a_2 x_i + b_2)]$, where $f(\_) = \tanh(\_)$, which is approximately linear in its central range, and saturates at $+/-1$ for extreme values of its input, $(\_)$ (see Fig. 24.1, bottom). This is an example of a "neural network" with one input, x_i, a single layer of hidden units, and only one output, y'_i. It

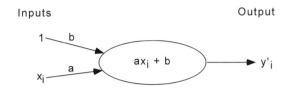

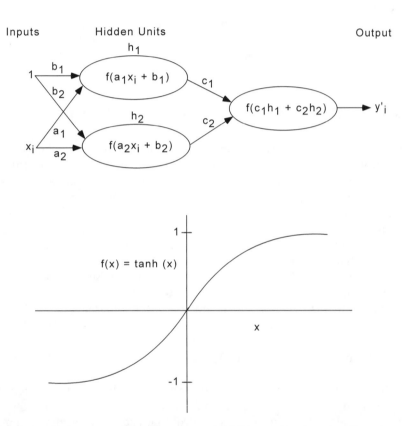

FIG. 24.1.   (top) Incremental linear regression, (middle) a combination of nonlinear functions to form a "neural network," and (bottom) a nonlinear function, $f(x) = \tanh(x)$.

can be used in a similar manner to the simple linear scheme in Fig. 24.1 (top) to approximate more complicated relations between x_i and y_i. The neural network is more elaborate than the simple linear model partly because it involves three sets of linear weighting constants, (a_1, b_1), (a_2, b_2), and (c_1, c_2), and partly because it uses a function $f(\_)$, which saturates at extreme values of its input, $(\_)$. However, the goal of this network is still to approximate the mathematical relation between x_i and y_i, and the general approach is the same. Namely, for each x_i, calculate y'_i, observe $E_i = (y_i - y'_i)$, and adjust the various parameters, $a_1, a_2, b_1, b_2, c_1, c_2$ in proportion to their sensitivities toward E_i^2. For example,

$$\partial E_i^2 / \partial c_1 = 2E_i \partial E_i / \partial c_1$$
$$c_1 \rightarrow c_1 - p(\partial E_i^2 / \partial c_1) \tag{3}$$

This method is still an example of gradient descent, which is also referred to as *back propagation* in the context of neural networks.

When there are multiple input variables, x_{1i}, x_{2i}, and so on, and multiple output variables, y_{1i}, y_{2i}, and so on, the corresponding network is more complicated. However, the general goal remains the same, namely, to approximate the mathematical relation between x_{1i}, x_{2i}, and so on, and y_{1i}, y_{2i}, and so on, by gradual adjustment of parameters (for more extended discussions of neural networks see Dayhoff, 1990; Hertz et al., 1991; Rumelhart & McClelland, 1986).

NEURAL NETWORKS AND DYNAMIC SYSTEMS

From an abstract perspective, a dynamic system is a mathematical relation between a set of input variables and a set of output variables. For some systems, this relation can be derived from basic principles of physics, chemistry, physiology, and so forth. For other systems, the processes relating input and output may be poorly understood. In these latter cases, an approximate relation between input and output can be derived from neural networks. Namely, the neural network can observe input–output pairs and adjust its parameters to approximate the poorly understood system. One might say that the neural network learns the structure of the dynamic system.

Neural networks can also be used to learn a control strategy. For example, if one is trying to implement feedforward control of a system, then it is necessary to develop a control strategy that is the approximate inverse of the dynamic system. Namely, the mapping

Desired Output → (Control Strategy) → Control → (Dynamic System) → Output

implies that the combination of the control strategy and the dynamic system must be approximately unity or an identity mapping. Namely, the control strategy, which transforms the desired output into the control that is exerted on the system, anticipates the transformation imposed by the dynamic system, which transforms the control into the actual system output. In order for the actual output to match the desired output, the control strategy essentially must act like the inverse of the dynamic system or (1/dynamic system). Therefore, a neural network could be used to learn the control strategy by gradually adjusting the parameters of the neural network so that the discrepancy between the actual output and the desired output is minimized (e.g., Jordan, 1996; Nguyen & Widrow, 1990).

Although this scheme sounds straightforward, there are complications. In order to adjust the parameters in the control strategy via gradient descent, it is necessary to determine the sensitivities of E_i^2 = (Desired Output$_i$ – Output$_i$)^2, to perturbations of the neural network parameters. However, in order to determine these sensitivities, it is necessary to know the entire mathematical relation that transforms desired output into actual output. The structure of the control strategy can be a known neural network. However, for a complex system, the mathematical description of the dynamic system may be unknown. This problem can be solved by training another neural network to mimic the dynamic system. This other network can be trained from observed input–output pairs of the actual system. With an approximate network model of the dynamic system, a mathematical representation of the full transformation from desired output to output is now available. Therefore, the control strategy can now be trained to be an effective inverse of the dynamic system, and thus provide feedforward control (see Jordan, 1996, for a more extended discussion).

Thus, in order to train an inverse model of the dynamic system for the control strategy, it is useful to have a "forward model" of the dynamic system itself. As noted by Jordan and Rumelhart (1992), "direct inverse" modeling can be attempted without a forward model by using observed input–output pairs from the dynamic system to train a neural network to produce the input when given the output. However, for nonlinear systems, this approach may produce systematic errors due to complications arising from system redundancy. Therefore, it is often better to learn a forward model as part of the process of learning to control the unknown system. These issues regarding the use of forward and inverse models parallel different approaches to adaptive control (Jordan, 1996; see chaps. 20 and 27).

BACKING UP A TRUCK

The problem of backing up a truck with a trailer to a particular position on a loading dock is a difficult, nonlinear control problem. Following an approach like that in the previous section, Nguyen and Widrow (1990) designed a neural network feedback controller for this system. Before reviewing their work, it is useful to consider how the nonlinear structure that makes this control problem difficult arises from the physical configuration of a truck and trailer.

As noted in a previous chapter, an automobile will travel in a circle if its steering wheel angle is held constant. If the front tire angle is θ, then the front tire closest to the center of the circular path will travel on a circle with radius $L/\sin\theta$, where L is the effective length of the automobile (Fig. 16.1). The corresponding rear tire will travel on a circle with a smaller radius, $(L/\sin\theta)\cos\theta$. If the front tire has a velocity with magnitude V, then the rear tire will have a velocity with magnitude $V\cos\theta$. Also, the rate of change of the car heading angle, ψ_c, will be $(V/L)\sin\theta$ (Equation 1, chap. 16). If the "car" is instead a truck cab, then the same approximate relations can be applied.

Suppose a trailer is attached to the back of the truck cab at a rotating pivot point. The front of the trailer is supported by this point, and the trailer only has a rear set of wheels. Let the heading angle of the truck cab be ψ_c, and the heading angle of the trailer be ψ_t. The heading angle of the truck cab relative to the trailer (i.e., the "relative cab angle"), will then be $\Delta\psi = \psi_c - \psi_t$. Because of the pivot point coupling of the truck cab to the trailer, the motion of the rear of the truck cab, $V\cos\theta$, will be transmitted to the trailer in two ways: (a) The component of the truck cab motion that is parallel to the longitudinal axis of the trailer, $V\cos\theta\cos(\Delta\psi)$, will cause the trailer to move back-

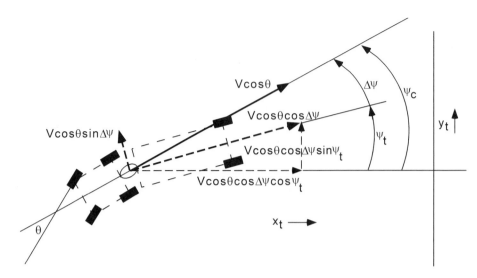

FIG. 24.2. Geometric analysis of the velocity components of a truck cab (dashed square) and trailer (dashed rectangle). A negative (leftward) tire angle (θ) is shown.

ward. (b) The component of the truck cab motion that is perpendicular to the trailer's longitudinal axis, $V\cos\theta\,\sin(\Delta\psi)$, will cause the trailer to rotate (heavy dashed arrows in Fig. 24.2). Because this rotation is about the rear wheels, the rate of change of the trailer heading angle will be $-V\cos\theta\,\sin(\Delta\psi)/L_t$, where L_t is the effective length of the trailer (Higgins & Goodman, 1994). Additionally, the longitudinal velocity of the trailer, $V\cos\theta\cos(\Delta\psi)$, can be partitioned into vertical and horizontal components, dy_t/dt and dx_t/dt by multiplying by $\sin\psi_t$ and $\cos\psi_t$, respectively (thin dashed arrows in Fig. 24.2). Summarizing these relations:

	Small angle approximations
$d\psi_c/dt = (V/L)\sin\theta$	$d\psi_c/dt = (V/L)\theta$
$d\psi_t/dt = -V\cos\theta\,\sin(\Delta\psi)/L_t$	$d\psi_t/dt = -(V/L_t)(\Delta\psi)$
$dy_t/dt = V\cos\theta\,\cos(\Delta\psi)\,\sin\psi_t$	$dy_t/dt = V\psi_t$
$dx_t/dt = V\cos\theta\,\cos(\Delta\psi)\,\cos\psi_t$	$dx_t/dt = V$

$$(4)$$

For small angles, the approximations cos(small angle) = 1 and sin(small angle) = small angle in radians can be used. The system to be controlled can then be represented with a linear approximation as in Fig. 24.3. Notice that the overall transformation from tire angle to trailer vertical position involves three integrations (i.e., this is a third-order system). Also note that the middle integration involves an unstable, positive feedback loop. Namely, a constant truck cab heading angle, ψ_c, will lead to an exponentially increasing or decreasing trailer heading angle, ψ_t (i.e., the trailer has a tendency to "jack knife" in a direction depending on the initial value of $\Delta\psi$). Based on this approximate analysis, it becomes easier to appreciate why backing up a truck and trailer is a difficult control task. If only the position of the back of the trailer were visible, then this system would probably be uncontrollable in the presence of pertur-

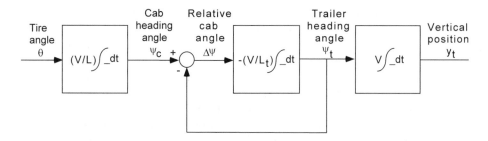

FIG. 24.3.   Approximate linear relation between the angle of the front tires of the truck cab and the vertical position of the rear of the trailer (refer to Fig. 24.2).

bations. However, the truck cab angle and trailer angle are also visible, which would be expected to facilitate control.

Nguyen and Widrow (1990) created a two-layer neural network that learned a feedback control strategy to perform the truck and trailer docking maneuver. They did not constrain the angle to be small, and therefore dealt with the full nonlinear equations.[1] The problem is to start at various initial distances and orientations of the truck and trailer, and to have the end of the trailer arrive at a particular docking location. Additionally, the longitudinal axis of the trailer should be perpendicular to the docking surface when the target location is reached. Following procedures similar to those discussed in the previous section, Nguyen and Widrow (1990) first had one neural network learn to mimic the dynamic response of the truck and trailer based on their response to various steering commands at different initial angles. Then a second neural net was trained to dock the truck and trailer by utilizing this neural network mimic of the actual dynamic system. The controller operated on the actual truck and trailer dynamics until it reached the dock. Then by using the neural network mimic of the truck and trailer, it was possible to calculate the sensitivity of the terminal docking error to the various controller parameters and thereby make appropriate adjustments.[2]

Nguyen and Widrow's (1990) controller had 26 units comparable to the nonlinear functions, f, in Fig. 24.1 (middle and bottom) and 200 weighting parameters. The network required thousands of trials to be successful at this control problem. Whereas this elaborate structure was shown to be sufficient to perform this task, simpler structures may also be successful. For example, Jenkins and Yuhas (1993) implemented a

---

[1]There are some minor differences between their discrete implementation of the truck and trailer equations and the continuous equations presented here (see Anderson & Miller, 1990, p. 493).

[2]One interesting detail is that error information was not available at every step in the control process, but was available only when the rear of the trailer reached the dock. In other words, a desired output was not specified at each instant in order to permit learning of feedforward control as discussed in the previous section. In calculating the sensitivities of terminal error to the controller parameters, Nguyen and Widrow (1990) treated each time step in the simulation as though a separate copy of the controller and mimic neural networks received state information from previous copies of these networks. Thus, there was effectively a chain of alternating controller and mimic networks corresponding to the whole sequence of actions. Sensitivities to the terminal error were calculated for each stage in the action sequence at the end of each trial, and the net adjustment to parameter weights was based on the adjustments indicated across the action sequence.

successful controller for this task with only three nonlinear elements and five weighting parameters. The upper left nonlinear function in Fig. 24.4 implements two springlike aspects of the control. First, it attempts to keep the trailer perpendicular to the dock, which corresponds to keeping the trailer heading angle near zero. Second, it attempts to keep the cab aligned with the trailer, which corresponds to keeping the difference between the cab heading angle and trailer heading angle in Fig. 24.3 (i.e., the relative cab angle, $\Delta\psi$) near zero. The weights are adjusted to give priority to the first alignment and to provide sufficient damping to avoid prolonged oscillations from the springlike behaviors. A third goal is to have the vertical position of the back of the trailer approximately equal to the target position, which corresponds to y_t equal to zero in Fig. 24.3. This latter goal is implemented with the nonlinear function at the bottom left of Fig. 24.4. The relative weighting of the first two goals and the third goal in determining the front tire angle is achieved with the third nonlinear function at the right of Fig. 24.4.

An important aspect of the Jenkins and Yuhas approach to this control problem was to decompose the state space of variables that could potentially influence the tire angle. Namely, a possible description of the truck cab and trailer is in terms of the two-dimensional position of the back of the trailer, (x_t, y_t), the trailer heading angle, the cab heading angle, and the front tire angle. A control strategy can be represented as a surface in a five-dimensional space with each dimension corresponding to one of these variables. The surface specifies the appropriate front tire angle for each combination of the other four variables. The neural network implements a possible control surface, which is gradually adjusted by gradient descent to improve system performance. Higher dimensional surfaces are generally adjusted more slowly on the basis of performance feedback.

The Jenkins and Yuhas approach essentially decomposes the problem to consider three one-dimensional problems of driving the trailer heading angle, the relative cab angle, and vertical position to their desired values. The horizontal position of the trailer can be ignored if the trailer does not start too close to the dock. A priority structure is established in the weighting parameters so that trailer heading angle generally takes priority over relative cab angle. The trailer will thus rotate until it is al-

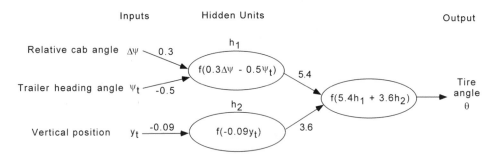

FIG. 24.4. A simple neural network controller for backing up a trailer. The ellipses represent nonlinear functions like those in Fig. 24.1 (bottom). From "A simplified neural-network solution through problem decomposition: The case of the truck backer-upper" by R. E. Jenkins and B. P. Yuhas, 1993, *IEEE Transactions on Neural Networks, 4*, pp. 718–720. Copyright 1993 by IEEE. Adapted by permission.

most properly aligned with the loading dock. Then its effect on tire angle will be diminished, and the subsequent springlike reduction of relative cab angle will contribute to maintaining the trailer's alignment.

Although Jenkins and Yuhas adjusted the five weights in Fig. 24.4 by hand rather than relying on a self-learning algorithm, their result demonstrates that structurally simpler controllers are possible for this problem. This step is important if this approach is to provide potential models of human performance with a limited number of interpretable parameters. On the other hand, the contrast between the Nguyen and Widrow (1990) approach to this problem and the Jenkins and Yuhas approach is similar to the contrast between computational models of arm movements involving forward and inverse models (e.g., Jordan, 1996; Kawato, 1996) and simpler spring models (e.g., Feldman, 1986; Polit & Bizzi, 1979). The Jenkins and Yuhas approach is a relatively simple composition of springlike elements that reduces the complexity of the coordination task. It performs well for the present task. However, for more difficult maneuvers, such as parallel parking between a dock and another truck, this simple technique may not be sufficient (Battiti & Tecchiolli, 1995).

As noted by Jang (1992, 1993), gradient descent can be used to adjust fuzzy control systems as well as neural network control systems. Therefore, the general approach used by Nguyen and Widrow (1990) could be implemented via a fuzzy controller as well. Jang (1992, 1993) demonstrated this technique by designing a self-learning fuzzy controller for balancing an inverted pendulum as well as forward model estimators in a number of different dynamic contexts. Higgins and Goodman (1994) implemented a nonlearning fuzzy controller for the truck and trailer problem.[3] Jenkins and Yuhas (1993) compared their decomposition of the truck and trailer problem to the implementation of fuzzy rules.

IMPLICATIONS FOR HUMAN PERFORMANCE

The emphasis in this area has so far been on demonstrating that a particular learning algorithm and control structure are sufficient to perform a given control task. A different question concerns whether such models will provide insight into human performance of these tasks. One important constraint in most human performance models is that there be a relatively small number of parameters, and they are behaviorally interpretable. The smaller neural net models (e.g., Jenkins & Yuhas, 1993) and similarly small fuzzy set models may have sufficient parsimony to meet this requirement at least for some tasks. Although some would argue that neural network and fuzzy control structures are a replacement for analytic modeling, small-scale models with these architectures may yield a good deal of analytic insight. More complex models with many parameters may also be used to describe human performance; however,

---

[3]Demonstration programs for personal computers for a truck cab backing up have been created by Beale and Demuth (1994) and by HyperLogic Corporation (in Kosko, 1992). The latter is based on a simplified version of the problem in Kong and Kosko (1992).

the problem of parameter identification is more difficult and typically requires larger numbers of observations.

Whether or not gradient descent will prove to be a useful model of human learning of control tasks is another interesting question. There are many alternative learning algorithms (e.g., Battiti & Tecchiolli, 1995; Mazzoni, Andersen, & Jordan, 1991; Miller, Sutton, & Werbos, 1990) that are more efficient and/or biologically plausible than gradient descent. Testing them empirically against human learning data is an intriguing challenge. The existence of learning control models has thus opened many new possibilities for human performance research.

REFERENCES

Ancerson, C. W., & Miller, W. T., III. (1990). A challenging set of control problems. In W. T. Millder, III, R. S. Sutton, & P. J. Werbos (Eds.), *Neural networks for control* (pp. 475–510). Cambridge, MA: MIT Press.

Astrom, K. J., & McAvoy, T. J. (1992). Intelligent control: An overview and evaluation. In D. A. White & D. A. Sofge (Eds.), *Handbook of intelligent control: Neural, fuzzy, and adaptive approaches* (pp. 3–34). New York: Van Nostrand Reinhold.

Baker, W. L., & Farrell, J. A. (1992). An introduction to connectionist learning control systems. In D. A. White & D. A. Sofge (Eds.), *Handbook of intelligent control: Neural, fuzzy, and adaptive approaches* (pp. 35–63). New York: Van Nostrand Reinhold.

Battiti, R., & Tecchiolli, G. (1995). Training neural nets with the reactive tabu search. *IEEE Transactions on Neural Networks, 6,* 1185–1200.

Beale, M., & Demuth, H. (1994). *Fuzzy systems toolbox for use with MATLAB* (pp. 5-2 – 5-15). Boston: PWS Publishing.

Dayhoff, J. (1990). *Neural network architectures.* New York: Van Nostrand Reinhold.

Feldman, A. G. (1986). Once more on the equilibrium-point hypothesis (λ model) for motor control. *Journal of Motor Behavior, 18,* 17–54.

Hertz, J., Krogh, A., & Palmer, R. G. (1991). *Introduction to the theory of neural computation.* Reading, MA: Addison-Wesley.

Higgins, C. M., & Goodman, R. M. (1994). Fuzzy rule-based networks for control. *IEEE Transactions on Fuzzy Systems, 2,* 82–88.

Jang, J. R. (1992). Self-learning fuzzy controllers based on temporal back propagation. *IEEE Transactions on Neural Networks, 3,* 714–723.

Jang, J. R. (1993). ANFIS: Adaptive-network-based fuzzy inference system. *IEEE Transactions on Systems, Man, and Cybernetics, 23,* 665–685.

Jenkins, R. E., & Yuhas, B. P. (1993). A simplified neural-network solution through problem decomposition: The case of the truck backer-upper. *IEEE Transactions on Neural Networks, 4,* 718–720.

Jordan, M. I. (1996). Computational aspects of motor control and motor learning. In H. Heuer & S. W. Keele (Eds.), *Handbook of perception and action: Vol. 2. Motor skills* (pp. 71–120). San Diego: Academic Press.

Jordan, M. I., & Rumelhart, D. E. (1992). Forward models: Supervised learning with a distal teacher. *Cognitive Science, 16,* 307–354.

Kawato, M. (1996). Trajectory formation in arm movements: Minimization principles and procedures. In H. N. Zelaznik (Ed.), *Advances in motor learning and control* (pp. 225–259). Champaign, IL: Human Kinetics.

Kong, S., & Kosko, B. (1992). Adaptive fuzzy systems for backing up a truck-and-trailer. *IEEE Transactions on Neural Networks, 3,* 211–223.

Kosko, B. (1992). *Neural networks and fuzzy systems.* Englewood Cliffs, NJ: Prentice-Hall.

Mazzoni, P., Andersen, R. A., & Jordan, M. I. (1991). A more biologically plausible learning rule for neural networks. *Proceedings of the National Academy of Sciences, 88,* 4433–4437.

Miller, W. T., III, Sutton, R. S., & Werbos, P. J. (Eds.). (1990). *Neural networks for control.* Cambridge, MA: MIT Press.

Nguyen, D., & Widrow, B. (1990). The truck backer-upper: An example of self-learning in neural networks. In W. T. Miller, III, R. S. Sutton, & P. J. Werbos (Eds.), *Neural networks for control* (pp. 287–299). Cambridge, MA: MIT Press.

Polit, A., & Bizzi, E. (1979). Characteristics of motor programs underlying arm movements in monkeys. *Journal of Neurophysiology*, *42*, 183–194.

Rumelhart, D. E., & McClelland, J. L. (Eds.). (1986). *Parallel distributed processing* (Vol. 1). Cambridge, MA: MIT Press.

Some Parallels Between Decision Making and Manual Control

> *The study of decision making in a dynamic, real time context, relocates the study of decision making and makes it part of the study of action, rather than the study of choice. The problem of decision making, as seen in this framework, is a matter of directing and maintaining the continuous flow of behavior towards some set of goals rather than as a set of discrete episodes involving choice dilemmas.*
>
> — Brehmer (1990, p. 26)

Although the close relation between control problems and decision problems has been known for some time (e.g., Bellman, 1961), control theory has had a relatively small impact on behavioral research on decision making. Controlling a dynamic system can be considered as making a sequence of interrelated decisions as to what control value(s) to execute from one instant to the next (e.g., Brehmer, 1990). The decision at one point in time interacts with previous decisions to determine the state of the system for which the next control decision is required. This scenario differs from much of the research on human decision making, which very often deals with single isolated decisions that do not impact the circumstances surrounding subsequent decisions (Hogarth, 1981). Even given such differences, there are a number of common themes within behavioral decision-making research and manual control research that are worth reviewing. This chapter considers Bayesian, regression, and heuristic approaches to decision making and decision boundaries in dynamic contexts in order to explore some of these parallels.

BAYESIAN ANALYSIS OF DECISION MAKING

A much investigated decision problem is to require a person to estimate the probabilities of two mutually exclusive situations based on prior information and additional observations. For example, consider an example of medical diagnosis studied by Christensen-Szalanski and Beach (1982):

1. Disease K is present in 7,000 out of 100,00 people.
2. There exists a test for disease K. The test is positive in 80% of persons who have disease K. The test is negative in 80% of persons who do not have disease K.
3. If a person receives a positive test result, what is the probability that the person has disease K?

One way of solving this problem is to use the odds form of Bayes' Theorem:

$$\frac{\text{Prior}}{\text{Odds}} \times \frac{\text{Likelihood}}{\text{Ratio}} = \frac{\text{Posterior}}{\text{Odds}}$$

$$\frac{P(K)}{P(\sim K)} \times \frac{P(+/K)}{P(+/\sim K)} = \frac{P(K/+)}{P(\sim K/+)}$$

$$\frac{.07}{.93} \times \frac{.80}{.20} = \frac{.056}{.186} = \frac{.23}{.77} \tag{1}$$

Equivalently, transforming the various fractions to have 1 in the denominator:

$$\frac{.075}{1} \times \frac{4}{1} = \frac{.30}{1} \tag{2}$$

$P(K)$ is the prior probability (prior to any medical test) of having the disease, $P(\sim K)$ is the prior probability of not having the disease, and $P(K)/P(\sim K)$ is the prior odds that a person has disease K. Odds are simply the ratio of the probabilities of two events. In this case, the ratio is .07/.93, which is equivalent to .075/1 or ".075 to one." In other words, few people have the disease.

The likelihood ratio indicates how much the positive test result favors having disease K versus not having disease K. $P(+/K)$ is the probability of a positive test result if a person has disease K, and $P(+/\sim K)$ is the probability of a positive test result if a person does not have disease K. Their ratio is the likelihood ratio for the positive test result. In this case, the positive test result is four times more likely when a person has disease K. The implication of Bayes' Theorem is that, given this likelihood ratio, the posterior odds should favor disease K four times more strongly than the prior odds. Namely, $P(K/+)$ is the probability of a person having disease K if there is a positive test result, and $P(\sim K/+)$ is the probability of a person not having disease K if there is a positive test result. Their ratio is the posterior odds (i.e., the odds after the test has been conducted). Whereas the prior odds were .075/1, the posterior odds are .30/1, which favors disease K four times as much. These posterior odds are equivalent to a probability of .23 of having disease K, and a probability of .77 of not having disease K (i.e., .30/1 = .23/.77). The probabilities of these two mutually exclusive events sum to 1.0.

Note that even though the positive test result favors having disease *K* by 4/1, the posterior odds are still much less than 1/1. Given that the prior odds are strongly against having disease *K*, it takes very strong evidence (i.e., a very high likelihood ratio) to make the posterior odds favor having disease *K*. If there were other independent medical tests, Bayes' Theorem could be used iteratively to update the posterior odds. Namely, the likelihood ratio associated with each of these tests could be successively multiplied times the current posterior odds to obtain new posterior odds that incorporated the evidential impact of the test. Together the net impact of a series of medical tests might then be sufficient to favor disease *K*. In the experiment by Christensen-Szalanski and Beach (1982), there was only one medical test.

When this problem was presented as listed earlier, people were not very accurate in estimating the posterior odds. The median estimate (50th percentile) of the posterior probability of disease *K* was .60, and the most common estimate was .80. It appeared that people were ignoring the prior odds and concentrating on the likelihood ratio, a phenomenon often referred to as *base-rate neglect* (e.g., Kahneman & Tversky, 1973).

In contrast, when people were shown a sequence of 100 slides that paired *K* or ~*K* with a positive or negative test result, they performed much better. The slides accurately reflected the relative frequencies with which *K* and ~*K* would be paired with positive or negative tests. After seeing the slides, people gave a median estimate of .15 of the probability of disease K given a positive test result, which is much closer to the correct answer of .23. Subsequent researchers have also found benefits for presenting data in frequency format even if the frequencies are not experienced on a trial-by-trial basis (e.g., Cosmides & Tooby, 1996; Gigerenzer & Hoffrage, 1995). However, for complex situations involving multiple cues, experiencing the cues and outcomes trial-by-trial may be insufficient to overcome base-rate neglect (e.g., see Kahneman & Tversky, 1996, for a review).

In the context of manual control, the primary use of Bayesian models is in the Kalman filter, which is a Bayesian estimator. Unlike the previous example in which there were only two states of interest (i.e., a person has disease *K* or does not have disease *K*), manual control tasks typically involve systems whose states are characterized by continuous variables. As noted in the optimal control model of Kleinman, Baron, and Levison (1971; chap. 17), one of the subtasks of a tracking task is estimating the present state of the dynamic system based on prior observations and the most current observation. The integration of this information by forming a weighted average with weightings inversely proportional to uncertainty (see chap. 18) is an optimal Bayesian estimation procedure. It combines prior information with the latest observation.

Given that information about a dynamic system is experienced over time in a manual control task rather than being presented as abstract probabilities in symbolic format, human performance hopefully would approximate Bayesian optimal performance. The success of the optimal control model in approximating human describing functions is some evidence supporting this expectation (chap. 17). Monitoring tasks that do not involve active control provide another test of whether people approximate the performance of a Kalman filter. For example, Gai and Curry (1976) used a Kalman filter to model a person's anticipations of the behavior of a second-order dynamic system perturbed by a noisy input signal. The deviations between the actual system performance and the anticipated performance were transformed into likelihood ratios favoring either the normal system state or each of two failure modes. Successive likelihood ratios were iteratively multiplied together (e.g., see Wald, 1947), which is equivalent to having 1:1 prior odds and iteratively calculating the Bayesian posterior odds favoring each of the two failure modes relative to the normal mode. In that the successive likelihood ratios stochastically vary across multiple observations, the successively updated posterior odds will exhibit a "random walk" until reaching a sufficiently extreme critical level at which the observer feels confident in identifying the occurrence of a system fault. Reaching a critical level of posterior odds corresponding to a failure mode was assumed to result in an overt response. Thus, the model can be interpreted as having two Bayesian elements—the Kalman filter for generating short-term dynamic anticipations and the Bayesian evaluation of successive deviations from those expectations (also called "residuals"). The model was successful in predicting how long it would take a human monitor to detect various step and ramplike changes in the mean of the noisy input signal. Gai and Curry (1976) also explored the application of this model to the detection of aircraft instrument failures.

An interesting aspect of the Gai and Curry model was that it was necessary to limit how certain it became that the system was in its normal state. The problem is that if a monitor multiplied successive likelihood ratios to update the posterior odds over a long period, then the model could (correctly) become very certain that a normal state of affairs existed. However, when a failure occurred, it would take a great deal of new evidence to overcome the accumulated odds favoring the normal state. Observing the system in a normal state for a long period of time would essentially create very extreme "prior odds" against which to evaluate new evidence of a system failure (i.e., a situation somewhat analogous to the medical example at the beginning of the chapter). In other words, the nonstationarity of the environment (i.e., its changing nature) made it beneficial to be "conservative." This term, often used by Edwards (1968) and his colleagues, means not becoming as extreme in one's posterior odds as would be dictated by Bayes' Theorem. In the present example, a monitor would react slowly to changes in the environment if they behaved according to Bayes' Theorem and simply incorporated the impact of successive likelihood ratios.[1] Navon (1978) similarly argued that data unreliability and the nonindependence of successive observations in many naturalistic environments would also encourage conservative Bayesian behavior. Testing a person in a stationary environment with highly reliable and independent likelihood ratios may yield conservative, suboptimal performance (i.e., insufficiently extreme posterior odds; Edwards, 1968). However, such environments may be atypical (Edwards & von Winterfeldt, 1986; Gigerenzer, 1996), and conservatism may be beneficial in nonstationary environments, especially when changes occur relatively infrequently.

---

[1]Hogarth (1981) argued that conservatism might prevent too rapid a response to new information indicating a large change. However, if large changes do occur rarely, a Bayesian estimator would typically respond slowly to such changes because of extreme posterior odds developed from previous observations of a relatively constant environment over a long period of time. Conservatism in the form of an iteratively imposed limit or saturation function on subjective posterior odds (e.g., see DuCharme & Peterson, 1968) could increase the speed of response to a change.

REGRESSION MODELS OF DECISION MAKING

Another analytic tool for understanding human decision making is multiple regression (e.g., see Dawes & Corrigan, 1974; Dawes, Faust, & Meehl, 1989; Hammond & Adelman, 1976; Slovic & Lichtenstein, 1971). Multiple regression is a mathematical technique for predicting one variable from a weighted sum of observations. For example, suppose when applying to graduate school individuals are interested in how well they will be rated by the admissions committee. One approach is to look at the committee's past behavior. For example, Dawes (1971) found that the ratings, Y_i, of the admissions committee of a well-known graduate program in psychology could be reasonably well predicted by a weighted sum of undergraduate gradepoint average (*GPA*), graduate record examination score (*GRE*), and the quality of undergraduate institution (*QI*, rated on a scale from 1–6). The predicted rating of the admissions committee, Y_i', for student (i) was:

$$Y_i' = \beta_0 + \beta_1 GPA_i + \beta_2 GRE_i + \beta_3 QI_i \tag{3}$$

The constants β_0–β_3 were calculated from the mathematical procedures of multiple regression to minimize the sum of squared deviations between the predicted committee ratings, Y_i', and the actual committee ratings, Y_i [i.e., to minimize $\Sigma(Y_i - Y_i')^2$]. The coefficient of multiple correlation was .78, which indicates that the three predictor variables accounted for $.78^2 = .61$ of the variance in the Y_i scores.

Equation 3 is thus a simple linear model of the admissions committee, which indicates how strongly it weights certain aspects of a student's admissions materials. For example, a heavy weighting of the GRE score would correspond to a relatively large value for β_2. Some schools actually send prospective students an equation similar to Equation 3 so that they can estimate whether they should go to the effort of applying to their graduate program. Although the predicted rating is approximate, a very low predicted rating would suggest that the probability of gaining admission is low.

This type of model has been used to describe many different varieties of human decision-makers, ranging from clinical psychologists diagnosing neuroses and psychoses (see Dawes et al., 1989, for a review) to police and citizen judgments of the desirability of various types of bullets (Hammond & Adelman, 1976). Several important points have arisen from such modeling efforts:

1. *"Simple models or simple processes?"* (Goldberg, 1968). Such models are not necessarily indicative of the underlying cognitive processes used by decision-makers. In other words, decision-makers may not simply multiply scores by weighting constants and then add them up. Nevertheless, such simple models may do a good job of approximating the decision-makers' behavior at an input–output level of description.

2. *Bootstrapping.* When a criterion measurement is available by which the validity of a decision-maker's rating can be evaluated, the model of the judge (Equation 3) often makes better evaluations than the judge's own ratings. For example, the Y_i' obtained from Equation 3 has been found to correlate more highly with measures of success in graduate school than the actual ratings of the admissions committee, the Y_i

(e.g., Dawes & Corrigan, 1974). This phenomenon is termed *bootstrapping* (i.e., pulling oneself up by one's bootstraps). One explanation of bootstrapping is that a regression equation such as Equation 3 implements the human judge's decision strategy more consistently than the human judge, who is subject to fatigue, boredom, variations in attention, and other stochastic variations in performance (Dawes & Corrigan, 1974; Goldberg, 1970).

3. *Equal weighting and the question of optimality.* For comparison, alternate linear models can be created by weighting the important cues equally after they have been converted to *z* scores (i.e., each cue has a mean = 0 and standard deviation = 1). In predicting a criterion variable like success in graduate school, such models typically do as well or better than a model (as in Equation 3) that uses β weights calculated via multiple regression from the human judge's behavior. In other words, the human judge's effective weighting of cues is typically not optimal (e.g., Dawes & Corrigan, 1974). Optimal weights can be estimated by replacing the judge's behavior on the left side of Equation 3 with the criterion variable, and recalculating the β values. Very often, deviations from the optimal weights have only a small impact on the relative predictions of the criterion variable as assessed by an overall coefficient of correlation (Dawes & Corrigan, 1974; von Winterfeldt & Edwards, 1986).[2] Many decision environments are only weakly understood and a range of additive approximations do about equally well.

4. *Nonlinear extensions.* Attempts to add nonlinear terms to representations like Equation 3 typically increase the total variance accounted for in the judge's behavior by only a small amount. Furthermore, such nonlinear models are typically not markedly superior in predicting some additional criterion variable such as performance in graduate school (Camerer, 1981).

Linear describing functions of human tracking behavior are analogous to linear regression models of decision making. Namely, they describe an approximate linear relation between a set of perceptual cues, such as displayed positions and velocities, and a person's control movements (see chaps. 14 and 16; see also Busemeyer, 2001). Some of the same issues that have been raised regarding linear models of human decision-makers can be raised regarding linear models of human perceptual-motor control:

1. *Simple models or simple processes?* Linear describing functions such as the McRuer crossover model typically do a good job of approximating human performance in tracking tasks, often accounting for a large proportion of the variance in the person's control motions (e.g., see chaps. 14 and 19). The success of such models is very good for tasks in which people are making small corrections from a desired trajectory. The success of such models does not imply, however, that the underlying behavioral processes closely mimic the linear operations in the model.

2. *Bootstrapping.* If the control movements in a tracking task are analogous to the decisions of a human judge, then the input signal is analogous to the criterion vari-

---

[2]However, if there are moderate negative correlations among the predictor variables, then the sensitivity to the weights may be much higher (e.g., von Winterfeldt & Edwards, 1986).

able. Namely, one can ask how well the actual output of the control system matches the desired output (i.e., the input signal). Most researchers in perceptual-motor control have not explicitly addressed the question of whether a linear model based on the human performer can match the input signal more accurately than the people themselves, perhaps because the answer seems straightforward from an engineering perspective. The human describing function typically contains a time delay, and an automatic control system without such a delay could perform better than the human performer. Even if comparison is restricted to systems with such delays, a linear model based on the human performer would not have a remnant component. The remnant (chap. 19) corresponds to variability in the person's motions which is not correlated with the input signal. The linear model would typically be expected to perform better than the human in an experimental tracking task by eliminating such variability, which would be analogous to bootstrapping in decision making (e.g., Baron, 1994, chap. 17). However, see point 4 for possible exceptions.

3. *Equal weighting and the question of optimality.* Tracking tasks are typically more sensitive to cue weightings than many decision tasks, and a linear model that equally weighted cues (e.g., position and velocity) would generally not be expected to do as well as a linear model based on the human performer. For example, compensatory tracking with a position control system demands a heavy weighting of position cues to generate laglike behavior; compensatory tracking with an acceleration control system demands a heavy weighting of velocity cues to generate leadlike behavior. Both the weighting of cues and the form of their transformation into motion patterns strongly changes across these tasks.

A more general question — Does the human tracker weight cues optimally? — turns out to be a more difficult question than in the decision-making domain, because there are more possible criteria for optimality. Although a mean-squared error criterion is often specified by the experimenter, performers may adopt implicit criteria that correspond to a trade-off between accuracy and effort. The success of the optimal control model (Kleinman, Levison, & Baron, 1971) indicates that tracking performance can be described as optimizing such a criterion (see chaps. 15 and 17 for a more general discussion of this issue).

4. *Nonlinear extensions.* There have been numerous efforts to supplement linear describing function models of the human performer with nonlinearities such as thresholds and discrete sampling (e.g., Pew, 1974; Sheridan & Ferrell, 1974). For many stationary tracking tasks with a relatively unpredictable input and an explicit criterion of minimizing tracking error, such efforts have generally not added very much to the amount of variance in the person's control movements captured by the model. On the other hand, time optimal control tasks, time-plus-fuel optimal control tasks (e.g., Athans & Falb, 1966), and stationary tracking tasks that push the performer close to the limits of stable performance (e.g., Young & Meiry, 1965) do have significant nonlinear aspects of performance that are not well captured by linear describing functions (chap. 7). Stationary conditions that promote strongly nonlinear behavior seem to be more easily demonstrated in manual control than in decision making.[3]

---

[3]Nonstationary tasks that demand adaptive adjustment for good performance would typically require nonlinear models in both decision and manual control domains (e.g., chaps. 20, 21, and 27).

HEURISTICS AND BIASES

A heuristic is a simplified procedure that yields an approximation to a more precise, elaborate procedure. For example, the use of equal weighting of important cues instead of weights derived from a formal regression analysis is an example of a decision heuristic that generally works well. The reliance on likelihood ratios alone while ignoring prior probabilities is a simplifying procedure that will only approximate the Bayesian posterior odds if the prior odds are close to 1:1 or if the likelihood ratios overwhelm the impact of the priors by a large margin. In many situations, such behavior has been considered an example of a "representativeness heuristic" (Kahneman & Tversky, 1973, 1996; Tversky & Kahneman, 1974) by which people estimate a probability based on the similarity of an observation to a stereotypic characterization of the generating process and/or parent population.

An example of a bias is the tendency for people to be insufficiently sensitive to the degree of change in patterns that arise when successive events have a sequential multiplicative structure. For example, consider a problem of Bayesian inference in which it must be decided from which of two possible diseases a patient is suffering. If a sequence of independent tests each favor a particular disease, then the successive likelihood ratios multiply together and can produce very extreme posterior odds. For example, suppose a doctor performs six independent tests, and the outcome of each test favors Disease X over Disease Y by 5/1. Also, suppose Disease X and Disease Y have equal prior probabilities. Each successive test increases the posterior odds of Disease X rather than Disease Y by a factor of 5: 5/1, 25/1, 125/1, 625/1, 3,125/1, and finally 15,625/1 = $P(X/\text{six test results})/P(Y/\text{six test results})$. Because of the multiplicative nature of odds combination, the absolute rise in the successive posterior odds gets larger and larger in an accelerating fashion. People's subjective estimates of posterior odds in analogous problems are not extreme enough, and as already mentioned, this phenomenon has been labeled *conservatism* (Edwards, 1968).

Another context in which a sequential multiplicative structure arises is in the probability of successfully achieving some plan that depends on some number of independent subtasks all being done correctly. For example, if each of six subtasks has a .90 probability of being performed correctly, then the probability that the overall plan will succeed is $.90^6 = .53$. In other words, even though the probability of success is high for the individual subtasks, the overall probability of success is much lower. People's subjective estimations of such probabilities tend to be too optimistic (e.g., Cohen, Chesnick, & Haran, 1972) even though the series considered sequentially has a decelerating trend ($.9^1 = .90$; $.9^2 = .81$; $.9^3 = .73$; $.9^4 = .66$; $.9^5 = .59$; $.9^6 = .53$). People show very little sensitivity to the multiplicative manner in which the successive probabilities combine. Tversky and Kahneman (1974) more generally referred to this behavior as an example of an "anchoring and adjustment" heuristic in which people insufficiently adjust relative to an initial anchor (e.g., .90).

Although the multiplicative structure was implicit in the problem structure for both the Bayesian odds revision and the conjunctive probability, a similar bias is exhibited when the multiplicative relation is explicit. For example, Tversky and Kahneman (1974) asked high school students to quickly estimate the product $8 \times 7 \times 6 \times 5 \times 4 \times 3 \times 2 \times 1$ or $1 \times 2 \times 3 \times 4 \times 5 \times 6 \times 7 \times 8$. The time limit of 5 seconds precluded performing an exact mental calculation, but presumably allowed for mental multiplication of the first few numbers followed by some type of extrapolation. The median respective estimates of 2,250 and 512 were both considerably smaller than the correct answer, 40,320.

A fourth example of a multiplicative structure is exponential growth. For example, if $10,000 is invested at an annual real growth rate of 5% (after allowing for inflation and taxes), then the money will double in value after roughly 14 years ($1.05^{14} = 1.98$). Human populations in many countries throughout the world are presently growing at rates such that a doubling will occur in less than 50 years. When people are given tables or graphs that reflect rapidly growing processes (e.g., more than doubling at each successive measurement interval), their predictions of future values tend to be severe underestimates (Wagenaar & Sagaria, 1975). However, the pattern of underestimations does reflect a multiplicative structure.

Multiplicative growth also occurs in various control problems. People are able to stabilize an unstable process like the critical tracking task up to a certain level of multiplicative "growth" that is determined by their effective time delay (chap. 15). Uncontrolled, this process would exhibit an exponential runaway. For a fixed value of λ (Fig. 15.6) and no control, the system output (or displacement) would equal $x_0 e^{\lambda t}$, where x_0 is the output at time $t = 0$. Therefore, when t increases by 1 second, the output would be multiplied by a factor of $e^\lambda > 1$. In other words, an exponential function has a sequential multiplicative structure (e.g., Wagenaar & Sagaria, 1975). A successful control strategy for this system, because it acts incrementally, only requires that a person respond proportionally to deviations from a desired reference point (Jex, McDonnell, & Phatak, 1966). Hogarth (1981) made a similar point that sequential incremental decision making need not be as accurate at each stage as long-term decision making, and he noted the analogy to motor control.

In contrast, compensatory stabilization of a third-order system requires a person to make control movements that are proportional to the acceleration of the error signal, and people are not very successful at such tasks (Sheridan & Ferrell, 1974). Namely, in order for a person plus three integrators in series to resemble a single integrator as in the McRuer crossover model, a person would have to act as a double differentiator. The generally poor performance that occurs in such tasks implies that people are not very sensitive to instantaneous visual acceleration. This result is roughly analogous to people's insensitivity to accelerating patterns of growth in the estimation tasks noted earlier.

Another example of a control problem requiring stabilization of exponential growth, but not performed in real time, is the "Reader's Control Problem" (Thomas, 1962). This problem provides evidence for heuristic simplification relative to the ideal control pattern. The system to be controlled in this problem has a control variable, y_k, and an output, x_k, that were related by the equation:

$$x_{k+1} = 1.4x_k + 0.1y_k \qquad (4)$$

The experimental participant's task was to generate a sequence of 10 successive values for y_k so as to minimize the criterion function:

$$I = 0.1\left(\sum_{k=1}^{10} (x_k^2 + y_k^2) + x_{11}^2\right) \tag{5}$$

In other words, the participant's task was to keep the system output, x_k, small, while using as little control, y_k, as possible. The problem is analogous to controlling a potentially runaway dynamic system while conserving energy. If no control is exerted (i.e., $y_k = 0$), the system output will grow larger and larger by a multiplicative factor of 1.4 at each time step, k. The optimal strategy consists of using large values of y_k early in the control sequence in order to diminish x_k quickly, and then using smaller and smaller values of y_k to keep x_k small. This strategy is pictured in Fig. 25.1 (Thomas, 1962). As can be seen in this figure, the human participant's behavior was not nearly optimal. This participant exerted too little control early in the sequence, and the system output, x_k, quickly grew very large. Of course, Thomas included this single subject's behavior merely for the purpose of illustration. Within this framework, Thomas suggested that studies of human learning, sensitivity to system parameters, sensitivity to noise added to the system's behavior, and adaptive behavior would all be promising topics for psychological research.

Following this suggestion, Ray (1963) studied the behavior of 32 participants who repeatedly attempted to solve the Reader's Control Problem. As a participant made each control decision, y_k, he was told by the experimenter the resulting x_{k+1}, as well as the resulting increment in the criterion function $y_k^2 + x_{k+1}^2$. The numbers were kept in view of the participant for each sequence of 10 decisions, and the participant also had access to records of his performance on previous sequences. Each participant was studied until that person either exhibited negligible deviation from the optimal solution or had fixed on a suboptimal strategy that was used from trial to trial without change.

Ray found that only 15 of his 32 participants converged on the optimal solution. Of the remaining 17 participants, 10 were labeled over-controllers and 7 were labeled under-controllers, depending on whether their performance held x_k to lower or higher average levels than the optimal strategy. Perhaps the most interesting finding of this study was the marked individual differences among these 17 participants, which seem to result from the participants imposing simplifying, but objectively unnecessary constraints on their sequences of control decisions. For example, three participants made y_1 through y_{10} equal to a nonzero constant value. If this constraint on the control sequence is assumed, then the constant level of y_k that will minimize the criterion function is $y_k = -3.9$ for all k. All three participants came extremely close to this constant value. In other words, it appeared that these participants did achieve optimal solutions within the context of the added simplifying heuristic constraint that they placed on the structure of their control sequences. Another group of five participants seemed to fix their value for y_1 early in practice and then modify their responding for control decisions y_2 through y_{10}. Ray reported that all five of these participants obtained nearly optimal solutions given their initial constraint on y_1. However, their behavior was suboptimal relative to the smaller set of constraints provided by the objective structure of the Reader's Control Problem itself. Additional studies involving stochastic versions of this problem were conducted by Rapoport (1966a, 1966b, 1967).

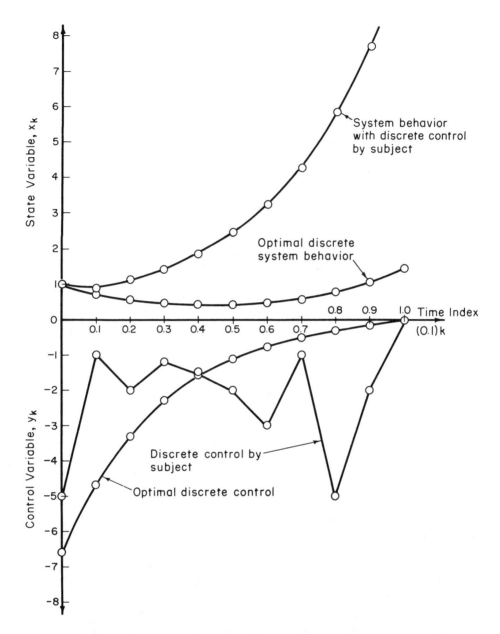

FIG. 25.1. The optimal pattern of control and the control pattern of an experimental subject for the Reader's Control Problem. From *Development of New Techniques for Analysis of Human Controller Dynamics* (p. 28) by R. E. Thomas, 1962, 6570th Aerospace Medical Research Laboratories Technical Documentary Report MRL-TDR-62-65. Wright-Patterson AFB, Ohio. Reprinted by permission.

The heuristic behavior in the Reader's Control Problem was judged against the optimal solution determined by dynamic programming, which is well suited to the discrete temporal structure of that problem (Bellman, 1961). This mathematical technique can be considered a very efficient way of recursively decomposing and searching the solution space for a best control. For continuous time problems, dynamic programming can sometimes be used; however, optimal control theory includes a wider variety of mathematical techniques as well (e.g., Athans & Falb, 1966; Bryson & Ho, 1969; White & Jordan, 1992). One example of a simple continuous optimal control problem is time-optimal control—namely, take a system from some initial state to a final state in as little time as possible (chap. 7). Optimal solutions to this type of problem require the use of the extreme settings of available control. For example, some taxi drivers in Manhattan seem to drive from stoplight to stoplight in near minimal time. To achieve this criterion, they depress the gas pedal as far as possible for as long as possible and then abruptly switch to depressing the brake pedal as far as possible to come to a screeching stop. This behavior does not minimize gas consumption or tire wear, and it does not maximize passenger comfort. It does minimize the time to go from one stop light to another.

In a laboratory investigation of a time-optimal control problem (Jagacinski, Burke, & Miller, 1977), participants were required to take a pendulumlike system, an undamped second-order feedback system (chaps. 6 and 11) or harmonic oscillator, from various initial positions and velocities to a well-marked stopping position. Participants were restricted to exerting either a constant rightward force or a constant leftward force, and they switched from one to the other by means of two pushbuttons.[4] The display showed the present position of the system and the desired stopping point. Participants were not told about the pendulumlike nature of the system dynamics. Instead, they were simply told that the task was to stop a "vehicle" at the indicated position in minimal time.

The initial position and velocity of the vehicle varied from trial to trial, but the vehicle always started to the left of the desired stopping point and initially had a rightward force applied to it. For initial conditions close to the desired stopping point (the origin in Fig. 25.2), only a single switch to a leftward force was necessary (e.g., if starting at point J, switch at point K). However, for starting positions farther from the origin (e.g., F), two switches, first to a leftward force (G) and then back to a rightward force (H), were necessary (see Athans & Falb, 1966, for additional detail).

Participants first practiced with initial conditions that required only a single switch to reach the origin (the inner lobe in Fig. 25.2). They exhibited an initial bias toward a monotonically increasing switching locus (Jagacinski et al., 1977); namely, the farther leftward the system was from the origin, the higher the velocity at which the leftward force was applied. This qualitative difference from optimal performance can be interpreted as participants' failure to understand the dynamics of the pendulumlike system (Jagacinski & Miller, 1978). Such behavior may well have resulted from participants' previous experience in stopping various types of vehicles. For example, in stopping an automobile or bicycle at a stop sign, the velocity at which people begin

---

[4]This dynamic problem is analogous to using a gas jet to stop an object in outer space from spinning about two of its three axes (Athans & Falb, 1966, pp. 569–570).

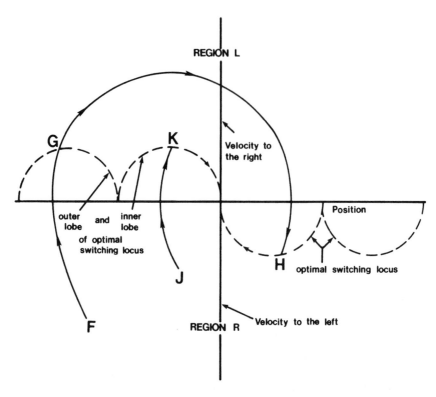

FIG. 25.2.   Time optimal switching loci (dashed lines) for an undamped harmonic oscillator. In Region R a rightward force is applied; in Region L a leftward force is applied. From "Use of schemata and acceleration information in stopping a pendulumlike system" by R. J. Jagacinski, M. W. Burke, and D. P. Miller, 1977, *Journal of Experimental Psychology*, 3, pp. 212–223. Copyright 1977 by APA. Reprinted by permission.

to apply the brakes typically increases with distance from the desired stopping point (i.e., the faster one is traveling, the longer distance it will take to stop, and hence the farther from the desired stopping point one begins braking). However, this pattern is inappropriate for the pendulumlike system, which requires application of a leftward force at very low velocities at the outer part of the inner lobe (Fig. 25.2).

Figure 25.3 shows participants' actual first switch to a leftward force after additional practice with initial conditions requiring both one and two switches to reach the origin (Jagacinski, Burke, & Miller, 1976). Although they had received more than 900 trials with initial conditions requiring a single switch and more than 300 trials with initial condition requiring two switches, the performance is far from optimal. The best performing participant (A) approximated optimal performance for initial conditions requiring a single switch, but did not exhibit a similar semicircular pattern for initial conditions requiring two switches. Although additional practice in this region of the phase plane might improve performance, it is interesting to note that engineers designing automatic control systems often use a zero velocity switching criterion instead of the more elaborate second semicircular pattern (outer lobe) for initial switches far from the stopping point. This heuristic switching criterion is easier to im-

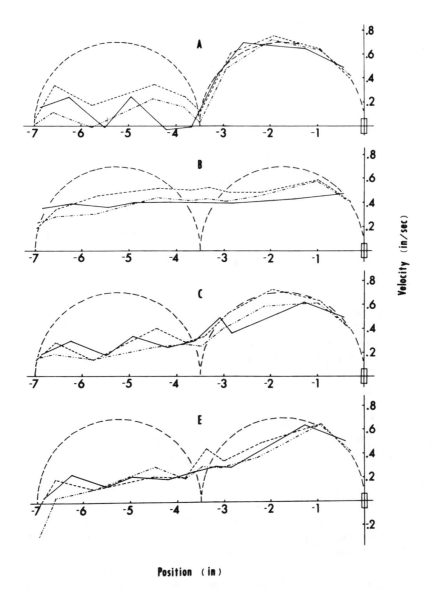

FIG. 25.3. First switches for four participants (A, B, C, E) in a time optimal control task with an undamped harmonic oscillator. The small dashed and dot-dashed switching loci are for well-practiced initial conditions, and the solid lines are for unpracticed initial conditions.

plement and increases the total time to reach the desired stopping point by less than 5% (Athans & Falb, 1966). Participant A comes close to mimicking this heuristic. The nearly constant velocity switching locus used by Participant B over much of the range of initial conditions is less efficient (see Athans & Falb, 1966, p. 588, for a closely related heuristic). Participants C and E exhibited intermediate switching loci that are also simplifications of the optimal switching locus. The switching loci in the time optimal control problem are the boundary between two values of the control variable (e.g., force). The loci can also be considered a decision policy—a rule for deciding which control to implement. The aforementioned experiment can thus be considered an example of heuristic decision making in a dynamic context.

In a hierarchical control structure (e.g., chaps. 7 and 27), loci in the phase plane can also demarcate boundaries between different styles of control. For example, in describing a helmsman's strategy for steering a supertanker, Veldhuysen and Stassen (1976) used two straight-line loci in the phase plane going through the origin to approximate boundaries between regions of active control and coasting (chap. 7). An optimal control policy for minimizing a weighted sum of maneuver time and the absolute value of steering control integrated over that time period would involve a state-space region of coasting surrounded by regions of maximum rightward and maximum leftward control values. The loci separating these regions would typically be curved for complex dynamics such as those of supertankers (e.g., Athans & Falb, 1966). The straight lines in Veldhuysen and Stassen's (1976) helmsman model can be interpreted as part of a heuristic approximation to this optimal control pattern.

Rasmussen (1996) and Flach and Warren (1995) suggested that decision aiding in dynamic control situations should involve making the boundaries between control regions more obvious to the controller (see chaps. 22 and 27). In the case of the undamped oscillator control problem (Fig. 25.2), Miller (1969) tested various phase plane displays, including one that explicitly showed the optimal switching locus. Participants performed better than with a simple position display, but still had difficulty in the cusp region where the two semicircles meet. It is apparently difficult to anticipate the intersection of the system trajectory with such an angular switching locus.

The present chapter has touched on just a few parallels between control theory and decision making. Other parallels include the distinction between open-loop (anticipatory) and closed-loop (error-based) decision making in supervisory control tasks. For example, Brehmer (1990) investigated this issue in the supervisory control of a fire-fighting commander, where significant communication delays made closed-loop decision making less efficient. Similarly, Klein's (1993) recognition-primed decision (RPD) model explicitly recognizes the dynamic, closed-loop nature of decision making in naturalistic settings. After rapid initial categorization of a time critical situation, the decision-maker may compare subsequent observations against expectations in the course of planning and executing various actions, and revise that categorization if a significant discrepancy is found. Another parallel is the theory of risk homeostasis (i.e., the tendency for users of a system to incur a relatively constant degree of risk despite the introduction of new safety features by the system designer; Wilde, 1976, 1985; see Rasmussen, 1996, for a generalization). There are a great many opportunities for increased interaction between researchers in behavioral decision making and dynamic control. For additional reviews of decision making in dynamic contexts, see Kerstholt and Raaijmakers (1997) and Busemeyer (2001).

REFERENCES

Athans, M., & Falb, P. L. (1966). *Optimal control*. New York: McGraw-Hill.
Baron, J. (1994). *Thinking and deciding*. New York: Cambridge University Press.
Bellman, R. (1961). *Adaptive control processes: A guided tour*. Princeton, NJ: Princeton University Press.

Brehmer, B. (1990). Strategies in real-time, dynamic decision making. In R. M. Hogarth (Ed.), *Insights in decision making: A tribute to Hillel J. Einhorn* (pp. 262–279). Chicago: University of Chicago Press.

Bryson, Jr., A. E., & Ho, Y. (1969). *Applied optimal control.* Waltham, MA: Blaisdell.

Busemeyer, J. (2001). Dynamic decision making. In N. J. Smelcer & P. B. Baltes (Eds.), *International encyclopedia of the social and behavioral sciences* (Vol. 6, pp. 3903–3908). Oxford, England: Elsevier Science.

Camerer, C. (1981). General conditions for the success of bootstrapping models. *Organizational Behavior and Human Performance, 27,* 411–422.

Christensen-Szalanski, J. J. J., & Beach, L. R. (1982). Experience and the base-rate fallacy. *Organizational Behavior and Human Performance, 29,* 270–278.

Cohen, J., Chesnick, E. I., & Haran, D. (1972). Confirmation of the inertial-ψ effect in sequential choice and decision. *British Journal of Psychology, 63,* 41–46.

Cosmides, L., & Tooby, J. (1966). Are humans good intuitive statisticians after all? Rethinking some conclusions from the literature on judgment under uncertainty. *Cognition, 58,* 1–73.

Dawes, R. M. (1971). A case study of graduate admissions: Application of three principles of human decision making. *American Psychologist, 26,* 180–188.

Dawes, R. M., & Corrigan, B. (1974). Linear models in decision making. *Psychological Bulletin, 81,* 97–106.

Dawes, R. M., Faust, D., & Meehl, P. E. (1989). Clinical versus actuarial judgment. *Science, 243,* 1668–1674.

DuCharme, W. M., & Peterson, C. R. (1968). Intuitive inference about normally distributed populations. *Journal of Experimental Psychology, 78,* 269–275.

Edwards, W. (1968). Conservatism in human information processing. In B. Kleinmuntz (Ed.), *Formal representation of human judgment* (pp. 17–52). New York: Wiley.

Edwards, W., & von Winterfeldt, D. (1986). On cognitive illusions and their implications. *Southern California Law Review, 59,* 410–451.

Flach, J. M., & Warren, R. (1995). Low-altitude flight. In P. Hancock, J. Flach, J. Caird, & K. Vicente (Eds.), *Local applications of the ecological approach to human–machine systems* (pp. 65–103). Hillsdale, NJ: Lawrence Erlbaum Associates.

Gai, E. G., & Curry, R. E. (1976). A model of the human observer in failure detection tasks. *IEEE Transactions on Systems, Man, and Cybernetics, SMC-6,* 85–94.

Gigerenzer, G. (1996). Rationality: Why social context matters. In P. Baltes & U. M. Staudinger (Eds.), *Interactive minds: Life-span perspectives on the social foundations of cognition* (pp. 319–346). Cambridge, England: Cambridge University Press.

Gigerenzer, G., & Hoffrage, U. (1995). How to improve Bayesian reasoning without instruction: Frequency formats. *Psychological Review, 102,* 684–704.

Goldberg, L. R. (1968). Simple models or simple processes? Some research on clinical judgments. *American Psychologist, 23,* 483–496.

Goldberg, L. R. (1970). Man versus model of man: A rationale, plus some evidence, for a method of improving on clinical inference. *Psychological Bulletin, 73,* 422–432.

Hammond, K. R., & Adelman, L. (1976). Science, values, and human judgment. *Science, 194,* 389–396.

Hogarth, R. M. (1981). Beyond discrete biases: Functional and dysfunctional aspects of judgmental heuristics. *Psychological Bulletin, 90,* 197–217.

Jagacinski, R. J., Burke, M. W., & Miller, D. P. (1976). Time optimal control of an undamped harmonic oscillator: Evidence for biases and schemata. In *Proceedings of the Twelfth Annual Conference on Manual Control* (NASA TM X-73, 170). University of Illinois, Urbana, IL.

Jagacinski, R. J., Burke, M. W., & Miller, D. P. (1977). Use of schemata and acceleration information in stopping a pendulumlike system. *Journal of Experimental Psychology, 3,* 212–223.

Jagacinski, R. J., & Miller, R. A. (1978). Describing the human operator's internal model of a dynamic system. *Human Factors, 20,* 425–433.

Jex, H. R., McDonnell, J. P., & Phatak, A. V. (1966). A "critical" tracking task for manual control research. *IEEE Transactions on Human Factors in Electronics, HFE-7,* 138–144.

Kahneman, D., & Tversky, A. (1973). On the psychology of prediction. *Psychological Review, 80,* 237–251.

Kahneman, D., & Tversky, A. (1996). On the reality of cognitive illusions. *Psychological Review, 103,* 582–591.

Kerstholt, J. H., & Raaijmakers, J. G. W. (1997). Decision making in dynamic task environments. In R. Ranyard, W. R. Crozier, & O. Svenson (Eds.), *Decision making: Cognitive models and explanations* (pp. 205–217). New York: Routledge.

Klein, G. A. (1993). A recognition-primed decision (RPD) model of rapid decision making. In G. A. Klein, J. Orsanu, R. Calderwood, & C. E. Zsambok (Eds.), *Decision making in action: Models and methods* (pp. 138–147). Norwood, NJ: Ablex.

Kleinman, D. L., Baron, S., & Levison, W. H. (1971). A control theoretic approach to manned-vehicle systems analysis. *IEEE Transactions on Automatic Control, 16,* 824–832.

Miller, D. C. (1969). *Behavioral sources of suboptimal human performance in discrete control tasks* (Engineering Projects Laboratory Tech. Rep. No. DSR 70283-9). Massachusetts Institute of Technology, Cambridge, MA.

Navon, D. (1978). The importance of being conservative: Some reflections on human Bayesian behavior. *British Journal of Mathematical and Statistical Psychology, 31,* 33–48.

Pew, R. W. (1974). Human perceptual-motor performance. In B. H. Kantowitz (Ed.), *Human information processing: Tutorials in performance and cognition* (pp. 1–39). New York: Wiley.

Rapoport, A. (1966a). A study of human control in a stochastic multistage decision task. *Behavioral Science, 11,* 18–32.

Rapoport, A. (1966b). A study of a multistage decision making task with an unknown duration. *Human Factors, 8,* 54–61.

Rapoport, A. (1967). Dynamic programming models for multistage decision-making tasks. *Journal of Mathematical Psychology, 4,* 48–71.

Rasmussen, J. (1996, August). *Risk management in a dynamic society: A modeling problem.* Keynote address presented at the conference on Human Interaction with Complex Systems, Dayton, OH.

Ray, H. W. (1963). *The application of dynamic programming to the study of multistage decision processes in the individual.* Unpublished doctoral dissertation, Ohio State University, Columbus, OH.

Sheridan, T. B., & Ferrell, R. (1974). *Man–machine systems.* Cambridge, MA: M.I.T. Press.

Slovic, P., & Lichtenstein, S. (1971). Comparison of Bayesian and regression approaches to the study of information processing in judgment. *Organizational Behavior and Human Performance, 6,* 649–744.

Thomas, R. E. (1962). *Development of new techniques for analysis of human controller dynamics.* (6570th Aerospace Medical Research Laboratories Technical Documentary Report, MRL-TDR-62-65). Wright-Patterson AFB, Ohio.

Tversky, A., & Kahneman, D. (1974). Judgment under uncertainty: Heuristics and biases. *Science, 185,* 1124–1131.

Veldhuysen, W., & Stassen, H. G. (1976). The internal model: What does it mean in human control. In T. B. Sheridan & G. Johannsen (Eds.), *Monitoring behavior and supervisory control* (pp. 157–171). New York: Plenum.

Von Winterfeldt, D., & Edwards, W. (1986). *Decision analysis and behavioral research.* New York: Cambridge University Press.

Wald, A. (1947). *Sequential analysis.* New York: Wiley.

Wagenar, W. A., & Sagaria, S. (1975). Misperception of exponential growth. *Perception and Psychophysics, 18,* 416–422.

White, D. A., & Jordan, M. I. (1992). Optimal control: A foundation for intelligent control. In D. A. White & D. A. Sofge (Eds.), *Handbook of intelligent control* (pp. 185–214). New York: Van Nostrand Reinhold.

Wilde, G. J. S. (1976). Social interaction patterns in driver behaviour: An introductory review. *Human Factors, 18,* 477–492.

Wilde, G. J. S. (1985). Assumptions necessary and unnecessary to risk homeostasis. *Ergonomics, 28,* 1531–1538.

Young, L. R., & Meiry, J. L. (1965). Bang-bang aspects of manual control in higher-order systems. *IEEE Transactions on Automatic Control, 6,* 336–340.

Designing Experiments with Control Theory in Mind

> *The experimental dialogue with nature discovered by modern science involves activity rather than passive observation. What must be done is to manipulate physical reality, to "stage" it in such a way that it conforms as closely as possible to a theoretical description. The phenomenon studied must be prepared and isolated until it approximates some* ideal *situation that may be physically unattainable but that conforms to the conceptual scheme adopted.*
>
> —Prigogine (1984, p. 41)

There are many books on experimental design in psychology. The strategies they suggest are often shaped by the statistical procedures used to analyze the data. For example, if the plan is to use analysis of variance, then it would be wise to consider a factorial arrangement of experimental conditions. Factor A might have three values, and Factor B might have two values, and all the combinations of these two factors will result in six different experimental conditions. The statistical procedures of analysis of variance can then be used to determine whether the measured pattern of results can be approximated as an additive combination of these two factors, or whether they interact in some nonadditive fashion. Often this style of analysis strongly shapes the way people theorize about the underlying behavioral mechanisms they are investigating, namely, analysis of variance functions as more than just a statistical procedure. The algebraic conception of additive and nonadditive factors becomes an integral part of the theorizing about some behavior (e.g., see Sternberg, 1969, for an excellent example).

Although this approach may be very fruitful for some problems, generally theories about the control and stability of dynamic behaviors are not theoretically tractable with simple algebraic structures. Loop structures, their interactions, and their stability properties typically require other forms of analysis, some of which have been introduced in this book. The goal of the present chapter is to summarize some of the experimental strategies that researchers have found helpful in understanding the structure of dynamic systems. They include opening a feedback loop, measuring stability boundaries, augmenting or weakening the stability of a dynamic system, measuring the response to perturbations and/or command inputs of various forms, and measuring the adaptive capability of a dynamic system. These experimental strategies do not argue against using statistical procedures, such as analysis of variance, as one possible means of analyzing experimental results. However, the intuitions behind these experimental strategies are based on control theory rather than typical statistical models.

OPENING THE LOOP

One strategy for demonstrating the closed-loop nature of a behavior is to prevent the feedback from circulating and thereby measure the open-loop response of a part of the system. In negative feedback loops, this manipulation disables some error-nulling mechanism and may result in an unstable, runaway sequence of behavior. In positive feedback loops, this manipulation may prevent a runaway sequence of behavior.

An interesting example of this manipulation comes from the study of eye movements. If a person is looking straight-ahead and then tries to look directly at a stationary target that is off to the side, then the resulting eye movement is a sudden jump or saccade. If the initial saccade misses the target, then a smaller secondary saccade occurs .15 to .30 s later (Young & Stark, 1963; see also Robinson, 1965). The delay between successive saccades is a basic constraint on this type of movement pattern, which can be demonstrated by opening the closed-loop relation between target angle and eye angle. Normally as the eye makes a saccadic movement toward the target, the difference between the angle of the eyes and the angle of the target is reduced. If the target is initially 15° to the right, and the eyes make a 13° saccade, then the target will be only 2° to the right of where the eyes are pointing after the saccade. Namely, the difference between the angle of the eyes and the angle of the target can be considered an error signal, which each saccade successively diminishes.

Young and Stark (1963) nullified this closed-loop relation by electronically measuring the angle of the eyes, and then moving the target the same number of degrees as the eyes rotated (Fig. 26.1). Now if the eyes made a 13° saccade to the right, so did the target, so that at the end of the saccade the target was still 15° to the right. The error signal remained constant, no longer diminished by the saccade, so effectively the feedback loop had been opened. In response to this constant error signal, the eyes exhibit a sequence of saccades, each separated by about .2 s, that gradually drives the target farther and farther to the right. In other words, saccadic eye movements act like a sampled data system rather than a continuous system. Opening the loop reveals the discrete, sampled nature of the successive saccadic responses more clearly.

A second example of opening the loop comes from the neural innervation of muscles. Alpha motor neurons have their cell bodies in the spinal chord, and the neurons extend out to individual muscles. The neural signals activate muscle fibers and cause them to contract. This contraction is a basic physiological mechanism for generating movement. The alpha motor neurons have short branches close to the spinal cord

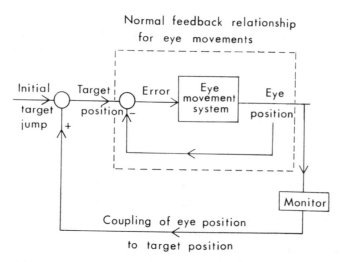

FIG. 26.1.   Opening the closed-loop relation between the angular position of the eyes and the angular position of a target (after Young, & Stark, 1963). From "A qualitative look at feedback control theory as a style of describing behavior" by R. J. Jagacinski, 1977, *Human Factors, 19*, pp. 331–347. Copyright 1977 by the Human Factors and Ergonomics Society. All rights reserved. Reprinted by permission.

that excite other neurons called Renshaw cells. The Renshaw cells in turn inhibit the activity of the alpha motor neurons (Fig. 26.2, Roberts, 1967; see McMahon, 1984, for a review). Thus, there is a negative feedback loop that modulates the activity of the alpha motor neurons. The inhibitory connections of the Renshaw cells to the alpha motor neurons can be blocked by either strychnine or tetanus toxin. The result of this chemical blocking is that the muscles go into convulsion (e.g., see Thompson, 1967, pp. 202–207). An interpretation of this effect is that the chemical blocking agents have effectively opened the feedback loop that normally modulates the response of the alpha motor neuron. If the gain of the feedback loop is normally $G/(1 + GH)$ (see chap. 2), then by blocking the feedback loop H is essentially set to zero, and the effective gain of the alpha motor neurons then becomes G. If G is considerably greater than $G/(1 + GH)$, then the muscles will be overstimulated, and convulsions will result (Roberts, 1967).

FIG. 26.2. Stretch receptors (middle far left) have excitatory connections to alpha motor neurons (lower left) in the spinal cord, which innervate muscles (lower far left). The alpha motor neurons also excite Renshaw cells (black, middle), which in turn inhibit the alpha motor neurons, and thus form a negative feedback loop. From *Neurophysiology of Postural Mechanisms* (p. 104) by T. D. M. Roberts, 1967, London: Butterworths. Reprinted by permission of Butterworth Heinemann.

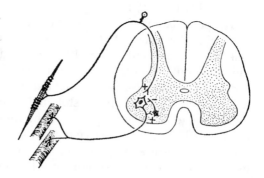

In sum, opening a feedback loop can be useful for showing that a particular behavior is associated with a closed-loop system and for examining the properties of a particular transducer by isolating it from its normal closed-loop environment.

MEASURING STABILITY

Given that an important property of closed-loop behavior is its stability, it is important to measure degrees of stability and conditions under which stability is lost. For example, the measurement of describing functions via Fourier analysis (chaps. 13 & 14) allows one to measure both gain margins and phase margins. Recall that stability of a single closed-loop typically requires that the open-loop amplitude ratio be less than 1.0 at the frequency for which the phase lag is 180°. Gain margins and phase margins are measures of whether the system just barely meets this criterion or meets it with something to spare. The gain margin is a measure of how much smaller than 1.0 the amplitude ratio is at the frequency for which the phase lag is 180°. Similarly, the phase margin is a measure of how close to 180° the phase lag is at the frequency for which the amplitude ratio is 1.0. If the gain margin or the phase margin is close to zero, then the system is just barely stable.

External perturbations will have a much larger effect on the performance of a system that is almost unstable in comparison with a system that has larger stability margins. Therefore, another way of characterizing a system's stability is to measure its response to various perturbations. For example, the perturbations could be continuous, and one could measure the transfer function characterizing a person's response to these perturbations. If, however, the person is simultaneously responding to a command signal, then the measurement problem is to distinguish between these two responses. One solution to this problem is to have the command input and the disturbance each consist of a sum of sine waves, but at different, interleaved frequencies. Separate describing functions can then be calculated for the command input and for the disturbance (e.g., Jex, Magdaleno, & Junker, 1978). This technique can similarly be used to simultaneously measure a performer's describing function to two different perceptual variables. For example, Johnson and Phatak (1990) used this technique to test which of several visual cues were used in a helicopter hovering task (chap. 22; see also Flach, 1990, for a review).

The perturbations could also be discrete, and then the measurement problem is to determine the transient response associated with the perturbation. For example, Dijkstra, Schoner, and Gielen (1994) induced postural sway in human observers who saw a computer simulation of a surface that moved back and forth sinusoidally. The postural sway was in-phase with the surface movement, until the experimenters suddenly introduced a 180° phase shift in the simulated surface motion pattern. The temporal changes in relative phase as the human observer gradually regained an in-phase relation with the surface motion was used to derive a stability measure for the coupling of posture to the surface motion. Additional discussions of phase resetting in perturbed oscillatory systems can be found in Glass and Mackey (1988).

DESTABILIZING A SYSTEM

A more qualitative way of testing the stability of a system is to introduce additional delays and note the deterioration and/or loss of stability. Pure time delays reduce the phase margin by introducing additional phase lags in a system's response without directly changing the amplitude ratio. One could measure how much delay can be introduced into a system before it became unstable (e.g., Jagacinski, 1977). Similarly, additional degrees of lag could be introduced into a system to destabilize it and/or to meet some other performance criterion such as a certain level of tracking error (Dukes & Sun, 1971). Either of these approaches demands gradually adjusting a delay or a lag parameter, and the adjustment must be done carefully so as not to introduce additional dynamic perturbations into the system (Dukes & Sun, 1971).

A similar manipulation is to introduce additional gains into the loop to reduce the gain margin. For example, the critical tracking task is an example in which the gain of a feedback loop is gradually increased until a performer loses control (chap. 15). As another example, the artificial coupling between eye position and target position in Fig. 26.1 can be used to introduce a large negative feedback gain. Namely, if a target is initially to the right, and a person shifts gaze to the right, then for a suitably large gain the target would shift to the left and be even farther from the center of gaze as it was initially. As the person vainly tries to gaze at the target, both the eyes and target make larger and larger excursions to the right and left. In other words, the system has been destabilized (Young & Stark, 1963). Even though participants in such an experiment may try to make smaller saccades to compensate for the unusual negative feedback, for a sufficiently large gain a sequence of alternating right and left saccades of increasing magnitude occurs nevertheless (Robinson, 1965).

Introduction of both additional gains and time delays could be used to map out stability boundaries for systems for which it may be difficult to measure instantaneous performance, but for which qualitative distinctions between stable and unstable performance can be measured (e.g., Jagacinski, 1977).

AUGMENTING STABILITY

The opposite approach to destabilizing a system is to augment its stability by various means. For example, the opposite of introducing greater delays into a system is to introduce greater anticipation and thereby reduce effective delays. Local anticipation can be introduced by making various derivatives of signals more easily perceptible. For example, display quickening replaces a signal with a weighted sum of the original signal and its velocity, acceleration, and so on (Birmingham & Taylor, 1954; Frost, 1972; chap. 9; see Fig. 26.3). Systems with many integrations that would be difficult or even uncontrollable without such a display are then within the range of human performance. Quickening has been applied to vehicular control on land, air, and sea (see Sheridan & Ferrell, 1974).

One difference between moving base and fixed base simulators is that the velocity and acceleration sensitivity of the inner ear can be utilized for greater local anticipa-

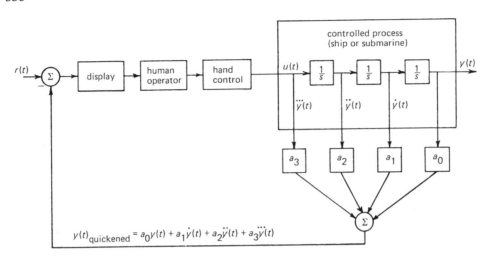

FIG. 26.3. An example of a quickened display for a simplified simulation of a ship or submarine. From *Man–Machine Systems: Information, Control, and Decision Models of Human Performance* (p. 269) by T. B. Sheridan and W. R. Ferrell, 1974, Cambridge, MA: MIT Press. Reprinted by permission.

tion or lead in a moving base simulator. This additional sensing capability often results in superior performance relative to a fixed base (nonmoving) simulation of the same system (e.g., Jex et al., 1978; Shirley & Young, 1968). Therefore, moving base simulator performance may more closely resemble performance with the actual system. However, it should be noted that with regard to using a simulator for training purposes, better performance in a moving base simulator does not necessarily imply that it will result in superior transfer to the actual system (e.g., Jacobs & Roscoe, 1980).

Another example of a multimodal display to improve anticipation is the utilization of variations in the proprioceptive feel of a control stick. For example, Herzog (1968) investigated manual control in which the torque (or force) applied by the human performer to the control stick was the control variable that was input into a second-order dynamic system. The task was compensatory tracking with a visual display. Herzog used a torque motor in conjunction with position, velocity, and acceleration sensors to adjust electronically the effective mass, damping, and springiness of the control stick (see Repperger, 1991, for other methods of varying the feel of a control stick). The relation between the applied force and the resulting control stick position was adjusted so that it approximately matched the relation between the applied force and the second-order dynamic system output. In other words, the control stick position provided an estimate of the control system output. The tracking task was thus made analogous to the task of moving one's arm under a dynamic loading (chap. 20) in response to a visual error signal, and performance was superior to performance of the same task with a typical position control stick. If the dynamic system to be controlled was more complex, such as a third-order system, then the control stick could be modified to simulate the higher order dynamics, or it could provide a second-order approximation to only part of the system dynamics. Even the limited dynamic information provided by this latter technique permitted a performer to exert much better control than with a typical position control stick. In some sense, Herzog's technique provided the performer with a proprioceptive model of the dynamic system to be controlled (Pew, 1970).

Detailed considerations of the relation between force and motion also arise in the design of haptic displays to simulate contact with the surface of an object in virtual environments. Although the control problems discussed herein have emphasized the achievement of spatiotemporal trajectories, an important class of control problem involves the simultaneous control of spatiotemporal trajectory and force in different geometric dimensions (see Lewis, Abdallah, & Dawson, 1993, for a review). For example, in shaving with a razor, one might simultaneously control the path of the razor across the skin and the force with which it touches the skin. Analyses of such problems mathematically characterize both a control strategy and the response of the environment in terms of *impedances* and *admittances* that are dynamic relations between force and motion (position, velocity, acceleration). Impedances transform motions into forces; admittances transform forces into motions. Interactions between a person or robot and a contacted object may be modeled as a combination of an admittance and an impedance (Anderson & Spong, 1988; Hogan, 1985). For example, if a man is controlling the force of a razor against his face, he may sense disturbances from the desired force or pressure and generate motions to control them. The face could be modeled as receiving these motions as an input, and generating reactive forces which the person in turn responds to in a closed-loop manner. The person acts as an admittance (transforming force disturbances into motions), and the face acts as an impedance (transforming these motions into forces) in a reciprocal closed-loop interaction. If the skin has a characteristic springiness and damping, then these properties will be represented in the mathematical characterization of its impedance. In a complementary example, if a person is sensing disturbances from a desired motion trajectory and generating forces to control them, then the person acts as an impedance. The environment can then be modeled as an admittance having force as its input and motion as its output, which is in turn fed back to the person. The person might respond as different impedances (or admittances) in order to achieve satisfactory disturbance nulling for different environmental dynamics or different task constraints. These various behaviors might occur even in different geometric dimensions of a single task such as shaving, which involves both force and spatial trajectory control (Anderson & Spong, 1988; Hogan, 1985; Lewis et al., 1993). This style of analysis can be important in understanding the stability of movement control in physical environments involving surface contact and in the design of haptic displays to enhance human performance in virtual simulations of surface contact and object manipulation (Adams & Hannaford, 1999; Minsky, Ouh-young, Steele, Brooks, & Behensky, 1990).

Long-range anticipation and planning can be augmented by introducing long-term predictive displays that present the extrapolated trajectory of the system (e.g., Jensen, 1981; Kelley, 1968; Roscoe, Corl, & Jensen, 1981). Such displays are particularly useful in slowly responding systems with complex dynamics. The range of possible trajectories corresponding to the extreme right, left, and center positions of the control stick (Fig. 26.4) and/or the extrapolated trajectory if the present control stick

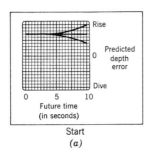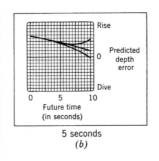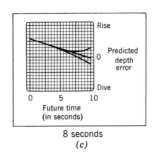

FIG. 26.4.   Predicted depth error for a submarine if the diving planes are set at their extreme settings (upper and lower traces) or at a center position (middle trace). From *Manual and Automatic Control* (p. 225) by C. R. Kelley, 1968, New York: Wiley. Reproduced by arrangement with the author.

position is maintained can be displayed. Such augmented anticipation can markedly improve performance with difficult systems such as a submarine or aircraft. Furthermore, if the dynamics vary over time, the simulation model used to generate the predicted trajectories can be modified accordingly (Kelley, 1968).

In addition to altering the display of information, it is sometimes possible to vary the dynamics of the control system. For example, Birmingham and Taylor (1954) were interested in systems used to continuously track slowly moving targets as might be encountered in military gunnery. Such systems often have one or more integrators so that the human performer can generate a continuous motion of the cursor without having to perform a prolonged constant velocity or accelerating movement of the control stick as would be necessary with a position control. Such a system can be made more responsive by having the output be proportional to a sum of the position, velocity, and possibly higher derivatives of the output of the final integrator. As a simple example, the weighted sum of position and velocity would add short-term linear anticipation or lead to the system and is sometimes termed *rate aiding* (Fig. 26.5). A more generic term is simply *aiding* (Birmingham & Taylor, 1954; Chernikoff & Taylor, 1957; chap. 9).

MEASURING ADAPTATION

Another experimental strategy is to vary the dynamic characteristics of the control problem during an experimental trial and see how well the human performer can adapt. In other words, given that human performers are not characterized by a single describing function, but vary their control characteristics depending on the particular control problem (chaps. 14, 15, and 20), it is important to measure dynamic aspects of the adaptive process. The need for such adaptation could arise from a hardware or software failure in a control system, from a change in environmental conditions, or from a change in the goal of the controller.

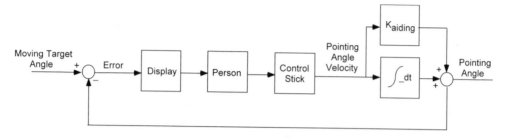

FIG. 26.5.   A simple rate-aided control system for following a moving target. From "A design philosophy for man–machine control systems" by H. P. Birmingham and F. V. Taylor, 1954, *Proceedings of the IRE, 42*, pp. 1748–1758. Copyright 1954 BY IEEE. Adapted by permission.

Miller (1965; see Elkind & Miller, 1966) introduced a sudden discrete change in the magnitude and/or polarity of the gain of a velocity control system. The performer was required to release a button when a system change was detected, and to then compensate for the change. Miller was able to identify a relation between change in control stick position and change in tracking error rate that predicted when the performer detected the change in the system. In other words, the system change might go undetected for several seconds if the performer's control actions did not result in an unexpected change in the tracking error rate. Additionally, Miller measured how the performer's gain varied over the few seconds after detection (i.e., the time course of the performer's adaptive adjustment; Fig. 26.6).

The change in a control system parameter can be continuous rather than abrupt. For example, the lateral handling characteristics of various vehicles (e.g., automobiles) can change continuously as a function of vehicle speed and environmental sur-

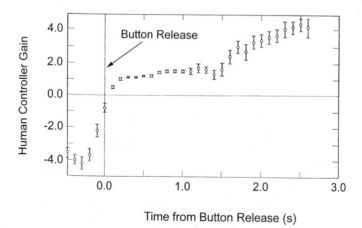

Time from Button Release (s)

FIG. 26.6.   Ensemble average estimates of a human controller's gain when there is a sudden change in the polarity of a velocity control system. The performer was required to release a button when a system change was detected, and then to compensate for the change. From "Process of adaptation by the human controller" by J. I. Elkind and D. C. Miller, 1966, *Proceedings of the Second Annual NASA-University Conference on Manual Control*, NASA SP-128, pp. 47–63. MIT, Cambridge, MA. Adapted by permission.

face conditions and require adaptive changes in control by the human performer. Delp and Crossman (1972) examined adaptive changes in the describing function of a performer controlling a second-order system that had sinusoidally varying natural frequency and damping. Describing functions were calculated over successive, partially overlapping, 15-second intervals so that their change over time within a trial could be examined. Delp and Crossman also calculated the describing function of the performer at various fixed values of natural frequency and damping as well. This performance also represents adaptive changes in the describing function, but without time pressure. This baseline pattern of changes was compared with the changes in the describing function across a single trial having a time-varying control system. For sinusoidal variations in the control system parameters having a period of 100 seconds, the human performer was almost the same as for the fixed trials in terms of magnitude of change. However, there was a substantial temporal delay. For sinusoidal control system changes with a period as short as 20 seconds, the performer's adaptive changes were greatly attenuated in amplitude and lagged behind the control system changes even further. Based on this pattern of results, Delp and Crossman approximated the dynamic response of the performer's adaptive adjustment of describing function parameters with a first-order lag plus a time delay.

OVERVIEW

The present chapter has briefly reviewed a number of experimental strategies in understanding the structure and stability of dynamic systems. For additional discussions of experimental strategies, refer to Glass and Mackey (1988), McFarland (1971), Milsum (1966), Pew (1974), Powers (1989), Robinson (1965), and Toates (1975).

REFERENCES

Adams, R. J., & Hannaford, B. (1999). Stable haptic interaction with virtual environments. *IEEE Transactions on Robotics and Automation, 13,* 465–474.

Anderson, R. J., & Spong, M. W. (1988). Hybrid impedance control of robotic manipulators. *IEEE Journal of Robotics and Automation, 4,* 549–556.

Birmingham, H. P., & Taylor, F. V. (1954). A design philosophy for man–machine control systems. *Proceedings of the IRE, 42,* 1748–1758.

Chernikoff, R., & Taylor, F. V. (1957). Effects of course frequency and aided time constant on pursuit and compensatory tracking. *Journal of Experimental Psychology, 53,* 285–292.

Delp, P., & Crossman, E. R. F. W. (1972). Transfer characteristics of human adaptive response to time-varying plant dynamics. In *Proceedings of the Eighth Annual Conference on Manual Control* (Air Force Flight Dynamics Laboratory Tech. Rep. No. AFFDL-TR-72-92, pp. 245–256). University of Michigan, Ann Arbor, MI.

Dijkstra, T., Schoner, G., & Gielen, S. (1994). Temporal stability of the action-perception cycle for postural control in a moving visual environment. *Experimental Brain Research, 97,* 477–486.

Dukes, T. A., & Sun, P. B. (1971). A performance measure for manual control systems. In *Proceedings of the Seventh Annual Conference on Manual Control* (NASA SP-281, pp. 257–263). University of Southern California, Los Angeles, CA.

Elkind, J. I., & Miller, D. C. (1966). Process of adaptation by the human controller. In *Proceedings of the Second Annual NASA-University Conference on Manual Control* (NASA SP-128, pp. 47–63). MIT, Cambridge, MA.

Flach, J. (1990). Control with an eye for perception: Precursors to an active psychophysics. *Ecological Psychology, 2,* 83–111.

Frost, G. (1972). Man–machine dynamics. In H. P. Van Cott & R. G. Kincade (Eds.), *Human engineering guide to equipment design* (pp. 227–309). Washington, DC: American Institutes for Research.

Herzog, J. H. (1968). Manual control using the matched manipulator control technique. *IEEE Transactions on Man–Machine Systems, 9,* 56–60.

Hogan, N. (1985). Impedance control: An approach to manipulation — Parts I, II, III. *Journal of Dynamic Systems, Measurement, and Control, 107,* 1–24.

Glass, L., & Mackey, M. C. (1988). *From clocks to chaos: The rhythms of life.* Princeton, NJ: Princeton University Press.

Jacobs, R. S., & Roscoe, S. N. (1980). Simulator cockpit motion and the transfer of flight training. In S. N. Roscoe (Ed.), *Aviation psychology* (pp. 204–216). Ames, IA: Iowa State University Press.

Jagacinski, R. J. (1977). A qualitative look at feedback control theory as a style of describing behavior. *Human Factors, 19,* 331–347.

Jensen, R. S. (1981). Prediction and quickening in perspective flight displays for curved landing approaches. *Human Factors, 23,* 355–364.

Jex, H. R., Magdaleno, R. E., & Junker, A. M. (1978). Roll tracking effects of G-vector tilt and various types of motion washout. In *Proceedings of the Fourteenth Annual Conference on Manual Control* (NASA CP-2060, pp. 463–502). University of Southern California, Los Angeles, CA.

Johnson, W. W., & Phatak, A. V. (1990). Modeling the pilot in visually controlled flight. *IEEE Control Systems Magazine, 10*(4), 24–26.

Kelley, C. R. (1968). *Manual and automatic control.* New York: Wiley.

Lewis, F. L., Abdallah, C. T., & Dawson, D. M. (1993). *Control of robot manipulators* (pp. 259–368). New York: MacMillan.

McFarland, D. J. (1971). *Feedback mechanisms in animal behaviour.* London: Academic Press.

McMahon, T. A. (1984). *Muscles, reflexes, and locomotion.* Princeton, NJ: Princeton University Press.

Miller, D. C. (1965). *A model for the adaptive response of the human controller to sudden changes in controlled process dynamics.* Unpublished master's thesis, Department of Mechanical Engineering, MIT.

Milsum, J. H. (1966). *Biological control systems analysis.* New York: McGraw-Hill.

Minsky, M., Ouh-young, M., Steele, O., Brooks, Jr., F. P., & Behensky, M. (1990). Feeling and seeing: Issues in force display. *Computer Graphics, 24,* 235–243.

Pew, R. W. (1970). Toward a process-oriented theory of human skilled performance. *Journal of Motor Behavior, 11,* 8–24.

Powers, W. T. (1989). *Living control systems.* Gravel Switch, KY: The Control Systems Group.

Prigogine, I. (1984). *Order out of chaos.* New York: Bantam.

Repperger, D. W. (1991). Active force reflection devices in teleoperation. *IEEE Control Systems Magazine, 11*(1), 52–56.

Roberts, T. D. M. (1967). *Neurophysiology of postural mechanisms.* London: Butterworths.

Robinson, D. A. (1965). The mechanics of human smooth pursuit eye movement. *Journal of Physiology, 180,* 569–591.

Roscoe, S. N., Corl, L., & Jensen, R. S. (1981). Flight display dynamics revisited. *Human Factors, 23,* 341–353.

Sheridan, T. B., & Ferrell, W. R. (1974). *Man–machine systems: Information, control, and decision models of human performance.* Cambridge, MA: MIT Press.

Shirley, R. S., & Young, L. R. (1968). Motion cues in man–vehicle control. In *Proceedings of the Fourth Annual NASA-University Conference on Manual Control* (NASA SP-192, pp. 435–445). University of Michigan, Ann Arbor, MI.

Sternberg, S. (1969). The discovery of processing stages: Extensions of Donders' method. In W. G. Koster (Ed.), *Attention and Performance II, Acta Psychologica* (Vol. 30, pp. 276–315). Amsterdam: North Holland.

Thompson, R. F. (1967). *Foundations of physiological psychology.* New York: Harper & Row.

Toates, F. M. (1975). *Control theory in biology and experimental psychology.* London: Hutchinson Educational.

Young, L. R., & Stark, L. (1963). Variable feedback experiments testing a sampled data model for eye tracking movements. *IEEE Transactions on Human Factors in Electronics, HFE-4,* 38–51.

Adaptation and Design

The main reason why humans are retained in systems that are primarily controlled by intelligent computers is to handle "non-design" emergencies. In short, operators are there because system designers cannot foresee all possible scenarios of failure and hence are not able to provide automatic safety devices for every contingency.
—Reason (1990, p. 182)

The smart machine, as it turns out, requires smart people to operate it as well as to maintain and support it.
—Rochlin (1997, p. 146)

The history of aviation provides a nice example of innovation that stems from formulating a problem within a control framework and of the central role played by the human in completing the design of many systems. A number of authors have observed that it was a focus on the control problem that led to the success of the Wright brothers. For example, Freedman (1991) noted that pioneers like Lilienthal and Chanute designed wings capable of lifting a person into the air. Langley showed that it was possible to build an engine–propeller combination that could propel a set of wings through the air. So, what was the contribution of the Wright brothers to flight? According to Freedman:

> The Wrights were surprised that the problem of balance and control had received so little attention. Lilienthal had attempted to balance his gliders by resorting to acrobatic body movements, swinging his torso and thrashing his legs. Langley's model aerodromes were capable of simple straight-line flights but could not be steered or maneuvered. His goal was to get a man into the air first and work out a control system later.
>
> Wilbur and Orville had other ideas. It seemed to them that an effective means of controlling an aircraft was the key to successful flight. What was needed was a control

system that an airborne pilot could operate, a system that would keep a flying machine balanced and on course as it climbed and descended, or as it turned and circled in the air. Like bicycling, flying required balance in motion. (p. 29)

The Wright brothers provided three controls (Fig. 27.1). One hand controlled an elevator, the other controlled a rudder. The third control was wing-warping (which in today's aircraft would be accomplished with ailerons). The wing-warping was controlled by a saddle-type device at the pilot's waist. This allowed the pilot to bank the aircraft by shifting his hips. Freedman (1991) observed that modern aircraft are controlled in an analogous manner: "A modern plane 'warps' its wings in order to turn or level off by moving the ailerons on the rear edges of the wings. It makes smooth banking turns with the aid of a moveable rudder. And it noses up or down by means of an elevator (usually located at the rear of the plane)" (p. 64).

The Wright brother story is relevant in two ways. First, it illustrates the importance of control considerations to innovation. Second, it illustrates that the "human" can be a critical element within many control systems. Thus, a background in control theory can be an important framework for attacking design questions, particularly when designing human–machine systems.

A central question to consider is how to utilize the human operator to achieve a stable control solution. Again, consider the Wright brothers' solution to the flight problem. At the time of their historic efforts, many people were concerned about whether the pilot should be given control over the lateral axis. Most designers were considering only passive solutions to lateral stability. The Wright brothers were unique in their choice to give the pilot active control over lateral stability. This proved to be critical to the ultimate design solution. However, it would be a mistake to blindly follow the lead of the Wright brothers to conclude that humans are always a good solution to a control problem. Advanced automated systems can often solve inner (faster) loop control problems more consistently and reliably than the human operator. In fact, the Wright brothers won the Collier Trophy in 1914 for the design of an automatic stabilizing system that would keep the plane straight and level without pilot intervention. This was motivated in part by the numerous accidents that plagued early aviation. However, despite increasingly capable automatic control systems, the pilot still plays a vital role in aviation. Billings (1997) provided a good description of the supervisory role of pilots in aircraft with advanced automated systems:

> The pilot must understand the functioning (and peculiarities) of an additional aircraft subsystem, remember how to operate it, and decide when to use it and which of its capabilities to utilize in a given set of circumstances. When it is in use, its operation must be monitored to ensure that it is functioning properly. If it begins to malfunction, the pilot must be aware of what it is supposed to be doing so he or she can take over its functions. Finally, the pilot must consider whether the failure impacts in any way the accomplishment of the mission and whether replanning is necessary; if so, the replanning must be done either alone or in communication with company resources. (pp. 80–81)

It is clear that automation is not eliminating the human as an important element in complex technological systems like aircraft. But it is displacing the human from being directly involved with the inner loop control. In these systems, the humans tend

Hand Controls for Rudder and Elevator

Wright Flyer III

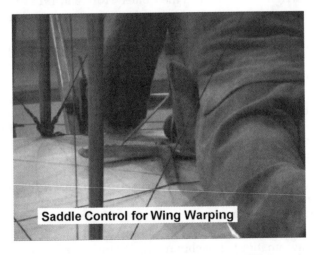

Saddle Control for Wing Warping

FIG. 27.1.   Three views of one of the first practical aircraft, the 1905 Wright Flyer III on display in the Wright Brothers Aviation Center, Carillon Historical Park, Dayton, Ohio. Photos were taken by Jens Rasmussen. Used by permission.

to function more as supervisors or managers. As managers, the humans' role is to monitor and tune the automation so that the overall process remains in control (i.e., remains stable). The human's role is to "adapt" the automatic control system. In some sense, the human's role is to "complete the design" to insure that the system does not become unstable due to situations that were not anticipated in the design of the automatic control systems.

Thus, this chapter first considers adaptive control. An adaptive control system is a system that "redesigns" itself. The chapter concludes with a brief discussion of cognitive systems engineering as a framework for thinking about the design of human–machine systems.

ADAPTIVE CONTROL

An important issue for manual control in natural work environments is whether the controlled process is stationary. That is, are the dynamics of the process fixed or do they change? For example, time delays in many systems can be variable. The delay in someone's shower may be variable. There may be long initial delays (to establish the flow of warm water), but once the warm water is flowing the effective time delay for later adjustments may be less. Variable time delays are common in many technologies. The delay of real-time computer graphics systems, such as those involved in virtual reality systems, may vary continuously as a function of the complexity of the computations. Communication delays for unmanned air vehicles or remote space systems may change as a result of motions and orientations of sensors, message characteristics, distance between the controller and the vehicle, and electromagnetic interference.

Another example of a nonstationary response is found in aviation. Sastry and Bodson (1989) described the problem that arises from the fact that the dynamic properties of aircraft can change as a function of speed and altitude:

> Research in adaptive control has a long and vigorous history. In the 1950s, it was motivated by the problem of designing autopilots for aircraft operating at a wide range of speeds and altitudes. While the object of a good fixed-gain controller was to build an autopilot which was insensitive to these (large) parameter variations, it was frequently observed that a single constant gain controller would not suffice. Consequently, gain scheduling based on some auxiliary measurements of airspeed was adopted. (p. 2)

The complexity of the adaptation problem is illustrated by Sastry and Bodson's (1989) description of an adaptive control system designed for the CH-47 helicopter:

> The flight envelope of the helicopter was divided into *ninety* flight conditions corresponding to thirty discretized horizontal flight velocities and three vertical velocities. Ninety controllers were designed, corresponding to each flight condition, and a linear interpolation between these controllers (linear in the horizontal and vertical flight velocities) was programmed onto a flight computer. Airspeed sensors modified the control

scheme of the helicopter in flight, and the effectiveness of the design was corroborated by simulation. (p. 5)

Note that although the design of adaptive controls is a relatively new field within control theory, the problem of nonstationarity has long been recognized. In many domains, there is a high degree of context sensitivity associated with control solutions. A solution that is exactly right for one context may be exactly wrong for another. Before the invention of "adaptive auto-pilots," the solution to this adaptive control problem was a well-trained pilot. Pilots recognized the need for adaptation with the term *situation awareness*. A *loss of situation awareness* described a situation where a pilot was insensitive to context changes and thus failed to adapt appropriately. A pilot with good situation awareness is a pilot who is well tuned to the changing context and who is able to adapt his control appropriately. Recently, a number of human factors' researchers have also adopted the term situation awareness to characterize the coupling between human operators and their work environments (e.g., Endsley, 1995; Flach, 1995).

Nonstationary systems raise the challenge of adaptive control. That is, the controller must perform at two levels. At one level, the controller must apply a current control law to reduce error. At another level, the controller must critique and evaluate the adequacy of the current control law and, if necessary, revise or replace it with a more appropriate law. Adaptive control systems are systems capable of adjusting control strategies in response to changing contexts. Several common styles of adaptation developed by control engineers are gain scheduling, direct and indirect adaptive control, and model-reference adaptive control (e.g., Ioannou & Sun, 1996).

The example of the CH-47 helicopter illustrates gain scheduling. With this style of adaptation, a fixed number of different control solutions are planned in advance. For example, instead of designing a fixed gain controller, several controllers with various gains are designed. These gains are chosen so that each will give optimal (or at least satisfactory) control for different contexts. The gain-scheduling controller then monitors the context (e.g., measures altitude and air speed) and activates the controller expected to be most appropriate for the context. Thus, a control with one set of gains may be active at low altitudes, and a different controller may take control at higher altitudes. Note that gain scheduling does not monitor system performance as a basis for its adaptive adjustments, and it can therefore be considered "open-loop" with regard to how well the system is performing. The number of different gain settings may vary as a function of the type of interpolation technique that is used. Fuzzy sets may be especially appropriate for this type of interpolation (Jang & Gulley, 1995, p. 2–55).

Although gain scheduling modifies the parameters of a single control structure, more drastic changes in the control process could also be implemented in different contexts. Switching among qualitatively different control strategies has sometimes been implemented as a production system that categorizes the performance at a higher level of abstraction than the individual control algorithms. Such structures have been useful in describing human performance in various target acquisition and continuous tracking tasks (e.g., see Jagacinski, Plamondon, & Miller, 1987, for a review). This kind of adaptation, where the mode of control is changed based on some dimension of the situation, is also observed in more complex task settings. For example, a surgeon might switch from a minimally invasive mode of action to an open surgery mode on observing a gangrenous gall bladder (Flach & Dominguez, in press).

A similar technique for switching among different control strategies can also be a solution to problems of controlling stationary, but highly nonlinear systems. The behavior of highly nonlinear systems can be represented in terms of qualitatively different behaviors that occur in various regions of their state space. Such behaviors may consist of stable points or oscillations that the system is attracted to once it is in a local neighborhood in the state space. Similarly, there may be unstable points and oscillatory regions that push the behavior of the system into more distant regions of the state space. Control strategies for such systems may consist of moving from one such neighborhood to another in an orderly manner, with different specialized control techniques in each different neighborhood. Design tools for implementing this general approach are under development and may prove useful for nonlinear design problems (e.g., Zhao, 1994).

Unlike gain scheduling, the method of direct adaptive control adjusts the parameters of the control system based on the difference between the desired response and the actual response. This difference can function as an error signal for the direct adaptive adjustment of controller parameters. For example, the gradient of this mean-squared difference signal can be calculated with respect to the various control parameters and used to guide their adjustment (i.e., "gradient descent"; chap. 15). This technique is an example of direct adaptive control. However, gradient descent does not guarantee stable adaptive control. Therefore, other composite functions involving the difference signal and parameter estimation errors are sought whose minization does guarantee stable adaptive control. Such functions are examples of *Lyapunov functions*, and their use in this manner (as described in chap. 20) is another example of direct adaptive control.

Indirect adaptive control attempts to solve the *dual control problem* in real time. The dual control problem is the two part, simultaneous problem of system identification (observation problem) and system control. An indirect adaptive controller is designed to observe the controlled process and construct a model of the process (chap. 20). This model is constantly being updated and adjusted based on continuous observation. The control algorithm is then adjusted so that it is appropriate for the current model of the system. The adjustment is indirect in the sense that the model is a mediator between new information about the process (e.g., unexpected response to control) and the changes to the control algorithm.

Note that the observation problem and control problem are in competition at one level and are collaborating at another level. Weinberg and Weinberg (1979) identified this as the *Fundamental Regulator Paradox*:

> The lesson is easiest to see in terms of an experience common to anyone who has ever driven on an icy road. The driver is trying to keep the car from skidding. To know how much steering is required, she must have some inkling of the road's slickness. But if she succeeds in completely preventing skids, she has no idea how slippery the road really is.
>
> Good drivers, experienced on icy roads, will intentionally test the steering from time to time by "jiggling" to cause a small amount of skidding. By this technique they intentionally sacrifice the perfect regulation they know they cannot attain in any case. In re-

turn, they receive information that will enable them to do a more reliable, though less perfect job. (p. 251)

Regardless of whether the adaptive adjustment of the controller parameters is direct or indirect, another aspect of the control problem is how to define the desired response. The desired response could simply be an exogenous input signal (e.g., the exact center of the driving lane of a winding road). On the other hand, the desired response might be specified in terms of a "model system" (i.e., follow the road in the manner of the famous race car driver Richard Petty). This latter path would involve deliberate deviations from the center of the driving lane, particularly on curves, and it would be an example of *model reference adaptive control*. Model reference adaptive controllers include a normative model for the desired system response to an exogenous input signal.

As another example, Jackson (1969) implemented an adaptive control algorithm (gradient descent) to make an adjustable crossover model match the performance of another crossover model with fixed parameters (chap. 15). The response of the crossover model with fixed parameters served as a model of the desired response to the common input signal. In subsequent experiments, a human performing a stationary tracking task was substituted for the fixed parameter crossover model. The adaptive adjustment of the simulated crossover model to match this human "model" system provided a means of "fitting" crossover parameters to human performance.

Human performers are also capable of model reference adaptive control. For example, Rupp (1974) showed that performers could mimic a fixed parameter crossover model when their deviations from the target parameters were explicitly displayed (chap. 15). Even without such displays, performers may have internal representations of performance analogous to the response of the crossover model, which are used to guide adaptive adjustments of their own control. The human's ability to adapt to changes in the order of control illustrated in Fig. 14.6 may reflect this style of adaptation. The human may effectively adjust the lead and lag components of their performance so that the human–machine system responds in a way that is consistent with the crossover model (e.g., McRuer & Jex, 1967; chap. 14). Surgeons also have clear expectations about how a surgery should proceed (in effect, a model reference). For example, surgeons have clear expectations about the anatomy and about the time course for a normal operation. They report that if they are not able to identify important structures of the anatomy (e.g., cystic duct, cystic artery, common bile duct, hepatic artery) within the first 30 minutes in a laparoscopic surgery, then they should convert to an open surgical procedure so that they can have more direct access to the anatomy. This type of surgical discipline is considered to be an important factor in avoiding common bile duct injuries, which are far more likely to happen in the laparoscopic (minimally invasive) mode of operation (Flach & Dominguez, in press).

Sometimes the best "reference" for control is provided by the behavior of a skilled human performer who is able to control a nonlinear, poorly understood system. In such instances, it would be helpful if skilled individuals could verbalize their control technique. However, very often much of their skill is difficult to verbalize. One tool that has proven helpful in this regard is the language of fuzzy sets (chap. 23). Namely, by using approximate verbal categories to describe both the people's actions and the situations that elicit those actions, it may be possible to capture a good deal of their control technique in a form that is useful to a system designer. The approximate description of the people's control strategy may be used to implement an automatic controller that can be subsequently adaptively adjusted on the basis of actual performance relative to the human model or some other standard (e.g., Jang, 1993). Given that the step of formally modeling the physical system is bypassed in this approach, there is typically a stronger reliance on physical testing of the automatic controller in a very wide variety of circumstances to be sure that unanticipated instabilities do not result.

It is important for designers to recognize when human performers may use different styles of adaptation requiring different types of support. For gain scheduling and other techniques for switching among preplanned control strategies, performers need unambiguous information about the context to specify which control is appropriate. For model reference adaptation, the behavior of a normative model should be integrated within representations so that the human can "see" discrepancies between actual system behavior and the normative expectations. To support direct adaptive control, it may be possible to display a multidimensional function of differences from the desired response and parameter estimates that can act as a Lyapunov fuction to guide the adaptation by the human operator. To support indirect adaptive control, the interface should provide a safe window for experiments (e.g., jiggling the wheel) and hypothesis testing, to support operators in solving the observation problem without serious threat to the control task.

It is important to note that adaptive control is a nonlinear process. Thus, stability problems should be approached cautiously and iterative testing and validation is essential. Even when stationary linear modeling is feasible, that does not eliminate the need for extensive testing of the control system to be sure that aspects of the system dynamics not captured in the formal model do not dominate performance. For example, in the design of the auxiliary manual control system for the Saturn V booster rocket used in early space missions, initial analyses of the control system neglected the possibility that unsteadiness in the motions of the human might excite high frequency bending movements inherent in the physical dynamics of the rocket (Denery & Creer, 1969; Jex, 1972). Once these effects were discovered in a simulator, it was possible to formally model them and to design appropriate filtering of the person's movements to avoid them.

Another cautionary point is that control engineering tools can be used inappropriately to produce performance that is "over-optimized." Namely, given a stationary description of the control environment, it is possible to design a system that will be highly effective for that particular description. However, if the description is incomplete or some aspect of the environment unexpectedly changes, then the behavior of the control system may drastically deteriorate and actually be poorer than a less finely tuned, but more "robust," control system (e.g., Astrom, 1970). Whereas adaptive control systems may overcome many of these problems, there are often complex dynamic processes that are only partly understood. The problem of creating control systems that are both adaptive to environmental change and robust under conditions where the dynamics are not accurately modeled is discussed in more advanced texts (e.g., Ioannou & Sun, 1996).

AN EXPANDED VIEW OF THE HUMAN OPERATOR

The classical cybernetic view of human performance is the *servomechanism*. This view of the human as a simple feedback device does not explicitly recognize the adaptive nature of human performance. For this reason, the servo metaphor often falls short. It is important to recognize that the servomechanism is an artifact of control theory. It is certainly an important artifact—so important that it has become an icon for control theory, but it is not control theory. Control theory allows engineers to design stable servomechanisms: to choose which state variables to measure and feed back in order to compute an adequate error signal, and to choose the gains that will result in stable regulation of that error. In a very real sense, the control designer might be a better metaphor for the kinds of problems that animals are faced with as they attempt to adapt to changing environments: to learn what variables to attend to and to discover the appropriate mapping between perception and action.

Figure 27.2 is an attempt to make the adaptive aspect of human performance more explicit. The supervisory loop could allow different styles of adaptation. Note that the supervisory loop is not a special augmentation needed only to model the human role in complex technological systems. The influence of this adaptation can be seen even in simple compensatory tracking tasks, where the human has to adapt a control strategy to accommodate the system dynamics (as seen in the discussion of the cross-over model). Note that there are two display boxes and two dynamic world models. This duplication is included to suggest that there may be important qualitative differences between the kinds of information and reasoning that occurs in the two loops. These qualitative differences might be reflected in Rasmussen's (1986) constructs of knowledge-based, rule-based, and skill-based behavior. This distinction is discussed more carefully later on. Note that in Fig. 27.2 the arrows between the human ele-

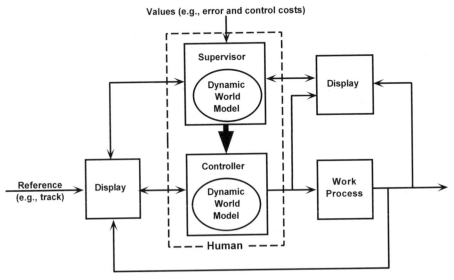

FIG. 27.2.   This alternative to the classical servomechanism view explicitly acknowledges the adaptive nature of human performance.

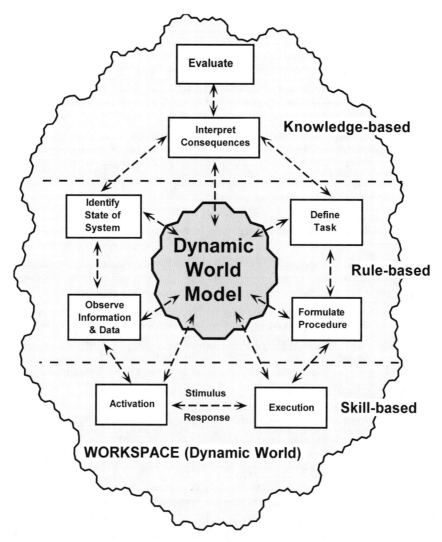

FIG. 27.3.   The information processing system is represented as a dynamic coordination where stimulus and response are intimately coupled, and where the links between perception and action are less constrained by structure than they are emergent properties of the dynamic. From "Cognitive engineering: Designing for situation awareness" by J. M. Flach and J. Rasmussen, 2000, in *Cognitive Engineering in the Aviation Domain* (p. 166), Hillsdale, NJ: Lawrence Erlbaum Associates. Copyright 2000 by Lawrence Erlbaum Associates.

ments and the displays are bidirectional. This is meant to reflect the active nature of the search for information. The human is not a passive receptor of information, but rather is seeking answers to specific questions. Thus, for example, different styles of adaptive control would require different kinds of information from the display within the outer loop. Also, the actor can play an important role in constructing the display representation, by externally or internally configuring the information. The term *display* is used in its broadest sense to include any available source of information (natural or artifactual). The arrow between the supervisor and the controller is distinct from the other arrows. This is meant to indicate that the supervisor operates on the transfer function within the control element, as opposed to the normal convention where the arrows represent signals that are operated on by the transfer function within the box. Finally, there are two inputs. One input from the left is intended to reflect the "reference" for the controller. This is where the system wants to be. The other input, from the top, is intended to reflect "values" as might be reflected in the cost functional of an optimal control model. In some sense, both the reference and value inputs are aspects of the "goal." But again, there are important qualitative differences between the goal of tracking a specific path and the goal of managing the associated resources. And these differences have important implications for how these dimensions fit within a control framework.

Figure 27.3 is an alternative representation from Flach and Rasmussen (2000) that also tries to capture the adaptive nature of human performance. This diagram includes traditional stages of information processing. However, the diagram has been reorganized to emphasize aspects of cognition that have not been represented well in traditional diagrams of information processing. Traditional images of information processing have used a communication channel metaphor that emphasizes the sequence of transformations with a fixed precedence relation among the processing stages. Most of these representations include a feedback link, but this link is represented as peripheral to the communication channel and the link has largely been ignored within cognitive research programs. In traditional information-processing diagrams, stimulus and response are represented as distinct and distant entities peripheral to the information-processing stream.

Figure 27.3 emphasizes the circular, as opposed to linear, flow of information. In this circular flow, there is an intimate link between perception and action. Thus, stimulus and response become the same line. Neisser (1976) recognized this circular flow of information in his attempts to make cognitive psychology more ecologically relevant. Recently, work on situation awareness has resulted in a deeper appreciation of Neisser's insights (e.g., Adams, Tenney, & Pew, 1995; Smith & Hancock, 1995).

The left half of the cycle represents different levels of the observation problem. The right half of the cycle represents different levels of the control problem. The arrows in the center reflect the different ways that observations (perception) can be coupled with control (action). Figure 27.3 emphasizes that there is no fixed precedence relation among the processing stages (i.e., there is a flexible coupling between perception and action). Rather, the cognitive system is an adaptive system capable of reorganizing and coordinating processes to reflect the changing constraints and resources within the task environment. The internal set of arrows symbolizes potential links between all nodes in the system. Note that these links are not *necessary* connections, but *potential* connections.

The stippled region in Fig. 27.3 represents the workspace. The processing loop is contained within the workspace and the workspace is illustrated as a substrate that extends within the processing loop. This is an attempt to illustrate that cognition is situated within an environmental context. Thus, the links between processing stages are often the artifacts within the workspace. Hutchins' (1995) analysis of navigation provided strong evidence that computations are distributed over humans and artifacts and coordination is achieved within the workspace (as opposed to being accomplished exclusively within the head; see also Zhang & Norman, 1994). The knowledge states shown in Rasmussen's (1986) decision ladder have not been included for simplicity. However, the presence of knowledge states is implied and these knowledge states can exist both in the head (i.e., knowledge of standard procedures, physical principles, etc.) and in the environment (i.e., instructions in a manual, checklists, written notes, graphical interfaces, etc.).

Because of the flexibility of this system and the ability to shunt from one pathway to another, processing in this system is not absolutely constrained by "channel capacity" or the availability of "energetic type resources." The fundamental constraint is the ability to attune and synchronize to the sources of regularity and constraint within the work domain. Rasmussen (1986) characterized this in terms of a dynamic world model. This internal model reflects the knowledge (both explicit and implicit) that the cognitive agent has about the invariant properties of the work environment. This attunement is constrained by the qualitatively different demands that govern the flow of information within this parallel, distributed network. Rasmussen (1986) distinguished three qualitatively different types of attunement within distributed cognitive systems: skill based, rule based, and knowledge based.

The different levels of processing (skill based, rule based, and knowledge based) reflect the different ways that the adaptive control problem can be solved. Depending on factors such as the level of skill of the operator and the constraints within a work environment, different processing paths can be utilized. The kinds of information required on the observer side of the loop may be quite different in different situations. For example, in a routine laparoscopic surgery, skilled surgeons may operate at a skill-based level in which they fluently respond to the familiar signals that allow them to maneuver within the abdomen and to accomplish their goals with little demand on higher levels of cognitive processing. However, the observation of unusual or unexpected conditions may involve other paths. The perception/action loop may be closed using preestablished conventions in a manner analogous to gain scheduling. That is, a sign (an inflamed gallbladder) may trigger a switch in the task (change from a minimally invasive mode of surgery to an open mode of surgery). Or the sign might simply change the parameters of the execution style (move with more care and deliberation within the minimally invasive mode—lower gain). Alternatively, surgeons who are faced with unusual anatomy may have to engage in problem solving in order to resolve ambiguities. They may initiate exploratory actions (tracing along a structure) in order to identify the state of the situation. Is this really the cystic duct or is it the common bile duct? And, if the ambiguity cannot be resolved within a reasonable time frame, surgeons must decide whether to continue laparoscopically or to convert to an open surgical procedure. Note that these paths can all be operating in parallel. Skill-based processes may be guiding the instruments as they trace the anatomy to solve a knowledge-based problem, whereas a rule-based process is ready to generate an interrupt if a time limit is exceeded.

The major point here is that the path through this flexible adaptive control system is not "determined" by the cognitive architecture and it is not "determined" by the environmental task, but it is "shaped" by both. Thus, in studying cognitive processes, it is important to keep an open mind to the wide range of possible solutions. Also, it is important to frame the problem of cognition in a way that respects the role of both the internal computational constraints and the external ecological constraints. Thus, an important goal of this book is not to provide a specific model of cognitive processing, but to help students to appreciate the range of possibilities. This book is not intended to provide new answers to the problem of cognition, but to stimulate a wider range of questions (and to provide some tools that may help researchers to manage the data that result).

The control language may also help to cut through some of the rhetoric that can hinder communication among researchers. For example, there is the curious notion of "direct" perception. Classical models of perception tend to segment the observation side of the control problem from the action side of the problem. The models that result tend to be analogous to "indirect" models of adaptation. Thus, these classical theories typically include discussions of an "internal model" as a necessary bridge between perception and action. Direct theories of perception tend to reject the "internal model" as an unnecessary constraint. This approach tends to look for direct links (e.g., optical invariants) between perception and action. These direct links may be analogous to the Lyapunov functions that link perception and action in direct adaptive control systems. Note that "direct" and "indirect" reflect different styles for implementing an adaptive control system, but at a functional input–output level of description these different solutions will often be isomorphic. Perhaps, if the arguments were framed within the language of adaptive control, then the different theoretical camps might be able to better appreciate the common ground and be able to frame empirical tests that address critical differences between the different models of adaptation.

Summary

Figures 27.2 and 27.3 are intended to help people to appreciate that the dynamics of human information processing are far more complex than the dynamics of a simple servomechanism or of a communication channel. However, the failure of these simple metaphors should not be mistaken as a failure of the control theoretic framework. In fact, it is because of this increased complexity that an investment in control theory is essential. The tools of control theory will be important for building theoretical frameworks that do justice to the complex nature of human performance. Two figures are presented, because neither figure fully captures the intuitions this chapter hopes to convey. The goal is to help people see beyond the icons associated with the cybernetic hypothesis and to appreciate both the richness of human performance and the richness of control theory. This richness was recognized by Pew (1974) in a review of human perceptual-motor performance:

> We should think of a continuum of levels of control and feedback, that the signal comparator operates at different levels at different times, and can even operate at different levels at the same time. What we observe in human skilled behavior is the rich intermingling of these various levels of control as a function of the task demands, the state of learning of the subject, and the constraints imposed on the task and the subject by the environment. The job of the researcher is different, depending on the level of analysis in which he is interested, but a general theory of skill acquisition will only result from con-

sideration of all the ramifications of this kind of multilevel process-oriented description of skilled performance. (p. 36)

COGNITIVE SYSTEMS ENGINEERING

The central issue is to consider the functional abstraction underlying control theory and to understand the implications of different control strategies on system behavior and design requirements. (Rasmussen, Pejtersen, & Goodstein, 1994, p. 8)

In complex systems, sometimes a set of controlled variables will have progressively longer time scales characterizing their influence on system performance. Then the design issues can be approximately decomposed into a set of nested control loops, and the design of the control system for each loop can be considered separately (e.g., Hess, 1989). However, information technology seems to be leading to systems where the couplings across the nested loops are increasingly complex. Cognitive systems engineering (CSE) (Rasmussen, 1986; Rasmussen et al., 1994; Vicente, 1999) has emerged as one framework for parsing these complex systems so that meaningful design decisions can be made. CSE involves decomposing complex problems along two dimensions (abstraction and part–whole relations). These two dimensions are illustrated in Fig. 27.4. One of the primary insights of the CSE approach is that domain experts tend to decompose systems along the diagonal of this two-dimensional space. Moving along the diagonal from the top down provides insights about the functional rationale of the system (the reasons why some states and paths through the workspace are more desirable than others). Moving along the diagonal from the bottom up provides insights about the causal relations within the system (options for how to

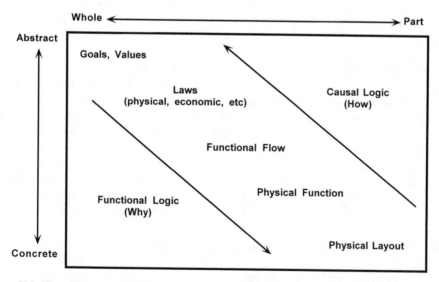

FIG. 27.4. The diagonal of the abstraction versus decomposition space provides important insights for understanding why and how a control system might function.

get from one state to another). This is fundamentally a control problem—to understand how information can be utilized so that a physical system can be constrained to behave in a way consistent with the design goals.

The two diagonals illustrate how control systems are different from simple physical systems (e.g., how an aviation system is different than a rock). To describe the trajectory of a rock, it is sufficient to know the momentary forces on the rock. It is not necessary to ask about what the rock wants to accomplish (although early physics did invoke intentional constraints to explain the fact that rocks consistently returned to earth). Today, a rock would be modeled as a purely causal system. Its path is completely determined by the forces acting on it. However, to predict the trajectory of an airplane, the intention of the pilot (or the design logic of an automatic control system) becomes a very important consideration. In this system, the causal constraints are organized in service to the functional goals. Thus, although the behavior of the aircraft (like the rock) is consistent with physical laws, it does not seem to be determined by simple physical laws, at least not in the same way that behavior of the rock is determined. Eventually, it may be possible to reduce the intentions of the pilot to causal laws. However, today this is not possible. So, the essence of control theory is to study the coordination between causal (physical) and functional (intentional) constraints. Another way to think about it, is that the rock is constrained by forces, but the aircraft is constrained by both force and information. The aircraft system can alter its course as a function of information that is fed back and evaluated relative to some reference or value system. The information constraints flow from the top down the diagonal, whereas the force constraints flow from the bottom up the diagonal.

The diagonal of the abstraction/decomposition space in Fig. 27.4 illustrates the type of reasoning essential to the design of any control system (whether for a simple temperature regulation system, an aircraft cockpit, or a nuclear power control room). In design the goal is typically to harness physical laws in service to some functional objectives. Thus, it is important to consider the goals for the system. Why is it being built? What goals are to be accomplished? How is performance to be scored (i.e., what are the values)? For simple systems, the goals are fairly obvious. The goal of a temperature regulation system is to control temperature. However, even in the design of simple control systems, other goals may constrain the solution (e.g., goals related to the cost of operation, environmental impact). For more complex systems (e.g., aviation or nuclear power), multiple goals must be considered (e.g., transport passengers and packages, generate a profit, maximize safety) and these goals often come into conflict. The trade-offs that result from competition among goals must be evaluated against some value function. How much safety/risk can be afforded? Consideration of values is explicitly dealt with in optimal control as the cost functions. In control texts, the goals and values tend to be "givens" for a control problem. The students' job is to derive a control law to satisfy the "given" constraints. But in design, correctly identifying the relevant goals and values is often a critical aspect of the problem.

Another level for consideration in designing a control system is to determine a model of the controlled process. This is represented in Fig. 27.4 as the global laws. Thus, before an autopilot can be designed, it is necessary to have a model of the "process dynamics." In this case, it is necessary to specify the aerodynamic properties of the vehicle to be controlled. Without a satisfactory model of the process, it is not possible to identify the state variables that need to be fed back in order to control the process. The question concerns how to tell whether the system is satisfying the functional goals. What are the dimensions of error that must be corrected by control actions? For example, throughout, this book has discussed the fact that controlling inertial systems typically requires feedback about both position and velocity. Thus, there is no one-to-one mapping between distance from a stop sign and the point where braking should be initiated. To know when to begin braking, the controller needs to have feedback about both distance and speed. At higher speeds, braking should occur at greater distances from the intersection. This reasoning reflects the physical laws of inertia (Newton's second law). The Wright brothers' first aircraft design required very careful studies of the physics of wings and lift. Through careful experimentation and analysis, they discovered that the constants that were generally used to compute lift at that time were in error. This had important implications for the camber and surface areas of the wings.

The designer must also consider the appropriate functional organization for the control system. Will the system be open and/or closed-loop? Will feedforward be utilized in addition to feedback? What parameters can be lumped together in a single control loop? This is the level where block diagrams become useful tools for visualizing the functional organization. The Wright brothers originally decomposed the flight problem into two control loops. Lateral (wing warping) and yaw (rudder) were yoked to a single control in the first aircraft, and a second control allowed manipulation of pitch (elevators). However, in their later designs, the Wrights provided independent controls for lateral (wing-warping) and yaw (rudder) control. This configuration remains the standard control organization in most modern aircraft.

Consideration must also be given to allocation of functions to specific types of physical systems. For example, the Wright brothers' automatic stabilization system was implemented "using a pendulum to control the wing warping and a horizontal vane to operate the elevator, both working through servomotors powered by a wind-driven generator" (Crouch, 1989, p. 459). Within a year after the Wrights won the Collier trophy, Sperry presented an alternative design in which the mechanical vanes and pendulum were replaced by gyroscopes. Sperry's solution proved to be the superior solution: "Not only did it form the basis for all subsequent automatic stability systems, it opened an entire range of new possibilities. The enormously complex inertial navigation system that guided the first men to the Moon in 1969 was directly rooted in Sperry's automatic pilot of 1914" (Crouch, 1989, p. 460). It is at this level of analysis that decisions about using humans (e.g., pilots) or automatic control devices (e.g., autopilots) to fly the plane come into consideration.

An additional level of analysis concerns the details of the physical layout. In the design of an electronic control system, this level might be reflected in the detailed wiring diagrams showing how the system is laid out and what wires are connected to what. In human–machine systems, questions about the position of the human, the layout of the controls, and the arrangement of displays are considered at this level.

An alternative way of visualizing the different levels of abstraction is as a nested set of constraints. The higher levels of abstraction set the constraints that bound solutions at lower levels of abstraction. The higher levels of abstraction represent "global" constraints, and the lower levels of abstraction represent "local" constraints on solutions. At high levels of abstraction, there are generally a few global degrees of freedom to consider (e.g., efficiency vs. safety). At lower levels of abstraction, more degrees of freedom (e.g., the arrangements of all the components and wires) must be considered. However, decisions about higher constraints help to bound the possibilities that need to be considered at lower levels of abstraction. It is tempting to think of the design process as a top down analysis in which goals are specified, models are built, a functional organization is chosen, the components are identified, and the system is assembled. However, for complex systems, design is an iterative process in which a discovery at one level of analysis can change the way constraints at another level of analysis are viewed.

The goal of CSE design is to support humans so that they can function expertly in solving the adaptive control problems in dynamic work domains. This generally involves building representations that help the humans to explore the diagonal of the abstraction/decomposition space. Thus, effective interfaces will typically involve configural representations that illustrate the nesting of constraints from global goals (abstract functions at a relatively gross level of decomposition) to local actions (physical functions at a relatively detailed level of decomposition). One way to think about this is that the displays must support many paths by which the human operator can close the loop around both the control problem and the adaptation problem. Rasmussen et al. (1994) and Vicente (1999) are recommended for more details on the CSE approach.

CONCLUSION

The primary goal of this book is to help make the intuitions of control theory accessible to a broader audience. In particular, the book is directed at people who are interested in human performance (e.g., psychologists, movement scientists, and human factors engineers). This reflects certain interests, but it is also a domain that is particularly rich with examples to which almost any student of dynamic behavior can relate. The beliefs contained herein strongly emphasize that the path to the intuitions of control theory requires a mastery of the fundamental elements. If individuals want to become great athletes or great musicians, they must start with the fundamentals. They must learn the scales first before they can create symphonies. The same applies to control theory. The goal of this book is to help make some of the fundamentals of control accessible to those outside of, or recently entering, the engineering discipline. This book provides some scales and early exercises in control theory. With this start, and at least 10 years of intense practice, there is optimism that great symphonies may emerge. These symphonies will put the notes together in ways that cannot be imagined today. However, the new models and theories of the future certainly will be built from some of the elements presented in this book.

REFERENCES

Adams, M. J., Tenney, Y. J., & Pew, R. W. (1995). Situation awareness and the cognitive management of complex systems. *Human Factors, 37,* 85–104.

Astrom, K. J. (1970). *Introduction to stochastic control theory*. New York: Academic Press.

Billings, C. (1997). *Aviation automation: The search for a human-centered approach*. Hillsdale, NJ: Lawrence Erlbaum Associates.

Crouch, T. (1989). *The bishop's boys: A life of Wilbur and Orville Wright*. New York: Norton.

Denery, D. G., & Creer, B. Y. (1969). *Evaluation of a pilot describing function method applied to the manual control analysis of a large flexible booster* (NASA Technical Note D-5149). Moffett Field, CA: NASA.

Endsley, M. R. (1995). Toward a theory of situation awareness in dynamic systems. *Human Factors, 37*, 32–64.

Flach, J. M. (1995). Situation awareness: Proceed with caution. *Human Factors, 37*, 149–157.

Flach, J. M., & Dominguez, C. O. (in press). A meaning processing approach: Understanding situations and awareness. In M. Haas & L. Hettinger (Eds.), *Psychological issues in the design and use of virtual environments*. Hillsdale, NJ: Lawrence Erlbaum Associates.

Flach, J. M., & Rasmussen, J. (2000). Cognitive engineering: Designing for situation awareness. In N. Sarter & R. Amalberti (Eds.), *Cognitive engineering in the aviation domain* (pp. 153–179). Hillsdale, NJ: Lawrence Erlbaum Associates.

Freedman, R. (1991). *The Wright brothers. How they invented the airplane*. New York: Holiday House.

Hess, R. A. (1989). Feedback control models. In G. Salvendy (Ed.), *Handbook of human factors* (pp. 1212–1242). New York: Wiley.

Hutchins, E. (1995). *Cognition in the wild*. Cambridge, MA: MIT Press.

Ioannou, P. A., & Sun, J. (1996). *Robust adaptive control*. Upper Saddle River, NJ: Prentice-Hall.

Jackson, G. A. (1969). A method for the direct measurement of crossover model parameters. *IEEE Transactions on Man–Machine Systems, MMS-10*, 27–33.

Jagacinski, R. J., Plamondon, B. D., & Miller, R. A. (1987). Describing the human operator at two levels of abstraction. In P. A. Hancock (Ed.), *Human factors psychology* (pp. 199–248). New York: North Holland.

Jang, J. R. (1993). ANFIS: Adaptive-network-based fuzzy inference system. *IEEE Transactions on Systems, Man, and Cybernetics, 23*, 665–685.

Jang, J. S. R., & Gulley, N. (1995). *Fuzzy logic toolbox for use with MATLAB*. Natick, MA: The Math Works.

Jex, H. R. (1972). Problems in modeling man-machine control behavior in biodynamic environments. In *Proceedings of the Seventh Annual Conference on Manual Control* (NASA SP-281, pp. 3–13). University of Southern California, Los Angeles, CA.

McRuer, D. T., & Jex, H. R. (1967). A review of quasi-linear pilot models. *IEEE Transactions on Human Factors in Electronics, HFE-8*, 231–249.

Neisser, U. (1976). *Cognition and reality: Principles and implications of cognitive psychology*. San Francisco: Freeman.

Pew, R. W. (1974). Human perceptual motor performance. In B. H. Kantowitz (Ed.), *Human information processing: Tutorials in performance and cognition* (pp. 1–39). Hillsdale, NJ: Lawrence Erlbaum Associates.

Rasmussen, J. (1986). *Information processing and human-machine interaction: An approach to cognitive engineering*. New York: North Holland.

Rasmussen, J., Pejtersen, A. M., & Goodstein, L. P. (1994). *Cognitive systems engineering*. New York: Wiley.

Reason, J. (1990). *Human error*. Cambridge, MA: Cambridge University Press.

Rochlin, G. (1997). *Trapped in the net*. Princeton, NJ: Princeton University Press.

Rupp, G. L. (1974). *Operator control of crossover model parameters*. Unpublished doctoral dissertation, University of Michigan, Ann Arbor, MI.

Sastry, S., & Bodson, M. (1989). *Adaptive control: Stability, convergence, and robustness*. Englewood Cliffs, NJ: Prentice-Hall.

Smith, K., & Hancock, P. A. (1995). Situational awareness is adaptive, externally directed consciousness. *Human Factors, 37*, 137–148.

Vicente, K. J. (1999). *Cognitive work analysis*. Hillsdale, NJ: Lawrence Erlbaum Associates.

Weinberg, G. M., & Weinberg, D. (1979). *On the design of stable systems*. New York: Wiley.

Zhang, J., & Norman, D. A. (1994). Representations in distributed cognitive tasks. *Cognitive Science, 18*, 87–122.

Zhao, F. (1994). Extracting and representing qualitative behaviors of complex systems in phase space. *Artificial Intelligence, 69*, 51–92.

Appendix: Interactive Demonstrations

In order to comprehend the dynamics of even simple systems, it is a good idea to interact with simulations of those systems rather than only study them on the static page of a book. The student can then represent the dynamic phenomenon in a number of ways — verbal description (categorical), differential equation (algebraic), simulation language (procedural), and perceived movement patterns (geometric, topological). Going from one form of representation to another provides a deeper understanding.

Here are a few examples of some demonstrations that may be helpful. In the style of the rest of this book, the list is by no means exhaustive, but simply illustrative of possibilities.

1. Interactive demonstrations of system dynamics are being developed on a website constructed by P. J. Stappers in consultation with the authors of this book (*http://studiolab.io.tudelft.nl/controltheory/*). They illustrate some of the concepts discussed in this book. Demonstrations include step tracking (target acquisition) and continuous sine wave tracking tasks. The software allows the user to specify the plant dynamics (position, velocity, and acceleration dynamics; see chap. 9). Generally it is instructive for a student to try out a particular control system in both target acquisition and continuous tracking tasks in order to have a broader conception of its dynamic behavior. Feedback is provided in the form of time histories.

2. The Manual Control Lab (MCL) provides real-time demonstrations of stationary target acquisition, compensatory tracking, and pursuit tracking in one or two dimensions. A menu system permits a choice of position, first-order lag, velocity, second-order lag, acceleration, or third-order dynamics (chaps. 4, 6, and 9). For the target acquisition task, one can choose a range of target distances and widths (chap. 3). After each trial, the MCL system displays reaction time and movement time, time

histories of control movement and system output position and velocity, and a state space trajectory of the system output (chap. 7). As the system dynamics change from position to velocity to acceleration controls, one can see the shape of the control movement for target acquisition change from a step to a pulse to a double pulse (chap. 9). The step correlates with system output position, the pulse correlates with system output velocity, and the double pulse correlates with system output acceleration or change of velocity vs. time.

The input for the compensatory and pursuit tracking tasks is a sum of up to 10 sine waves, which has sufficient complexity to be relatively unpredictable to the human performer. A menu system permits selection of both the frequencies and amplitudes. A typical input might consist of several low frequency sine waves with large amplitudes and a set of higher frequency sine waves with much smaller amplitudes. The frequency range of the larger sine waves determines the effective bandwidth of the input signal. The smaller amplitude, higher frequency sine waves permit Fourier analysis of the tracker's performance (chap. 12) without making the overall speed of the input seem too fast. After each trial, the MCL system provides a time history of the input signal, control movement, system output, and error along with root mean squared error and other statistics. Also provided is rapid calculation of the amplitude ratios and phase shifts of the control system (plant), the human tracker, and the human plus plant (open-loop describing functions). This performance feedback permits one to see how well the crossover model approximates the person plus plant and permits visual estimation of the crossover frequency and the phase margin (chaps. 13 and 14). Also, by changing the dynamics from position to velocity to acceleration control systems, one can see the describing function for the person change from lag-like, to gain-like, to lead-like, while the general form of the describing function for the person plus plant stays relatively invariant in the region of the crossover frequency (Fig. 14.6).

Tracking with an acceleration control system may be somewhat easier with a joystick than a mouse. Both are supported by the MCL system. Quickening the display (chaps. 9 and 26) is often necessary for naive trackers with the acceleration control. For more advanced trackers a time delay can be added to the dynamics to make them more challenging.

Vendor: ESI Technology
2969 Piney Pointe Drive
St. Louis, MO 63129
Telephone: 314-846-1525
E-mail: bobtodd@i1.net (Note: The symbol after the "i" is a "one".)
Web-page: www.i1.net/~bobtodd/mcl.htm

Computer system: PC compatible running Microsoft(R) MS-DOS(R) Version 3.x or later or Windows 9x DOS window

3. The Computer Aided Systems Human Engineering: Performance Visualization System (CASHE: PVS) provides a set of 11 demonstrations including various tests of visual and auditory sensitivity as well as manual control. There are also links to relevant entries in the Engineering Data Compendium edited by Boff and Lincoln (1988), which summarizes behavioral studies of the variables included in the simulation.

Links are also available for system designers to military standards (MIL-STD-1472D, 1992).

The manual control demonstration is organized around a conceptual flow chart of the tracking task. There are associated menus for changing the input signal (stationary targets or continuously moving targets), system dynamics (position, first-order lag, velocity, acceleration, and third-order controls, plus time delays and/or various degrees of aiding in the form of leads [chaps. 4, 9, and 26]), and display (compensatory or pursuit, plus quickening [chaps. 9 and 26], and a left/right compatibility option). The tracking is performed with a mouse.

For one-dimensional stationary target acquisition, up to three different adjustable amplitudes and widths can be factorially combined to generate nine targets (chap. 3). Performance feedback consists of reaction times, movement times, and time histories of the mouse movement and the system output position and velocity.

Compensatory and pursuit tracking of continuously moving targets can be performed in one or two dimensions (chap. 10). The continuous target in each axis is generated by summing up to 9 sine waves with adjustable frequencies and amplitudes. A sum of 9 sine waves is relatively unpredictable, whereas a single sine wave is much easier to predict and track (chap. 21). Performance feedback includes root mean squared error and control.

Engineering data compendium: Human perception and performance (vol. 1–4). (1988). K. R. Boff & J. E. Lincoln (Eds.). Wright-Patterson Air Force Base, Ohio: Armstrong Aerospace Medical Research Laboratory.

Vendor: Human Systems Information Analysis Center
2261 Monahan Way
Wright-Patterson AFB, OH 45433-7022
Telephone: 937-255-4842
E-mail: paul.cunningham@wpafb.af.mil
Web-page: iac.dtic.mil/hsiac
Computer system: MacIntosh computer running OS 7 or later

4. Tutsim is a simple programming language for simulating dynamic systems. This language is relatively easy for beginning students to learn. It only requires four different steps with simple syntax to implement a simulation:

1. Construct a numbered list of primitive elements, for example, gain, time delay, integrator, pulse, and indicate how they are connected, that is, which elements are inputs to other elements. Readers old enough to remember slide rules will note the similarity to programming an analog computer.

2. Specify the numerical parameters of the primitive elements, for example, the magnitude of the gain, the duration of the time delay, the initial output of the integrator, the start time, amplitude, and stop time of the pulse.

3. Indicate the temporal granularity, that is, the update rate, and the total duration of the simulation. The temporal granularity should be fine enough so that the plots of system performance develop over time at a comfortable rate on the

computer monitor. Faster computers require finer granularity to slow down the rate of plotting.

4. Indicate which variables should be plotted. For example, one can plot up to three variables vs. time to generate time histories or plot velocity versus position to generate a state-space trajectory (chap. 7). Additionally, one can successively plot a family of system behaviors on the same graph corresponding to quantitative variations in parameters, for example, gains, delays, and so on.

These four steps provide a relatively simple programming procedure, so that within a few hours someone new to dynamic simulation can implement and begin to explore on their own a first-order lag (chap. 4), the crossover model (chap. 14), a second-order system (chap. 6), and so on.

Vendor: Actuality Corporation
           805 West Middle Avenue
           Morgan Hill, CA 95037
           Telephone: 408-778-7773
           E-mail: tutsimpa@aol.com
Computer system: PC compatible running DOS 5 or later, Windows 9x, or Windows NT 4.0

5. MATLAB provides a very large and highly developed set of demonstrations and programming capabilities ranging from control system simulation to signal processing to fuzzy sets to many other topics beyond the scope of the present text. For example, the Fuzzy Logic Toolbox (Jang & Gulley, 1995) provides a simple introduction to fuzzy sets in terms of the problem of deciding how much to tip a waiter in a restaurant based on the quality of the food and service. The interactive program permits a student to manipulate the latter two variables and see the contributions of various fuzzy rules in determining the tip. A three-dimensional surface representation of the input-output behavior of the fuzzy system is also provided. The Fuzzy Logic Toolbox also provides a number of dynamic examples of fuzzy control, e.g., backing up a truck to a loading dock (chap. 23). Additional discussion of fuzzy control in the context of MATLAB can be found in Beale and Demuth (1994).

The University of Michigan and Carnegie Mellow University have developed a website which provides a set of control tutorials for MATLAB (*www.engin.umich. edu/group/ctm/*). Seven different control problems (e.g., balancing an inverted pendulum, aircraft pitch control) are used to show how different control theoretic representations can be used to provide insights into these tasks. This material is also available on CD-ROM through Prentice Hall. Additional demonstrations for MATLAB can be found in Frederick and Chow (2000) and Djaferis (1998), which also illustrate various feedback control design techniques in the context of engineering problems.

Beale, M. & Demuth, H. (1994). *Fuzzy systems toolbox for use with MATLAB*. Boston, MA: PWS.
Djaferis, T. E. (1998). *Automatic control: The power of feedback using MATLAB*. Boston: PWS.
Frederick, D. K., & Chow, J. H. (2000). *Feedback control problems using MATLAB and the Control Systems Toolbox*. Pacific Grove, CA: Brooks/Cole.

Jang, J.-S. R., & Gulley, N. (1995). *Fuzzy Logic Toolbox for use with MATLAB*. Natick, Massachusetts: The MathWorks.

> Vendor: The MathWorks
>   3 Apple Hill Drive
>   Natick, MA 01760
>   Telephone: 508-647-7000
>   Web page: www.mathworks.com
> Computer system: PC compatible running Windows 9x, Windows 2000, Windows NT
> 4.0 with service pack 5, or Linux. Other platforms are also available.

6. An elaboration of the fuzzy control of a truck backing up (chap. 23) can be found on the floppy disk in Kosko (1992). This version of the problem was implemented by HyperLogic Corporation (www.hyperlogic.com) and provides a display of the instantaneously active cells in a fuzzy state space in parallel with a display of the two-dimensional time history of the truck's position as it backs up to a loading dock. The fuzzy state space represents a set of rules that associates steering wheel angles with different combinations of truck position and heading angle. The instructor can selectively alter or even delete cells in the fuzzy state space to demonstrate their effects on the observed trajectories of the truck.

Kosko, B. (1992). *Neural networks and fuzzy systems*. Englewood Cliffs, New Jersey: Prentice Hall.

Computer system: PC compatible running DOS or Windows DOS window

7. The Control Systems Group (CSG) is an organization that examines Perceptual Control Theory as a general theory of human behavior. This group was inspired by the work and writings of William Powers (e.g., 1973, 1998). The CSG website includes several clever interactive demonstrations developed by Rick Marken (*www.ed. uiuc.edu/csg/*).

Powers, W. T. (1973). *Behavior: The control of perception*. New York: Aldine de Gruyter.
Powers, W. T. (1998). *Making sense of behavior: The meaning of control*. New Canaan, Connecticut: Benchmark.

8. Simple physical models of systems are sometimes excellent teaching aids. For example, William Powers (1998) describes an interactive demonstration using rubber bands to illustrate several aspects of control systems. The demonstrations involve linking two rubber bands so that the knot joining them becomes the system output. The linked bands can then be stretched over a table, and a target can be designated as a spot on the table. One person holds one of the rubber bands and acts as the controller. That person's task is to keep the knot aligned with the target spot. A second person simulates disturbances to the control system by moving a second rubber band so as to perturb the knot away from the target spot. Numerous variations on this task are possible. A third person might move the target spot. Also, additional rubber bands can be added in order to illustrate either multiple disturbances or multiple control linkages, that is, the degrees of freedom problem.

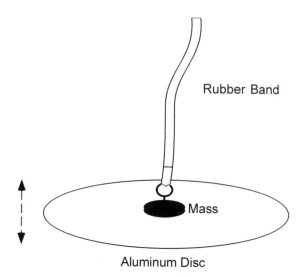

FIG. A.1.   A rubber band, a mass, and an aluminum disc form a second-order under-damped system.

A simple second-order system demonstration used by R. W. Pew consists of a thin 10-inch diameter, circular aluminum disc, a mass, and an 5-inch rubber band that is cut to make a single strand 10-inches long (measurements are approximate). A screw with an eye-ring is used to connect the mass and one end of the rubber band to the center of the aluminum disc (Fig. A.1). The rubber band provides springiness, and the aluminum disc moving through the air provides damping for this second-order, underdamped system. The demonstration proceeds by holding the free end of the rubber band in one's hand and bouncing the suspended mass up and down in a yo-yo like fashion. By moving one's hand vertically in a sinusoidal pattern at low, near resonant, and high frequencies, the amplitude ratios and phase shifts of the closed-loop frequency response can easily be demonstrated (chap. 13). The amplitude ratio of disc movement to hand movement is near 1.0 at low frequencies, greater than 1.0 near resonance, and much less than 1.0 at high frequencies. The phase lag is near 0 degrees at low frequencies, near 90 degrees near resonance, and near 180 degrees at high frequencies.

A toy car or a bicycle can be used to demonstrate lateral steering dynamics. The front tire(s) and steering wheel should be clearly visible, so one can demonstrate a proportional relation between steering wheel angle and front tire angle, single pulse steering control to change the heading angle, and double pulse steering control to perform a lane change maneuver (chaps. 9 and 16).

Powers, W. T. (1998). *Making sense of behavior: The meaning of control.* New Canaan, Connecticut: Benchmark.

The above demonstrations explore a few elementary dynamic systems. A wide range of additional examples and analyses can be found on the World Wide Web.

Author Index

Subject Index

A

Abstraction/decomposition space, 355–358
Adaptive control
  arm movement, 239–251
  compensatory tracking, 164–183, 339–340
  convergence, 173, 247–248, 347–348
  direct, 245–248, 347–349
  see also Gain scheduling
  see also Gradient descent
  indirect, 244–248, 347–349
  see also Lyapunov function
  model reference, 170–175, 348–349
  of sinusoidal patterns, 255–259
Admittance, 337
Aiding, 98–99, 338–339, 362
Aircraft
  air speed, 273–275
  altitude, 15, 272–275, 281–287
  banking, 343
  see also Design
  high order dynamics, 90–91, 98–99
  hovering, 284–287
  landing, 281–282
  level flight, 273
  see also Optical flow
  see also Pilot-involved oscillation
Ataxia, 13
Attractor, 260–265
Automobile, *see* Car

B

Ball hitting, 278–281
Ballistic movement
  see Discrete control
  see Nonproportional control
  see Optimal control, time optimal
Bandwidth, 23, 66, 153, 176–180, 361
Bang-bang control, 59–63, 68, *see also* Optimal
  control, time optimal
Base-rate neglect, 316
Bayes' Theorem, 207, 214, 314–317, 321, *see also*
  Kalman filter
Beat frequency, 128
Block diagram, 41–44
Bode plot, 137–157, *see also* Frequency response
Bootstrapping, 318–320

C

Car
  braking, 269–272, 275–279
  lane change, 189–190
  see also Optical flow
  parking, 294–299
  steering, 184–194, 365
  window control, 96
Choice reaction time, 18–20
Cognitive systems engineering, 355–358

375